Anthroposophy and Science

Peter Heusser

Anthroposophy and Science
An Introduction

Bibliographic Information published by the Deutsche Nationalbibliothek
The Deutsche Nationalbibliothek lists this publication in the Deutsche Nationalbibliografie; detailed bibliographic data is available in the internet at http://dnb.d-nb.de.

Library of Congress Cataloging-in-Publication Data
Names: Heusser, Peter, 1950- author.
Title: Anthroposophy and science : an introduction / Peter Heusser.
Other titles: Anthroposophische Medizin und Wissenschaft. English
Description: 1 [edition]. | New York : Peter Lang, 2016.
Identifiers: LCCN 2016014692 | ISBN 9783631672242
Subjects: LCSH: Anthroposophical therapy.
Classification: LCC RZ409.7 .H4813 2016 | DDC 615.5/3--dc23 LC record available at https://lccn.loc.gov/2016014692

Authorized translation of the first German language edition
Heusser, Anthroposophische Medizin und Wissenschaft
© 2011 by Schattauer GmbH, Stuttgart/Germany

ISBN 978-3-631-67224-2 (Print)
E-ISBN 978-3-653-06753-8 (E-Book)
DOI 10.3726/978-3-653-06753-8

© Peter Lang GmbH
Internationaler Verlag der Wissenschaften
Frankfurt am Main 2016
All rights reserved.
Peter Lang Edition is an Imprint of Peter Lang GmbH.

Peter Lang – Frankfurt am Main · Bern · Bruxelles · New York ·
Oxford · Warszawa · Wien

All parts of this publication are protected by copyright. Any utilisation outside the strict limits of the copyright law, without the permission of the publisher, is forbidden and liable to prosecution. This applies in particular to reproductions, translations, microfilming, and storage and processing in electronic retrieval systems.

This publication has been peer reviewed.

www.peterlang.com

Foreword

From the very first conversations with the author of this book, it soon became apparent that we shared a common attitude towards medicine: the struggle to understand and put into practice an integration of all effective measures when caring for people as individuals in both health and illness. In this book, Peter Heusser has made the unusual attempt to create the foundation for a complete scientific and empirical integration of the different medical approaches. The result goes far beyond the customary confrontation of the natural scientific and spiritually oriented ways of looking at medicine and, with his evidence-based logic, provides inspiration for a completely new insight into human nature for a medicine of the future. The rapid and skilful change of perspective which runs throughout the book testifies to a rare tolerance of apparently opposing ways of thinking and to a medical attitude which can only develop from a devoted involvement with patients. Through the view of the human being developed here the ancient but often paternalistically compromised medical attitude of "voluntas aegroti suprema lex" seems to me to be given new meaning and a new justification. But what fascinates me the most is that Peter Heusser's reflections encourage the demand for a scientific re-evaluation using suitable methods, methods which cover a massive span of the whole spectrum from atomic physics via biological diagnostic techniques in molecular medicine all the way to qualitative research in the arts and social sciences and finally to medical research in the healthcare sector. While working through the book the reader develops a multitude of hypotheses, making this book truly innovative and integrative in the real sense of the word. I hope that all those involved in healthcare provision will read this book, so that an integrative way of thinking in medicine can help to strengthen the practice of individualised medicine.

Witten, October 2010

    Univ.-Prof. Dr. med. Eckhart G. Hahn, MME (UniBe)
    Dean of the Faculty of Health
    Witten/Herdecke University

Preface to the first edition

In the biological and psychological sciences and in human medicine in particular, the time has come to think in a new and serious way about the nature of the human being. The successful explanation of the physical-material *conditions* for life, the soul and the spirit in human beings, has led to the belief that life, soul and spirit are *caused* by matter, lacking a reality of their own. The scientific view has indeed focused very successfully but very one-sidedly on the material level. However, this has led to the progressive loss of what is essentially human in the human being. For at the level of molecules, the difference between man, animal and other living creatures becomes blurred. And living creatures even share the elements of these molecules with the non-living world of matter. Where is the uniquely human now to be found? It cannot be found in the realm of substance, leading John Martin to state in 2000: "Perhaps the great problem of the next 100 years in biology will be *to understand what makes a human being a human being*" (Martin, 2000, 934).

In this book I have attempted to make a contribution to solving this problem and to show that it is possible to have a scientific view of mankind which not only recognises the reality of the material side of the human being, but the living, emotional and spiritual sides as well. This work was produced in 2009 under its German title *"Anthroposophische Medizin und Wissenschaft. Erkenntniswissenschaftliche und konzeptionelle Beiträge zu einer ganzheitlichen medizinischen Anthropologie"* as a habilitation thesis at the University of Witten/Herdecke. While engaged on its creation I received many valuable suggestions and support from various sides.

I wish to thank my medical colleagues Prof. Peter Matthiessen, Prof. Arndt Büssing and Dr. Friedrich Edelhäuser for constructive criticism, the development biologists Prof. Wolfgang Schad and PD Dr. Bernd Rosslenbroich, the geneticist Dr. Johannes Wirz, the physicist Dr. Stephan Baumgartner and the mathematician and epistemologist Dr. Renatus Ziegler. My thanks are also due to the Software AG foundation for the award of a habilitation grant and the Asta Blumfeld Stiftung, the Dr. Hauschka Stiftung, the Mahle Stiftung, the Iona Stichting, the Stichting Triodos, Weleda AG and the Gerhard Kienle Stiftung for contributions to the printing costs without which this book could not have been published. I am also very grateful to the Schattauer Verlag for being immediately willing to publish this work and in particular to the editor Marie Teltscher for her reliable work and her quick and friendly assistance.

My thanks go also to my secretary and assistant Marina Frieben for technical help and to my dear wife Ursula Heusser for her help in tracking down errors of style and typing mistakes and for her patience and constant support during the months when this work was in progress, in addition to many other professional duties. I therefore dedicate this book to her with my love.

Witten/Herdecke, September 2010

Peter Heusser

Preface to the 2016 English edition

This book is the updated and slightly expanded English version of my book which was published in 2011 by Schattauer Verlag, Stuttgart, under the title *"Anthroposophische Medizin und Wissenschaft. Beiträge zu einer integrativen medizinischen Anthropologie"*. This was my habilitation thesis presented to the University of Witten/Herdecke in 2009. The aim of this thesis was – and still is – to present a scientific explanation of the view of the human being underlying anthroposophical medicine, a view which is also capable of providing the conceptional basis for modern integrative medicine. The book starts with the epistemological explanation of science and the concept of reality developed by Rudolf Steiner (1861–1925) in his writings on the theory of knowledge and applied to both natural and spiritual science. This is followed by the systematic application of this concept of science and reality first to the understanding of substance in modern physics and chemistry, then to present-day issues in genetics, molecular biology and morphogenesis, to neurobiology, the mind-body problem, psychology, the philosophy of mind and also to the question of human free will. This results in a modern, scientifically founded holistic understanding of the human being which is capable of overcoming the reductionistic naturalism in these sciences and in medicine. The human being ceases to be viewed mechanistically as a molecular-genetic and neurobiological machine but instead as a differentiated being composed of body, life, soul and spirit. This scientifically founded understanding of the human being corresponds fully with that of anthroposophy. However, when reading this book it must be borne in mind that the anthropological medical concepts developed in it will only be comprehensible when their epistemological basis has been understood. For this reason, Chapter 2 is a prerequisite for the further chapters. This book also shows how the concept of anthroposophy already existed in the European history of science in the early and mid-19th century and how Rudolf Steiner then laid the methodological and epistemological basis for modern anthroposophy as spiritual science at the turn of the 20th century. He went on to develop the subject matter of this modern anthroposophy and, in cooperation with various experts, to take it all the way to practical applications in various cultural realms including medicine. Particular importance is attached to showing how anthroposophical spiritual scientific concepts and methods relate to the scientific research in modern medicine and how far research in this area has progressed and how much still needs to be done.

The general scientific principles of epistemology, physics, chemistry, genetics, morphogenesis, biology, neurobiology, psychology, philosophy of mind, anthropology, and anthroposophy presented in this book are developed in such a way that they are valid outside the field of medicine. They are then applied to medicine, simply because this is my own specialism. But they could just as well be applied to pedagogy, curative education and other specialist areas which deal with the human being as a whole. In fact it was primarily specialists from the pedagogical field who reacted with particular enthusiasm to the first edition of this book. Basically this book first provides an introduction to anthroposophy and its relationship to the sciences mentioned above, and then goes on to relate this to a practical area of application using medicine as an example. This updated English edition is therefore entitled: "Anthroposophy and Science. An Introduction".

In conclusion I should like to extend my heartfelt thanks to the translator of this book, Lynda Hepburn BSc MSc MITI, Edinburgh, for the energy, patience, care and reliability which she has brought to this translation. She has achieved a clarity and familiarity with the concentrated line of thought of the German original and has succeeded in retaining the often complex philosophical and scientific subject matter in her rendering into the English idiom, something that is no easy undertaking. I should also like to thank Lydia Garnitschnig, Vienna, for her very careful proofreading, Erich Colsman and the Barthels Feldhoff Stiftung as well as Ursula Piffaretti and the Stiftung zur Förderung anthroposophischer Institutionen for funding the printing costs, and Mrs. Marion Ulrich of Schattauer Verlag, Stuttgart, as well as Dr. Benjamin Kloss of the Peter Lang International Academic Publishing Group, Frankfurt am Main, for their willingness to print the book under licence at Peter Lang, enabling its distribution in the international English-speaking world. My thanks also go to Dr. Jürg Lehmann, Arlesheim, Basel, who enabled the funding of the translation in memory of the philosopher Werner A. Moser (1924–2003), Basel. To Werner Moser I owe my first and well-grounded introduction to the fundamentals of anthroposophical science 40 years ago, something which became the foundation of my later medical scientific career and therefore also of this book. I dedicate the book to him, in grateful memory.

Herdecke, January 2016

>Univ.-Prof. Dr. med. Peter Heusser, MME (UniBe)
>Gerhard Kienle Chair for Theory of Medicine, Integrative and Anthroposophical Medicine,
>Institute of Integrative Medicine
>Faculty of Health
>Witten/Herdecke University
>Germany

Translator's Note

German and English, like any pair of languages, do not have exact equivalents for a range of words, so in translation a solution has to be found to try to convey the author's meaning. Sometimes this is because the two terms do not have the same scope, one being a broader term while its possible translations are more specific and narrow. In this book, this applies particularly to the German words "Seele" and "Geist". While the English "soul" and "spirit" may seem obvious equivalents, this is not quite true and is certainly not the case for all instances of their use. In addition, the adjectives "seelisch" and "geistig" cannot be rendered into English by any simple single term.

In general, the German "Seele" has a much broader meaning than the English "soul" and is used far more frequently, including in technical contexts. In translating it here, I have sometimes used simply "soul" as for the comparison "Leib-Seele-Verhältnis = relationship between the body and soul", but references to "Seelenleben" have sometimes been rendered as "the emotional life" while for the "denkende Seele des Menschen" I have opted for the "thinking human mind" which sounds more acceptable than the "thinking soul". In yet other instances "Seele" is better translated by "psyche".

The adjective "seelisch" designates the whole array of soul or consciousness functions, not merely emotions, and different solutions have been used in different contexts.

While "Geist" has been translated as "spirit" in the majority of instances, this was not possible in some contexts. "Geist-Gehirn-Theorien" is better rendered as "mind/brain theories" and in yet other instances "intellect" may be nearer to the intended meaning. "Spirit" can easily be misunderstood (as can soul), though this book soon gives clear definitions of both to avoid misunderstandings.

While these two nouns and their adjectives constitute the most frequent difficulties, many other terms which occur less frequently in the text also present similar or related problems. It is worth bearing this in mind and, rather than always assuming the standard meaning of any term, paying careful attention to how it is applied in the context of the material in this book.

As I am obviously not the first translator to be faced with these issues, the readership is referred to two helpful discussions on the translation of anthroposophical terminology, one by Michael Wilson in *"The Philosophy of Freedom"* (Steiner, 2011, xiii–xxvii) and the final paragraph of Owen Barfield's introduction to his translation of Steiner's *"Von Seelenrätseln"*, *"The Case of Anthroposophy"* (Steiner, 1978a).

Edinburgh, January 2016

Lynda Hepburn BSc MSc MITI

Contents

1 Introduction 17
    1.1   Summary 24

2 Epistemology: Knowledge and truth 25
    2.1   The theory of knowledge or epistemology: knowledge of knowing 25
    2.2   Experience 30
    2.3   Thinking as a fact of experience 36
    2.4   Objective empirical knowledge 41
    2.5   Science and reality 52
    2.6   Objective ontological idealism 56
    2.7   Summary 60

3 Science and ontological idealism in physics and chemistry 61
    3.1   Empirical ontological idealism instead of reductionism 61
    3.2   Spirit in matter 78
    3.3   Emergence, self-organisation and causality in physics, chemistry and biochemistry 80
    3.4   Summary 96

4 Ontological idealism in biology 99
    4.1   Chemical explanation of life? Genes, genetic information and proteins 99
    4.2   Gene regulation. From a static to a dynamic concept of the gene 105
    4.3   Self-organisation and causality in biology and Goethe's archetype 109
    4.4   Organism or mechanism? 115
    4.5   Typus and morphogenesis 119

	4.6	Organic versus inorganic cognition 122
	4.7	Morphogenetic fields or morphogenetic substances? 129
	4.8	Causality and systems biology 141
	4.9	Biology beyond vitalism and mechanism 147
	4.10	Reflections on the thermodynamics of organic processes 150
	4.11	Life versus death: physical and etheric organisation 153
	4.12	Summary 161
5	**Neurobiology, psychology and philosophy of mind: the reality of the soul and spirit** 163	
	5.1	Neurobiology and the emergence of consciousness. Is the soul real? 163
	5.2	The problem of ontological monism and psychophysical causation 169
	5.3	Consciousness versus life: organic and psychological activity 172
	5.4	The soul as a formative principle. The "astral" organisation 177
	5.5	Spirit versus soul: the emergence of self-consciousness and self-determination 183
	5.6	Soul and spirit: Intelligence in animals and humans 189
	5.7	The importance of the spiritual factor in health and medical care 197
	5.8	The question of freedom 202
	5.9	A comprehensive basis for medical anthropology: body, life, soul and spirit 221
	5.10	Summary 228
6	**From anthropology to anthroposophy** 231	
	6.1	The question of the reality and cognition of the spiritual 231
	6.2	The anthropology and anthroposophy of I. P. V. Troxler 240

	6.3	The anthropology and anthroposophy of I. H. Fichte 243
	6.4	Summary..249

7 Anthroposophy as an empirical spiritual science 251
    7.1   The limits to knowledge and their transcendence 251

    7.2   Rudolf Steiner's empirical spiritual science 258

    7.3   Summary..269

8 Anthroposophical spiritual science and natural scientific medicine 271
    8.1   The fourfold image of man as a basis for medical anthropology, nosology and therapy...271

    8.2   Scientific examination of anthroposophical concepts and medical rationale...276

    8.3   Anthroposophical medicine and modern scientific medical research..281

    8.4   Concerning the extension of medical anthropology through anthroposophy 287

    8.5   The status of clinical scientific research in anthroposophical medicine 296

    8.6   Summary..315

Bibliography 319

1 Introduction

The international and increasingly expressed public desire for holistic or complementary methods in medicine – which includes anthroposophical medicine – is substantiated by numerous studies and is generally accepted as fact. There is a large overall number of different complementary medical methods on offer and a considerable volume of complementary medical preparations are sold, something which is also of economic importance. There is therefore a need for evidence-based scientific validation of the safety and efficacy of complementary methods. An increasing number of scientists and centres, including those in academic institutions, are currently working on this. As little as 20 years ago complementary medicine was mainly a matter of alternative practitioners and practising doctors; there were only a few inadequately trained researchers, only a few – almost all small – clinical facilities for complementary medicine (as is still the case!) and, in contrast to orthodox medical research which is largely supported by a massive pharmaceutical industry, funding by the manufacturers of complementary medicines was and is insignificant. The scientific evaluation of complementary medicine was (and is) often correspondingly inadequate, which continues to be a frequent cause of conflict between the different approaches.

However, this situation seems to be gradually improving. State research funding has contributed to systematic evaluations, such as the German Federal Ministry of Education and Research project "Unconventional Medical Therapies" (Matthiessen et al., 1992), the Swiss National Science Foundation's "National Research Project 34" (NFP 34) (Baumann & von Berlepsch, 1999), the Swiss national Complementary Medicine Evaluation Programme (PEK) (Melchart et al., 2005), and the numerous grants from the US American National Center for Alternative and Complementary Medicine (NCCAM)[1] at the National Institute of Health (http://nccam.nih.gov/) and the Office of Cancer Complementary and Alternative Medicine (OCCAM) at the National Cancer Institute (www.cancer.gov/CAM/). Chairs or professorships in complementary medicine have been established at universities, in German-speaking Europe for example in Witten/Herdecke, Berlin, Freiburg, Zurich and Bern[2]; and an increasing number of international peer-reviewed journals have appeared. All this has considerably improved the capacity and quality

---

1. In 2015 the name of the NCCAM was changed to "National Center for Complementary and Integrative Health" (NCCIH) (https://nccih.nih.gov. Viewed 03/12/2015).
2. For example, three professorships for anthroposophical medicine have been established at European Universities: in 2009 at Witten/Herdecke University (Chair

of research in complementary medicine, and it is clear that complementary medicine has learned a great deal from traditional conventional medicine in this process.

This change is also reflected in the nomenclature. While 20 years ago the preferred term was "alternative medicine", 10 years ago the more cooperative term "complementary medicine" was adopted and now the even more unifying expression "integrative medicine" is used. In the USA in particular, leading universities such as Harvard, Stanford, John Hopkins, the Universities of California, Texas and Michigan and others have quickly set up centres for "integrative medicine" and joined forces in the "Consortium of Academic Health Centers for Integrative Medicine" (CAHCIM)[3] which includes more than 50 centres. This consortium defines integrative medicine as follows: "Integrative Medicine is the practice of medicine that reaffirms the importance of the relationship between practitioner and patient, focuses on the whole person, is informed by evidence, and makes use of all appropriate therapeutic approaches, healthcare professionals and disciplines to achieve optimal health and healing" (www.imconsortium.org)[4]. A few years ago this trend also started in Europe with congresses (www.ecim-congress.org) and, since July 2009, an "Institute for Integrative Medicine" has been built up at Witten/Herdecke University – the first of its kind amongst German-speaking universities.

This has brought research on complementary or integrative medicine to a new and more profound stage. Up until now the focus was mainly a practical one on the scientific proof of safety and efficacy, to a certain extent irrespective of the nature of the various methods and their underlying concepts. The relevant research methods have been established and the efficacy and safety of many methods are gradually becoming better defined. However, the result is an "integration" of conventional and complementary medical methods and concepts, which are often very different from one another and indeed seem logically to be contradictory. While in conventional medicine, molecular biological events are viewed as the ultimate cause of physiological

---

for Theory of Medicine, Integrative and Anthroposophical Medicine), 2014 at the University of Bern and 2015 at the Charité Medical University in Berlin.

3 In 2015 the name of the Consortium was changed to "Academic Consortium for Integrative Medicine & Health" (https://www.imconsortium.org/. Viewed 03/12/2015).

4 The new 2015 definition reads: "Integrative medicine and health reaffirms the importance of the relationship between practitioner and patient, focuses on the whole person, is informed by evidence and makes use of all appropriate therapeutic and lifestyle approaches, healthcare professionals and disciplines to achieve optimal health and healing" (https://www.imconsortium.org/about/about-us.cfm. Viewed 03/12/2015).

and pathological occurrences in life and consciousness, in complementary medicine there are explanations for the same occurrences which are attributed to *non-material* causal factors, such as "chi" forces in Chinese medicine, "prana" in Ayurvedic medicine, "life force" in homoeopathy, "etheric" and "astral" forces in anthroposophical medicine, "information" in neural therapy and various "energetic" factors in other systems. And the therapeutic methods used correspond to these: tai chi is intended to harmonise the chi forces, curative eurythmy to work on the etheric body, homoeopathically potentised substances – which, at a dilution beyond the Avogadro limit, no longer contain any molecules – to stimulate life forces, etc. So conventional and complementary methods are used alongside each other without a real mutual understanding. This is more like an "aggregation" than an "integration".

A true integration going beyond the present-day situation would mean that these kind of therapies would not only be evaluated at conventional centres and accepted into mainstream medicine if they pass the "test", but that their nature and the concepts on which they are based would need to be understood, and broader concepts developed which include the conventional *and* the complementary, the material *and* the immaterial in one united perspective.

It cannot merely be assumed that the *conceptual integration* will consist of one day being able to abandon the complementary medical concepts in favour of the conventional ones and of explaining the corresponding therapies as being caused by molecular biological interactions. This is indeed attempted at times, but basically without success, because explanations like these are just as much of an evasion as the ones which explain the phenomenon of consciousness from the functioning of the brain. Consciousness with its *real experience of emotional and mental qualities* is not "explained" by getting to the bottom of the physical processes in the brain which are necessary for the *occurrence* of those emotional qualities. The reductionist thinking habit of the last 200 years has led conceptually to a type of negation of those experiences of a non-molecular, indeed non-material kind which every person has on a daily basis and which form a significant part of their *humanity*, that is, their emotional and intellectual life, achievement and suffering. However it is these reductionist thinking habits which significantly influence medical theory and ultimately medical practice. And this is at least partly responsible for the frequently voiced dissatisfaction of patients with conventional medicine, despite all the credit which is gladly accorded to it, precisely *because of* its ability to explain the physical basis of the human organism and to use this for the benefit of the patient. A representative study in 2002 by the Swiss Academy of Medical Sciences is an example which shows clearly where the deficits of conventional medicine were perceived to be: 69% of Swiss wanted *more humanity* in medicine, 58% wanted more alternative medicine but only 27% wanted more basic care and 21% more top level medicine; and in a direct comparison with the present-day situation, a more holistic view of the patient is seen as one of the most important requirements by far for the future (Leuenberger & Longchamp, 2002). Patients look to complementary medicine for what they miss in conventional medicine (Heusser, 2002a). It cannot be ruled out a priori that there may "be something" in the complementary concept, that human beings may in fact be more than their physical constitution, as patients in fact feel to be the case, and that in this regard conventional medicine can also learn something from complementary medicine. But this would only seem to be possible if a – so far non-existent – scientific debate at the level of anthropological medical concepts and their epistemological basis were to be held between orthodox and complementary medicine which would approach what already exists at the level of clinical studies. Only then would there be a prospect of the true integration which seems necessary if medicine is ultimately to become *unified*.

Although *anthroposophical medicine* is the most recent of what are referred to as complementary medical systems, it appears to offer a unique starting point for a debate of this kind. In its roughly 90 years of existence, it has achieved a high degree of integration between conventional and anthroposophical elements in both theory and practice, something which does not apply to any other complementary medical discipline. This integration is already present in its most fundamental basic principle, as expressed by Otto Wolff et al. (Wolff et al., 1990):

> Anthroposophical medicine is the spiritual scientific extension of natural scientific medicine. In judging health, illness and healing it relies on the physical laws which are determined by the natural sciences and accords equal value to the laws of life, soul and spirit in their mutual dependencies.

This basic principle is fulfilled through the fact that anthroposophical medicine requires a normal course of medical education at a university medical school followed by further general medical or specialist training and that the anthroposophical aspects are acquired through a regulated further training which, e.g. in Switzerland, leads to a recognised certificate of competence from the Swiss Medical Association FMH[5]. In addition, anthroposophical doctors and hospitals participate in the general hospital provision of their regions, and the services they provide in this framework are covered by the health insurance schemes in the usual way. Anthroposophical hospitals are also acknowledged as training centres for standard postgraduate medical training in various medical disciplines and in some cases in the training of medical students. Another aspect of this integration is that anthroposophical doctors' practices and hospitals are set up diagnostically and therapeutically

---

5   FMH: Foederatio Medicorum Helveticorum.

basically just like the purely mainstream ones and that clinical examination, laboratory tests, radiological diagnoses and any necessary conventional treatments are carried out just as normal. The difference from purely conventional medicine lies in the fact that the findings and symptoms in question are not interpreted purely in accordance with natural science and psychology in the customary manner of mainstream medical anthropology, but against the background of a scientifically expanded picture of the human being comprising body, life, soul and spirit. The medical history and diagnostic assessment may therefore contain additional elements and extra treatment methods may be used in accordance with this.

An additional factor is that anthroposophy sees itself as an empirical spiritual *science* which is similar to natural science in its own domain and which transfers the epistemological principle of natural science to the realm of spiritual perception. Unlike any of the other complementary medical disciplines, it has its roots in recent developments in European science, and applies itself to the active development of this on a spiritual level.

For these reasons anthroposophical medicine appears to be particularly suited to making a contribution to a scientific debate between orthodox and complementary medicine which, in the longer term, can aim towards the goal of making *medicine per se* into a holistic and, in this sense, conceptually integrative discipline.

The aim of this book is to make a contribution towards this goal. Using an epistemological basis following Steiner and Goethe, I shall first develop a scientific concept of substance which enables the matter in physics, chemistry and biochemistry to be thought of in such a way that it is not at odds with the spiritual but in a certain respect already contains it. Based on this I shall discuss fundamental biological concepts such as genes, genetic information, gene regulation, organic autoregulation, morphogenesis and biological systems so that it is possible to see the connection between modern biology, Goethe's concept of the typus and Steiner's concept of the etheric, and to distinguish between empirically justified and purely hypothetical organic concepts of a vitalistic nature. This will lead to an investigation of consciousness, the relationship between body and soul, the issue of psychophysical causation and finally the question of freedom of the human spirit. Amongst other things I shall explain why the human ability for freedom is not questioned despite the much-discussed readiness potential as defined by Benjamin Libet.

Working from this basis I shall establish a fourfold conception of the human being based on the conventional sciences, a conception in which the physical body, its specifically living aspect, the soul and the spirit will be defined by each of the different emergent laws and forces. This leads to an expansion of medical anthropology in terms of a differentiated integrative conception of the human being, corresponding to that of anthroposophy.

In what follows, anthroposophy will be discussed as *empirical spiritual science*. I shall briefly describe how this spiritual scientific anthroposophy appeared as a result of the scientific development in Central Europe and how it was then founded by Steiner. I shall go on to mention its application to medical anthropology and explain how the anthroposophical medical concepts which have arisen from this can be logically connected to modern scientific concepts, on the one hand to enable a holistic rational medicine which includes natural *and* spiritual aspects and on the other to allow – or rather demand – conventional empirical scientific testing of anthroposophical concepts. Finally, on this foundation, I will briefly review the status of the anthroposophically expanded medical anthropology which has developed since Steiner and the body of evidence related to the efficacy and safety of anthroposophical treatments as well as the methodological problems arising from them.

The basis of this entire undertaking is provided by the scientific foundation of anthroposophy, as laid down in Goethe's scientific method and Steiner's basic epistemological writings. As these are almost unknown nowadays and also completely unaccepted – or leastways ignored – by critics of anthroposophical medicine such as Franz Stratmann (Stratmann, 1988), Klaus Dietrich Bock (Bock, 1993), Thomas Dinger (Dinger, 1996), Robert Jütte (Jütte, 1996), Barbara Burkhard (Burkhard, 2000) and Helmut Zander (Zander, 2007), some of whom vehemently dispute the scientific method claimed by Steiner[6], in Chapter 2 I have quoted Steiner himself at length on the concepts of knowledge and reality. It is shown how this results in an empirical ontological objective idealism which does away with the need for reductionism and in its place recognises the phenomena and laws of each of the emerging levels of being in their own reality. The concept of reality and knowledge inherent in Goethe's and Steiner's works thus becomes a universal element which on the one hand distinguishes all the differing areas of being, such as matter, life, soul and spirit from one another and on the other combines them in a unified overall scientific view.

It is unavoidable that elements from general epistemological and ontological principles are sometimes repeated in the specialist chapters and are applied in a similar way using examples from the different sciences. This should serve

---

6 The same applies to Ernst & Schmacke (2015). Cf. also other commentaries on these critics in Sections 8.2, 8.3 and 8.5 as well as in the circumstantial treatise *"Anthroposophische Medizin und ihr integratives Paradigma"* by Michaela Glöckler, Matthias Girke und Harald Matthes (Glöckler et al., 2011) which was published after this book but also describes scientific fundamentals of the integrative concept of anthroposophical medicine.

to increase the internal coherence of the whole and facilitate the assimilation of potentially new points of view by the reader.

I have ventured this attempt to outline an epistemologically well-founded *overall scientific view* of differing scientific fields because the general lack of an integrative overall view is one of the main characteristics of our era of highly specialised individual sciences, not only to the benefit of our patients and our medical practices but also to their detriment. It will be shown that an overall view of this kind is fundamentally possible. This will provide a contribution to the scientific development of an integrative and therefore holistic medical anthropology which has become necessary in our time and which, according to the warning given by Gerhard Kienle (1923–1983), will need to be not merely the familiar natural scientific knowledge but, without doubt, "also a knowledge which encompasses the individuality of the human being" (Kienle, 1980).

What *cannot* be achieved within the bounds of this work is a *systematic* discussion of Steiner's theory of knowledge and the concept of reality based on it in relationship to the more recent scientific theories. With reference to this I would direct you to Helmut Kiene's *"Grundlinien einer essentialen Wissenschaftstheorie"* and the section on epistemology in Peter Schneider's *"Einführung in die Waldorfpädagogik"* (Kiene, 1984; Schneider, 1985)[7]. There is also no discussion of these publications by the critics mentioned above which is all the more surprising as their dispute with Steiner and anthroposophical medicine is primarily based on the question of scientific method.

This work is also unable to offer a thorough and systematic discussion of other major topics in relation to the prevailing current debate. This refers particularly to the topics of emergence, the concept of substance, biological concepts, the body/soul relationship, questions about the human spirit and freedom, the history of science and philosophy and the methodology of clinical studies. This would far exceed the limits of this book and each topic could easily form the subject of a comprehensive monograph. This could be seen as a weakness of this study, as in many instances the topic is not dealt with fully and completely and may therefore not fulfil some readers' need for discussion. Given the present-day status of a range of discussions in each of these fields, this is scarcely attainable in a book by a single author.

Starting from a sound epistemological foundation based on observation instead of mere theory, the aim here is to develop an internally consistent overall view of the different realms of relevance to medical anthropology, such as the concept of substance, biology, consciousness and the body/soul relationship, and spirit and the possibility of freedom. In addition, the author

---

[7] Addendum 2015: Jaap Sijmons' „*Phänomenologie und Idealismus. Struktur und Methode der Philosophie Rudolf Steiners*" (Sijmons, 2009).

aims to show how the further development of biological and psychological anthropology is possible in a spiritual scientific anthroposophy which complements anthropology, and how anthroposophical concepts in medicine can be verified and developed for medical practice through empirical scientific work. This also includes examples from the author's own experimental and clinical research.

At the same time, this is the first work in the medical field to attempt this kind of comprehensive anthropological/anthroposophical overall view on an epistemological basis which is internally logically consistent and in agreement with the empirical facts. It will provide an example of the way in which a conceptual integration between conventional and complementary medicine could be envisaged. This treatise also serves as a practical basis for the author's work as the holder of the Gerhard Kienle Chair of Medical Theory, Integrative and Anthroposophical Medicine at Witten/Herdecke University.

Note on the text: square brackets in quotations are additions or omissions by the author. Emphases by the author in text citations are marked with "ePH" (emphasis added PH) and footnotes in citations are those of the author.

1.1 Summary

A growing trend towards an "integrative medicine" is currently discernable. Up until now this integration has generally consisted of the methods used in complementary medicine being subject to practical testing in conventional clinical studies at academic centres and, if successful, being put into clinical practice. What is still lacking is a *conceptual integration* in which what at first appear to be the incompatible molecular biological causative explanations in orthodox medicine and the non-material/energetic explanations or methods in complementary medicine are brought onto a common basis of mutual understanding. Anthroposophical medicine provides particular opportunities for this kind of basis of understanding because, due to its very nature, it is based on an integration of natural and spiritual scientific elements in its development, theory and practice, and because, according to its view of itself as spiritual science, it corresponds to a continuation of the natural scientific principle of knowledge at a spiritual level of observation. Both natural and spiritual science are based on a common theory of knowledge.

2 Epistemology: Knowledge and truth

2.1 The theory of knowledge or epistemology: knowledge of knowing

In his first epistemological book, the *"Grundlinien einer Erkenntnistheorie der Goetheschen Weltanschauung"* from 1886 (*"The Science of Knowing, Outline of an Epistemology Implicit in the Goethean World View"*), following on from Goethe Steiner characterises the task of every science as follows (Steiner, 1988a, 18):

> It is ultimately true for all science what Goethe so aptly expressed with the words: 'In and for itself, theory is worth nothing, except insofar as it makes us believe in the interconnections of phenomena.' Through science we are always bringing separate facts of our experience into a connection with each other. In inorganic nature we see causes and effects as separate from each other, and we seek their connection in the appropriate sciences. In the organic world we perceive species and genera of organisms and try to determine their mutual relationships. In history we are confronted with the individual cultural epochs of humanity; we try to recognize the inner dependency of one stage of development upon the other. Thus each science has to work within a particular domain of phenomena in the sense of the Goethean principle articulated above. Each science has its own area in which it seeks the interconnections of phenomena.

This points to the fact that scientific knowledge always requires two elements: first, empirically given perceptual *phenomena* and second, the *theory*, i.e. thoughts, ideas, concepts which are used to explain the interconnection of the phenomena. While in the epistemological debate prior to Kant or Goethe, "rationalism" and "empiricism" disputed whether knowledge is derived from empiricism or reason, Immanuel Kant, the actual founder of the modern theory of knowledge, brought about a synthesis of these positions by pointing out (Schneider, 1998) that, "the cognition of the reality of things" requires *both*: the "perception" on the one hand and the "concept" on the other, as well as their "connection" in the form of a judgement (Kant, 1926, 266–273). Knowledge of things can be obtained neither from the pure experience, as postulated by empiricism, nor from the concept alone, but requires the synthesis of both. The concept is "ascribed as a predicate" to the perception in the cognitive judgement (Kant, 1926, 266). Karl Popper, who is still accorded considerable importance in the modern theory of knowledge, also views cognition as the relationship of our "theories" to the objects of "experience" gained through observation (Popper, 1976) and the same generally applies to the epistemology of the 20th century (Schneider, 1998), for example for

Donald Davison, when he mentions "our perceptions and thoughts" as the two elements required for cognition (Davidson, 1993).

Cognition accordingly appears as a product of the human subject which requires two interrelated elements: percept and concept. This is independent of the way in which these far from compatible views of knowledge from 200 years ago define cognition as such, and what relationship they assign its two elements – both of which are found in the subject – to an "objective reality". Whether, in the percept and concept, the subject remains within himself or whether these elements actually provide access to truth and to the things "in themselves" is irrelevant to the fact that percept and concept are both equally necessary for cognition.

From a scientific point of view, what cognition is and what relationship it or its elements have to reality can naturally not be determined a priori or arbitrarily, but must itself be the result of scientific research. In other words: to answer these questions, cognition needs to be directed *at the process of knowing itself*. This is Steiner's view which he presents in his basic writings, and this, as *knowledge of knowing*, is how he interprets epistemology or the theory of knowledge[8] (Steiner, 1988a, 18ff.).

For Steiner the science of knowledge is therefore the *fundamental science of all sciences*, because it takes as its object the instrument of science, cognition as such. Steiner built his natural and spiritual scientific views logically on the science of knowledge[9], first in his description of and following on from Goethe's scientific method in the above-mentioned *"Science of Knowing"*, and then from his own standpoint from a purely philosophical and epistemological angle in his doctoral thesis which he subsequently published in an expanded form in 1892 as *"Wahrheit und Wissenschaft. Vorspiel einer Philosophie der Freiheit"* (Steiner, 1980a) (translated as *"Truth and Knowledge"*, 1981a), and in 1894 with a greater emphasis on the psychology of cognition in his main philosophical work *"Die Philosophie der Freiheit; Grundzüge einer modernen Weltanschauung. Seelische Beobachtungsresultate nach naturwissenschaftlicher*

---

8 *Theory* of knowledge (or epistemology) here and as used by Steiner is synonymous with *science* of knowledge (Steiner, 1988a).

9 This has been convincingly presented in particular by Peter Schneider in his habilitation thesis for the chair of vocational pedagogy at the University of Paderborn, published as *"Einführung in die Waldorfpädagogik"* (Schneider, 1985). The general part of his presentation of the epistemological and methodological basis of anthroposophy as the foundation for Steiner/Waldorf educational theory can also be applied to the foundation of other applications of anthroposophy in practical areas of life such as anthroposophical medicine. The same is true of Renatus Ziegler's treatise on *"Erkenntniswissenschaft als Grundlage von Natur- und Geisteswissenschaft"* (*epistemology as a basis of natural science and spiritual science*) (Ziegler, 2014).

Methode" (Steiner, 1978b) (translated as *"The Philosophy of Freedom. The Basis for a Modern World Conception"*) (Steiner, 2011).

According to Steiner's view, the theory of knowledge can only turn science into a *critical* one by the knowledge of knowing bringing its inherent laws to light. Until then cognition remains unreflecting or "naïve" in the original, i.e. non pejorative sense of the word: while it is put into practice, it is still unaware of what the rules which govern this activity are (Steiner, 1981a, 48–49):

> In all our activities, two things must be taken into account: the activity itself, and our knowledge of its laws. We may be completely absorbed in the activity without worrying about its laws. The artist is in this position when he does not reflect about the laws according to which he creates, but *applies* them, using feeling and sensitivity. We may call him *naive*. It is possible, however, to observe oneself, and enquire into the laws inherent in one's own activity, thus abandoning the naive consciousness just described through knowing exactly the scope of and justification for what one does. This I shall call *critical*. I believe this definition comes nearest to the meaning of this concept as it has been used in philosophy, with greater or lesser clarity, ever since Kant. Critical reflection then is the opposite of the naive approach. A *critical* attitude is one that comes to grips with the laws of its own activity in order to discover their reliability and limits. Epistemology can only be a critical science. For its object is an essentially subjective activity of man: *cognition*, and it wishes to demonstrate the laws inherent in cognition. Thus everything "naive" must be excluded from this science. Its strength must lie in doing precisely what many thinkers, inclined more toward the practical doing of things, pride themselves that they have never done, namely, 'think about thinking'.

However, thinking about thinking or "knowledge of knowing" starts with *observation* of knowing, in fact with the observation of knowing as such, i.e. *our mental activity* and not with observation of the processes in the brain which are the material correlate of our mental activity. This has to be stressed because, under the influence of the reductionist mind/brain theories and the naturalistic and evolutionary theories of knowledge of the 20th century, the *actual*, i.e. *empirically apparent* difference between material states in the brain and the emotional and spiritual states of consciousness which correspond to them has become *ontologically* blurred, for the very reason that each state of consciousness is dependent on the state of the brain to which it corresponds. This is the view held by Gerhard Vollmer, a prime exponent of the evolutionary theory of knowledge in Germany (Vollmer, 1987, 90):

> These psychophysical findings suggest that every state of consciousness corresponds uniquely to one state of the brain or that there is actually only one state which is perceived differently – that is, psychologically or physiologically.

However, a correspondence or a causation or dependency does not equate with being ontically the same, and phenomenologically nothing remains but to observe the psychological or mental on the one hand and the physiological or physical on the other hand, initially *individually* in order to actually be able to recognise the nature of their *relationship* (cf. Chapter 5). In this respect Steiner's theory of knowledge is very radical, in contrast to many other epistemological movements, especially the newer ones.

In order to be able to think about thinking, it must be present as such, i.e. as a mental activity, as an object of thinking, in other words as an object of itself so to speak. This, however, presents a problem, as the *thinking* thinking is then no longer the same as the thinking or thought which is *to be* thought about. The view held by Steiner on this (Steiner, 2011, 34–35) is that:

> While I am thinking I pay no heed to my thinking, which is of my own making, but only to the *object* of my thinking, which is not of my making. [...] I can never observe my present thinking; I can only subsequently take my experiences of my thinking process as the object of fresh thinking. If I wanted to watch my present thinking, I should have to split myself into two persons, one to think, the other to observe this thinking. But this I cannot do. I can only accomplish it in two separate acts. The thinking to be observed is never that in which I am actually engaged, but another one.

Thinking or knowing can therefore only be made an object of knowledge after the event, i.e. in retrospect[10] when a later thought process is directed at already completed acts of thinking[11]. In this sense thinking is "the unobserved element

---

10 Only in this way, i.e. in reflecting on the "object" of the completed thinking process can e.g. the laws of formal logic first described by Aristotle be found, in other words the forms in which the thought processes take place in judgements and conclusions. In contrast to the science of logic which only concerns itself with the laws of *thinking*, the science of knowledge examines *cognition*, i.e. the activity which in the spirit of the above-cited statement by Goethe, creates the connection between the conceptual theory and the phenomena.

11 Here it might be objected that the later thinking which is actually taking place is in fact a different one from that completed earlier and observed in retrospect, so that the active thinking is always neglected. However, this does not apply to thinking itself, but only to the status of thinking. As a current activity this is, as it were, fluid or living, however as a completed activity it is fixed or "dead". But qualitatively it nevertheless remains thinking and follows its laws. And this is the point because: "[...] when I observe my own thinking, no such neglected element is present. For what now hovers in the background is once more just thinking itself. The object of observation is qualitatively identical with the activity directed upon it" (Steiner, 2011, 39). But how can we know that the thinking is actually a living thinking, when we can only observe it after the event, as its "corpse"? We know this because we have to produce it ourselves by our own activity if it is to be there in the first place. We experience it directly as a living activity in the process of creating it. "It is a percept in which the perceiver is himself active, and a self-activity which is at the same time perceived." (ibid., 216). This direct

in our ordinary mental and spiritual life". "This is just the peculiar nature of thinking, that the thinker forgets his thinking while actually engaged in it. What occupies his attention is not his thinking, but the object of his thinking, which he is observing" (Steiner, 2011, 34), and therefore, in comparison with the observation of other objects, the observation of thinking is "a kind of exceptional state" (ibid., 32).

This fact, that it is easy for cognition to remain the unacknowledged element of the knower and that the observation of thinking forms an exception to the normal state of consciousness, may be the reason why e.g. in a scientifically based training such as medicine, there is practically no mention of the science of knowledge[12]. The problem here is that cognition in the above-mentioned sense remains "naive" and that science is not actually a critical one, i.e. it cannot become conscious of its own principles and therefore runs the risk of not using these correctly.

For this reason Steiner places the science of knowledge at the start of all science which has to begin with the empirical observation of its activity just as natural science has to begin with the observation of empirically given facts. Steiner therefore intended his basic writings not as mere *theory* of knowledge but as a description of the concepts of intersubjective, testable inner empirical facts. This is also clear from the subtitle of *"The Philosophy of Freedom"*: "Some results of introspective observation following the methods of Natural Science". The discussion of the two elements of cognition, experience and thinking, which follows is also meant in this way.

---

perception or *experience* in which we are involved as a witness while we produce it actively ourselves must be distinguished from *observation*. Because observation is "contemplative comparison" (Steiner, 1978b, 43). And in the case of thinking this is only possible after it has been produced: "The reason why it is impossible to observe thinking in the actual moment of its occurrence is the very one which makes it possible for us to know it more immediately and more intimately than any other process in the world. Just because it is our own creation do we know the characteristic features of its course, the manner in which the process takes place" (Steiner, 2011, 35). However, "Kenntnis soll Erkenntnis warden" as Hegel said, which is probably best translated as "knowledge needs to become understood knowledge": by observing the thinking, that with which we are already familiar "naively" can be understood by means of critical consciousness.

12 This accords with the fact that, for example in an encyclopaedia written by experts in their field which is available throughout the world and constantly updated, in detailed sections such as "Cognition", "Theory of knowledge", "Epistemology", "Knowledge", "Science" and "Theory of science" no mention is made of the observation of thinking and cognition as a basis for the science of knowledge (http://en.wikipedia.org/wiki/. Viewed for these terms 16/12/2015).

2.2 Experience

Self-observation shows that cognition is sparked by given phenomena. Two things therefore need to be distinguished: the given phenomenon which stimulates the need for knowledge, and the knowledge itself. Steiner calls the objects "the content of experience to the extent that this is accessible to our observation" (Steiner, 1979a, 27). It should be stressed at this point that, following Volkelt, in *"The Science of Knowing"* Steiner understands *"experience"* to be the *percept* as also used in *"The Philosophy of Freedom"* and not, as the total extent of repeated earlier experiences, as could be implied by another usage of the word. Experience is therefore that which observation *finds* when cognition begins to be active (Steiner, 1988a, 20):

> What is experience? Everyone is conscious of the fact that his thinking is kindled in conflict with reality. The objects in space and in time approach us; we perceive a highly diversified outer world of manifold parts, and we experience a more or less richly developed inner world. The first form in which all this confronts us stands finished before us. We play no part in its coming about. Reality at first presents itself to our sensible and spiritual grasp as though springing from some beyond unknown to us. To begin with we can only let our gaze sweep across the manifoldness confronting us.
>
> This first activity of ours is grasping reality with our senses. We must hold onto what it thus presents us. For only this can be called pure experience.
>
> We feel the need right away to penetrate with organizing intellect the endless manifoldness of shapes, forces, colors, sounds, etc., that arises before us. We try to become clear about the mutual interdependencies of all the single entities confronting us. [...] But what results from such asking and seeking is no longer *pure experience*. It already has a twofold origin: experience and thinking.
>
> *Pure experience is the form of reality in which reality appears to us when we confront it to the complete exclusion of what we ourselves bring to it.*

What is meant here by "the complete exclusion of what we ourselves bring to it" – as is clear from the context – is the exclusion of the *thinking judging* self which, by producing connections, would like to introduce order, value and judgement into the given phenomena. Before exercising this thinking the experience is "pure" to the extent that it is still without the connection-forming concepts which result from critical thinking[13]. This does not exclude the possibility that this pure experience – in *epistemological* terms – is not at all pure in terms of the *psychology of cognition* and that, for example, it contains not only pure sense impressions but mental pictures or concepts which arise spontaneously along with the perception. In relation to the current

---

13 On the other hand "pure thinking" is unconnected to the world of phenomena, such as in pure mathematics or in pure logic of contents. Cf. Steiner (2011) or Hegel (1977).

process of cognition, even these mixed experiences – from a psychological perspective – are epistemologically still "pure" as long as they are not permeated by a cognitive *judgement*. This is what Steiner means. Pure experience in this sense is "the directly given" through which cognition first feels itself to be challenged (Steiner, 1981a, 52):

> *Before* our conceptual activity begins, the world-picture contains neither substance, quality nor cause and effect; distinctions between matter and spirit, body and soul, do not yet exist[14]. Furthermore, any other predicate must also be excluded from the world-picture at this stage. The picture can be considered neither as reality nor as appearance, neither subjective nor objective, neither as chance nor as necessity; whether it is "thing-in-itself", or a mere mental picture, cannot be decided at this stage.

For these kind of characterisations can only be made as the *result* of cognition[15].

For this reason, from an epistemological standpoint the world of percepts cannot be seen from the outset as a purely subjective construct of the human mode of perception arising solely in consciousness, as interpreted by the naturalistic theory of knowledge dominated by people such as Willard Quine in the 20th century and summarised by Norbert Schneider as follows: "That which is considered to be observation can now be explained by the stimulation of the sense receptors – wherever consciousness may be" (Schneider, 1998, 154). Steiner called this kind of approach "a preconception, existing since Kant" (Steiner, 1988a, 28–29):

> The preconception I mean is the view: It is already established from the very beginning that the whole world of perception, this endless manifoldness of colors and shapes, of sounds and warmth differentiations, etc., is nothing more than our subjective world of mental pictures (Vorstellungen), which exists only as long as we keep our senses open to what works in upon them from a world unknown to us. This view declares the entire world of phenomena to be a mental picture *inside* our individual consciousness, and on the foundation of this presupposition one then erects further assertions about the nature of our activity

---

14 What is meant is: *as the content of cognition* – not, for example, in terms of existence – "not yet created", as will appear in what follows.
15 Interestingly, the founder of modern Japanese philosophy, Kitaro Nishida (1870–1945), also acknowledges this definition of pure experience: "By *pure* I am referring to the state of experience just as it is without the least addition of deliberative discrimination. [...] The moment of seeing a color or hearing a sound, for example, is prior not only to the thought that the color or sound is the activity of an external object or that one is sensing it, but also to the judgment of what the color or sound might be. In this regard *pure* experience is identical with direct experience. When one directly experiences one's own state of consciousness, there is not yet a subject or an object" (Nishida, 1992, 3–4). Nishida had studied German idealism; the author does not know whether he was also familiar with Steiner.

of knowing. [...] This preconception is not a *fundamental truth* and is in *no way* qualified to stand at the *forefront* of the science of knowledge.

But do not misunderstand us. We do not wish to raise what would certainly be a vain protest against the *physiological* achievements of the present day. But what is entirely justified physiologically is still far from being qualified on that basis to be placed at the portals of epistemology. One may consider it to be an irrefutable physiological truth that only through the participation of our organism does the complex of sensations and perceptions arise that we have called experience. But the fact remains, nevertheless, that any such knowledge can only be the result of many considerations and investigations. This characterization – that our phenomenal world, in a *physiological sense*, is of a subjective nature – is already what thinking *determines* it to be, and has therefore absolutely nothing to do with the initial appearance of this world. This characterization already presupposes that *thinking* has been applied to *experience*. The examination of the relationship between these two factors of knowing activity must therefore precede this characterization.

This is only possible if *cognition itself*, the human being's very own mental activity, is made into the object of cognition.

It is not the task of epistemology to define the *content* of experience, but its *form*, i.e. the *concept* of experience (Steiner, 1988a, Chapter 4). The former is the task of the relevant branches of science, for example physiology, psychology, etc. The epistemological concept of pure experience therefore applies in like manner to *all the observed content of experience*, irrespective of whether this involves physical or psychological phenomena, whether sense perceptions, feelings, dream images or phantasies, recalled mental images or even well-developed concepts. In epistemological terms these elements are considered to be "pure experience" to the extent that they are available to consciousness to be cognised, and have *not yet been cognised* in themselves nor in their relationship to other elements. Steiner explains this using a passage from Johannes Volkelt, which he describes as an "excellent characterization of pure experience" (Steiner, 1988a, 25–26):

> [It] simply presents us with the pictures which, in a limited period of time, pass before our consciousness in a completely unconnected way. Volkelt says: "Now, for example, my consciousness has as its content the mental picture of having worked hard today; immediately joining itself to this is the content of a mental picture of being able, with good conscience, to take a walk; but suddenly there appears the perceptual picture of the door opening and of the mailman entering; the mailman appears, now sticking out his hand, now opening his mouth, now doing the reverse; at the same time, there join in with this content of perception of the mouth opening, all kinds of auditory impressions, among which comes the impression that it is starting to rain outside. The mailman disappears from my consciousness, and the mental pictures that now arise have as their content the sequence: picking up scissors, opening the letter, criticism

of illegible writing, visible images of the most diverse written figures, diverse imaginings and thoughts connected with them; scarcely is this sequence at an end than again there appears the mental picture of having worked hard and the perception, accompanied by ill humor, of the rain continuing; but both disappear from my consciousness, and there arises a mental picture with the content that a difficulty believed to have been resolved in the course of today's work was not resolved; entering at the same time are the mental pictures: freedom of will, empirical necessity, responsibility, value of virtue, absolute chance, incomprehensibility, etc.; these all join together with each other in the most varied and complicated way; and so it continues." Here we have depicted, within a certain limited period of time, what we really *experience*, the form of reality in which *thinking* plays no part at all.

This thinking has "absolutely no part" in pure experience which, *for current* cognition, has to *actively* find the connections between the given elements of perception, whereas the thoughts which are already contained in these elements have either been produced actively at an earlier point – i.e. *before* the current act of perception – (e.g. philosophical concepts of the above example), or appear spontaneously – in other words, without our conscious assistance – permeated by cognitive content, such as is the case of e.g. fantasies and recalled mental pictures, and also what are known as inherent or current mental pictures which arise with perceptions as cognitive content along with sensory impressions, i.e. in relation to these. In an epistemological sense all the thoughts and mental pictures which do not appear as a result of our active thinking are intrinsically "experience", even though they might have been brought into consciousness at an earlier point by this thinking activity[16]. It is a generally accepted fact of present-day cognitive psychology that observation in the latter sense is always experienced as "theory-laden" (Hacking, 1996, 285). In pure experience or perception as understood by cognitive psychology, the sensory impression would have to arise in consciousness *devoid of any inherent mental picture*. This is impossible, because clearly in all sense perception through the human physical and mental organisation – usually without the conscious participation of the originator – cognitive units are linked to sensory data, so that the diverse world of the senses, in which from

---

16 It is always a matter of the *current* cognition: the explanandum, the experience, must already be *present* for the act of cognition in question: how this experience came about is of no importance for *this* act of thinking. However, this act of thinking is indeed what produces the concept (explanans) and the connection of this concept to the experience. This – actively produced – concept can in epistemological terms, however, again become a – passively received – *experience* in a further act of cognition (in a reasoned judgement) because due to the earlier act of thinking it is already present. This explains why Volkelt also includes *concepts* as *content* of "pure *experience*" in the reference above.

a purely sensory perspective everything merges into everything else without any interruption, appears as varied separate *units*: leaves, trees, pebbles, etc., right from the first form in which it arises in consciousness. In the direct form in which they appear these "impure", in other words cognitive/sensory units[17] therefore present the elements of pure experience (in *epistemological* terms) whose specific content itself and whose lawful connection with other content must first be found through thinking.

So if cognitive and sensory elements as such are identified as the content of the experience this can only be the *result* of cognition. But this identification requires that the relevant elements have first been *observed*, and in a pure manner i.e. in the form of experience, without a defining judgement. Observation then shows that the percept in no way displays merely the cognitive elements belonging to the subject, but also the sensory ones which are experienced as being given to the subject from outside. This is apparent from the fact that the sensory impression is only present for the subject when the object of perception is actually present, whereas the mental picture also remains connected with the subject's inner being after the object has disappeared. Steiner expresses this as follows (Steiner, 2011, 56):

> When the tree disappears from my field of vision, an after-effect of this process remains in my consciousness – a picture of the tree. This picture has become associated with my self during my observation. My self has become enriched; its content has absorbed a new element. This element I call my *mental picture* of the tree.

This *empirical* distinction between the sensory and cognitive elements of observation is important for the question of whether perception gives the subject a relationship to an objective world or not. Under the influence of several movements, from Kant to modern constructivism such as advocated by Humberto Maturana for example, it has become common in epistemology to treat knowledge and especially the percept component of it as being determined by the human subject or even by their physical constitution (Schneider, 1998). However, the subjectivity of the experience would then also have to be proved by the *observation*. This is not the case for the sensory component of the experience, so the definition of what is observed as "subjective" or "construed" is empirically unsatisfactory.

---

17 This fact of sensory-cognitive units can be demonstrated well using what is known as pictographic ambiguity, in which *different* forms are "seen" *in the same* sensory material. This is because the sensory elements, e.g. those printed on a page are arranged in such a way that different mental pictures can logically be related to them. These types of phenomena are used in gestalt psychology.

This differentiation of the content of perception into a sensory and a cognitive component, the identification of what is perceived as subjective or objective, as given or construed, as physiologically or psychologically determined, as having evolved or been socially produced, can only ever be the result of specific empirical knowledge, in other words can only arise from perceptual psychology, sensory physiology, evolutionary biology and psychology, sociology, etc. However, all these particular sciences presuppose cognition and can therefore not be placed at the forefront of the science of knowledge. The fact that this is seldom noticed is one of the reasons for the vast range of diverging epistemological positions to which the 20th century has given rise (Schneider, 1998) and which almost all *come from* specific scientific fields.

Steiner's epistemology differs radically from these positions by systematically considering cognition as such *before* all the particular sciences, in other words the mental activity of the human being which is activated in all other knowledge but which cannot view itself in this process and therefore remains "the unobserved element in our ordinary mental and spiritual life" (Steiner, 2011, 34). It is the task of epistemology as a basic science to raise this element in an empirical manner to an object of cognition. Its goal is to find the "laws inherent in cognition" (Steiner, 1981a, 48), i.e. the "nature of knowledge itself" (Steiner, 1981a, 91). This should answer the question: "What significance is there in the mirroring of the outer world in human consciousness; what connection exists between our thinking about the objects of reality and these objects themselves?" (Steiner, 1988a, 19). Knowledge of knowing can only be achieved by first observing this activity of cognition, not the neurophysiological correlate of cognition, but cognition *as a mental activity*, so that through this observation its laws can be found. Hence, for the purpose of epistemology, the necessity arises of raising not just any kind of facts of experience to the object of observation but thinking itself which, for the purpose of cognition, creates the connections between the elements of "pure experience" (Steiner, 1988a, 22):

> *We must seek thinking among the facts of experience as just such a fact itself.* Only in this way will our world view have inner unity. It would lack this unity at once if we wanted to introduce a foreign element into it. We confront experience pure and simple and seek within it the element that sheds light upon itself and upon the rest of reality.

It therefore appears that that which first sparks thinking, pure experience, can only obtain its characterisations through this thinking. This is why no judgements about the content of experience can be made at the beginning of epistemology. It is first necessary to show to what degree thinking is justified in making these kind of judgements. And this in turn is only possible by an empirical examination of thinking itself.

2.3 Thinking as a fact of experience

In common with Goethe, Steiner pointed out that thinking as a fact of experience is a "higher experience within experience": "We find, within the unconnected chaos of experience, and indeed at first also as a fact of experience, an element that leads us out of unconnectedness. It is *thinking*" (Steiner, 1988a, 35). But, and this is the crucial point: "Even as a fact of experience within experience, thinking occupies an exceptional position" (ibid.). Because: "What, for the rest of experience, must first be brought from somewhere else – if it is applicable to experience at all – namely, *lawful interconnection*, is already present in thinking in its very first appearance. [...] *In thinking, what we must seek for with the rest of experience has itself become direct experience*" (ibid.). This statement is of fundamental importance as it enables the consistent *foundation of epistemology on the principle of experience*. A longer passage from Steiner is therefore cited which explains this point in greater detail (ibid., 36–37):

> With this the solution is given to a difficulty that will hardly be solved in any other way. That we stick to experience is a justified demand of science. But no less so is the demand that we seek out the inner lawfulness of experience. *This inner being itself must therefore appear at some place in experience as experience*. In this way experience is deepened with the help of experience itself. [...] The principle of experience, in its implications and actual significance, is usually misunderstood. In its most basic form it is the demand that we leave the objects of reality in the first form in which they appear and only in this way make them objects of science. This is a purely methodological principle. It expresses absolutely nothing about the content of what is experienced. If someone wanted to assert, as materialism does, that only the *perceptions of the senses* can be the object of science, then he could *not* base himself on this principle[18]. This principle does not pass any judgment as to whether the content is sense-perceptible or ideal. But if, in a particular case, this principle is to be applicable in the most basic form just mentioned, then, to be sure, it makes a presupposition. For, it demands that the objects, as they are experienced, already have a form that suffices for scientific endeavor. With respect to the experience of the outer senses [...] this is not the case. This occurs only with respect to thinking.
> *Only with respect to thinking can the principle of experience be applied in its most extreme sense.*

It is important for Steiner's epistemological position that, in this connection, he sees himself in complete accord with Goethe, whose attitude to cognition[19]

---

18 Even materialism makes use of *thinking* in order to explain sense perceptions.
19 This view of cognition held by Goethe, in which the content of a theory of nature corresponding to reality may not be arbitrary but only "experience of a higher nature", has been presented in greater detail by the author as part of the lecture

had become clear to him through the publication of Goethe's natural scientific writings (ibid., 37):

> A science of knowledge established in the sense of the Goethean world view lays its chief emphasis on the fact that it remains absolutely true to the principle of experience. No one recognized better than Goethe the total validity of this principle. He adhered to the principle altogether as strictly as we demanded earlier. All higher views on nature had to appear to him in no form other than as experience. They had to be "higher nature within nature".

But how can "the basic element of what is scientific: ideal lawfulness", even when this appears as experience, have any kind of relevance for objective reality when this element has first to be produced by the activation of thinking, in other words a definite, likewise experienceable accomplishment of the subject?

The answer to this seems possible when the difference between the content of thinking and the activity of thinking is clarified by means of inner observation. The *activity* of thinking is, of course, our own accomplishment, something we are totally certain of in that we produce each of its threads ourselves by our own activity[20]. But we do not bring about the *content* of thinking in the same way. Our thinking activity merely causes it to *appear*, we do not *create* it. This can be seen from the fact that thought content such as e.g. mathematics or geometry cannot be thought arbitrarily, but only as it actually is (ibid., 40–41):

> We definitely do not produce a thought-content as though, in this production, we were the ones who determined into which connections our thoughts are to enter. We only provide the opportunity for the thought-content to unfold itself *in accordance with its own nature*. We grasp thought *a* and thought *b* and give them the opportunity to enter into a lawful connection by bringing them into mutual interaction with each other. It is not our subjective organization that *determines* this particular connection between *a* and *b* [but the content of *a* and

---

series "Goethe's contribution to the renewal of the natural sciences" organised by him at the University of Bern to mark the 250th anniversary of Goethe's birth (Heusser, 2000c).

20 Cf. Section 2.1. It is also stressed here that in this work, if not noted otherwise, "thinking" refers solely to thinking created by the *personal activity of the thinker* and not to thoughts which arise in consciousness *by themselves*, such as associations, spontaneous ideas, recalled mental images, etc. Cf. also Steiner: "One should not confuse the 'having of thought-images' with the elaboration of thought by thinking. Thought-images may appear in the soul after the fashion of dreams, like vague intimations. But this is not *thinking*" (Steiner, 2011, 45). An example of the thinking meant here is performing mathematics or geometry, or thinking through Hegelian logic. Nothing can be understood of such subjects without *full inner activity*.

b is the *sole determining factor*]. [We have not the slightest influence on the way in which *a* and *b* relate] in precisely one particular way and no other. *The human spirit effects the joining of thought masses only in accordance with their content.* In thinking we therefore fulfil the principle of experience in its most basic form.

By way of comparison, something is experienced here in thinking similar to the sense perception when touching a rock face made of granite. Just as you have to accept all the corners and curves of this in the form in which they occur and in the way they follow one after another, so with the activity of thinking you need to move along the thought content as this actually is and as it links together objectively. For example, any attempt to divide a prime number by another number than itself or one, so that a whole number results, will be unsuccessful. It is apparent that the ideas contained in this attempt are countered *spiritually or intellectually* by a passive resistance just like the rock wall in the case of our senses: it can only be experienced as it actually is and not as we might want it to be[21].

This internal experience of resistance is an empirical proof of the objectivity of the thought content and therefore of the objective idealism or universal realism argued by Steiner (cf. Section 2.6). The thought content turns out empirically, i.e. in experience, to be a given, an internal "objectum". However, an object which is no longer *externally* present for thinking like the sense object for observation, but with which it is completely and *inwardly united* in the thinking act. Without this union, which is appropriately called "insight" in the German and English languages and "intuition" in the *intellectual* sense by Thomas Aquinas, Descartes and Steiner, thinking would be completely unable to understand its content[22]. This objectivity is the reason for the fact that, e.g. mathematical and geometrical content can be experienced as "laws" and logical connections must follow the thought content as a "necessity" and that these connections must be "intersubjectively verifiable", as science rightfully demands. Because the thought content is not *created* by the activity of thinking but merely *made manifest* in the consciousness of the subject. It exists objectively, independently of the individual subjects. But the subjects participate in it through thinking.

---

21 This is also the empirical argument against the theory that thought connections are only *conventions*. Incidentally, the convention theory actually tends to provide confirmation of what is said here because in a convention the many subjects have to agree on "one" thing (convenire: Latin for to meet). This "one" must then be a different object from all the subjects.

22 "I must work the thought through, must recreate its content, must experience it inwardly *right into its smallest parts* if it is to have any significance for me at all" (Steiner, 1988a, 39). It is the thought *activity* that, in experiencing the thought *content*, creates its *form*. Content and form create a unity.

The fact that mistakes in thinking can be made is no argument against the objectivity of the thought content. When the thinking activity proves to be too weak or not concentrated enough to follow the content, then incorrect connections can be made. However, the fact that the mistakes can be corrected and the correctness of mathematical or other logical thought connections verified intersubjectively, something which everyone knows from experience, actually confirms the principal objectivity of the thought content.

In terms of its existence as a "given" (i.e. its objectivity), the thought content is therefore comparable to the content of perception. However, there is a difference in that the content of perception appears as it were "passively", i.e. without our help as a phenomenon "given directly" to the senses from outside, whereas the thought content only appears as a given inside us and "actively", i.e. through the intermediary of our thinking activity (Steiner, 1981a, Chapters 4 and 5).

But thinking performs the same function towards the thought content as does observation towards the content of perception: In Steiner's terminology it is the *"organ of apprehension"* (Steiner, 1988a, 67): "Thinking is an organ of the human being that is called upon to observe something higher than what the senses offer" (Steiner, 1988a, 53–54), that being the content of ideas through which the knower wishes to understand the objective interrelationships of the phenomena. In this context Goethe also speaks of the "eyes of the mind", through which the scientist should be able to discover the inner law of that which the "eyes of the body" allow him to see externally (Goethe, 1950r). Hegel called the "pure knowledge" of the objective content of ideas of the world "intellectual intuition" ("intellektuelle Anschauung") (Hegel, 1979, 76) or "intellectual" or "inner intuition" ("Anschauen") (Hegel, 1977, 10, 35).

Steiner also used the term "intuition" – which actually means a *rational intellectual intuition* – for the same process, as already demanded for science by Descartes at the beginning of the scientific age[23], not the appearance of

---

23 As already shown by the author elsewhere (Heusser, 1999c), in his famous *"Regulae ad directionem ingenii"* (Rules for the Direction of the Mind, Descartes, 1954), René Descartes demanded a thinking for science that "attain a certitude equal to that of the demonstrations of Arithmetic and Geometry" (ibid., Rule 2). In this process "we ought to investigate what we can clearly and evidently intuit or deduce with certainty, and not what other people have thought or what we ourselves conjecture. For knowledge can be attained in no other way" (ibid., Rule 3). Intuition and deduction together make up the "intellectual activities" [intellectus nostri actiones] (ibid., Rule 3). "By intuition I mean, not the wavering assurance of the senses, or the deceitful judgment of a misconstructed imagination, but a conception, formed by unclouded mental

spontaneous ideas, dreamlike experiences or feelings commonly referred to by this expression, which arise without any thinking effort by the individual[24] (Volkamer et al., 1991).

> attention, so easy and distinct as to leave no room for doubt in regard to the thing we are understanding. It comes to the same thing if we say: It is an indubitable conception formed by an unclouded mental mind; one that originates solely from the light of reason [rationis], and is more certain even than deduction, because it is simpler [...]. Thus, anybody can see by mental intuition [animo] that he himself exists, that he thinks, that a triangle is bounded by just three lines, and a globe by a single surface, and so on; [...]" (ibid., Rule 3). By intuition, Descartes therefore understands purely intellectual "seeing" of the ideational content such as the mathematical content on the one hand, and also "seeing" of the thinking activity which causes these to appear on the other. In his *"Philosophy of Freedom"* Steiner likewise understands "intuition" to mean pure intellectual intuition: in the first instance cited above, intuition is "the form", in other words the thinking *activity*, in which the thought-content appears, but in subsequent places it is the thought *content* which has been produced and which explains the percept. On p. 123 the concept of intuition is expanded, thus including the *experience and knowledge* of the thinking activity: *"Intuition* is the conscious experience – in pure spirit – of a purely spiritual content. Only through an intuition can the essence of thinking be grasped" (Steiner, 2011, 123). Steiner speaks accordingly of "intuitive thinking" or "intuitively experienced thinking" and describes this as the basis of the *"Philosophy of Freedom"*: "The argument of this book is built upon intuitive thinking which may be experienced in a purely spiritual way and through which, in the act of knowing, every percept is placed in the world of reality" (ibid., 215). This expanded definition of intuition is also open in principle to *higher*, spiritual scientific intuition, which is not mentioned explicitly in *"The Philosophy of Freedom"* (cf. Chapter 7), but has nothing to do with forms of intuition *less than* completely conscious active thinking.

24 This "intuition" arising from itself as a direct given is, looked at epistemologically, merely *experience* like any other percept whose meaning must first be found in active thinking through the rational intuition indicated here. Evidence for this is provided by famous examples from the history of science such as the "intuitive" discovery of the ring structure theory of benzene by Kekulé. Kekulé recounts how he was travelling at night on the last bus through London and "sank into dreams" where he saw rotating atoms which moved and formed into groups, arranged themselves in chains and "in whirling dances". "I was awakened from my dreaming by the conductor shouting: 'Clapham Road', but I spent part of the night putting at least a sketch of this dream picture onto paper. This is how the structure theory came about". "Let us learn to dream, gentlemen, then perhaps we shall find the truth. [...] But let us beware of publishing our dreams before they have been *tested by the alert intellect*" (Volkamer, et al., 1991, 34, ePH). It is only *intellectual* intuition which makes this possible. *This* is what Steiner is referring to.

Thinking therefore has to find a *content* in itself[25] which is not apparent in the percept and the act of this finding is insight, intuition (Steiner, 2011, 79–80):

> The separate facts appear in their true significance, both in themselves and for the rest of the world, only when thinking spins its threads from one entity to another. This activity of thinking is one *full of content*. For it is only through a quite definite concrete content that I can know why the snail belongs to a lower level of organization than the lion. The mere appearance, the percept, gives me no content which could inform me as to the degree of perfection of the organization.
>
> Thinking offers this content to the percept, from man's world of concepts and ideas. In contrast to the content of percept which is given to us from without, the content of thinking appears inwardly. The form in which this first makes its appearance we will call *intuition*. *Intuition* is for thinking what *observation* is for the percept. Intuition and observation are the sources of our knowledge. An observed object of the world remains unintelligible to us until we have within ourselves the corresponding intuition which adds that part of the reality which is lacking in the percept. To anyone who is incapable of finding intuitions corresponding to the things, the full reality remains inaccessible.

According to this description, active thinking therefore grasps an objective thought content and relates this to a percept given by observation. This creation of relationships is the real cognitive achievement of thinking.

2.4 Objective empirical knowledge

According to Steiner's description, the result of the two preceding sections is that two forms of given facts have to be distinguished within the totality of experience: first the observation of what is given directly, which appears without the help of the subject for its sensory perception[26] and, second, what appears for the subject through the intercession of his thought activity or

---

25 It is not unnecessary to mention this because, according to Kant concepts without sensory contemplation are empty (cf. Note 37).

26 "Sensory perception" is not meant here only in a physical/sensory sense, i.e. with the help of the physical sense organs, but in an all-inclusive way, as expressed by Steiner in *"The Science of Knowing"*: "If we now wished to have a name for the first form in which we observe reality, we believe that the expression that fits the matter the very best is: *manifestation to the senses*. By *sense* we do not mean merely the outer senses, the mediators of the outer world, but rather *all* bodily and spiritual organs whatsoever that serve the perception of immediate facts. It is, indeed, quite usual in psychology to use the expression *inner sense* for the ability to perceive inner experiences. Let us use the word *manifestation*, however, simply to designate a thing perceptible to us or a perceptible process insofar as these appear in space or in time" (Steiner, 1988a, 32–33).

intuition, the concept. The experience is "pure" in as far as it is not yet linked to the concept. The concept is "pure" in as far as its content is not (yet) related to any content of experience i.e. to a perception. Examples of this are the mathematical concepts of *pure* mathematics *before* they are related to sensory facts in *applied* mathematics.

Observation of the pure concepts further shows that these are related internally to more concepts, are connected with them in a lawful manner in accordance with their content. For example, the concept of the cause is intrinsically related to that of the effect, the concept of the part to that of the whole, and vice versa. And "pure thinking" means nothing more than thinking about these pure *self-contained* conceptual relationships. In the act of thinking these are therefore contemplated internally, i.e. intellectually. Hegel calls this "intellectual" or "pure" intuition: "More than this. Pure and simple *intuition* is completely the same as pure and simple thought" (Hegel, 2001b, 54).

This inherent reference character of all ideas is the reason for the objectivity and intersubjective verifiability discussed above and at the same time the reason for the human ability to form logically consistent thought systems, i.e. a systematic unity in the multiplicity of thoughts (Steiner, 1988a, 47):

> How does our thinking manifest to us when looked at *for itself*? It is a *multiplicity* of thoughts woven together and organically connected in the most manifold ways. But when we have sufficiently penetrated this multiplicity from all directions, it simply constitutes a unity again, a harmony. All its parts relate to each other, are there for each other; one part modifies the other, restricts it, and so on. As soon as our spirit pictures two *corresponding* thoughts to itself, it notices at once that they actually flow together into one. Everywhere in our spirit's thought-realm it finds elements that belong together; *this* concept joins itself to *that* one, a third one elucidates or supports a fourth, and so on. Thus, for example, we find in our consciousness the thought-content 'organism'; when we scan our world of mental pictures, we hit upon a second thought-content: 'lawful development, growth'. It becomes clear to us at once that both these thought-contents belong together, that they merely represent two sides of one and the same thing. But this is how it is with our whole system of thoughts. All individual thoughts are parts of a great whole that we call our world of concepts.

On the one hand, to produce an ideational whole or system of thoughts is experienced as an elementary need (Steiner, 1988a, 47):

> If any *single* thought appears in my consciousness, I am not satisfied until it has been brought into harmony with the rest of my thinking. A separate concept like this, set off from the rest of my spiritual world, is altogether unbearable to me. I am indeed conscious of the fact that there exists an inwardly established harmony between all thoughts, that the world of thoughts is a unified one.

On the other hand there is the objectivity of the subject's sought-for ideal discussed above[27]. This is why the "thorough [...] harmony of all the concepts we have at our command", is experienced as *truth*: "If we have struggled through to where our whole thought-world bears a character of complete inner harmony, then through it the contentment our spirit demands becomes ours. *Then we feel ourselves to be in possession of the truth*" (ibid., 48)[28].

This internal logical agreement of concepts is determined on the basis of the *concept-judgement* in which a concept (as the subject of the judgement) is linked (copula) to a concept (as the predicate). (The other possible form of a judgment is a *percept-judgment*, in which the subject concerns a percept instead of a concept, ibid., 64). In this respect the development of all logically consistent scientific theory rests on concept-judgements. Popper even claims that the "objective knowledge or knowledge in the objective sense [...] consists of the logical content of our theories" (Popper, 1972, 73)[29]. But the

---

27 The objection that because of *the subject's need* for unity in his ideas, this is then itself subjective, is countered by Steiner as follows: "It is indeed correct that subjective reason has the need for unity. But this need is without any content; it is an empty *striving* for unity. If something confronts it that is absolutely lacking in any unified nature, it cannot itself produce this unity out of itself. If, on the other hand, a multiplicity confronts it that allows itself to be led back into an inner harmony, it then brings about this harmony. The world of concepts created by the intellect is just such a multiplicity" (ibid., 62).

28 Here, by "truth", Steiner initially means the "universal harmony of all the concepts *we have at our command*", keeping the concept of truth not for the harmony of *absolutely all* concepts which incidentally would be completely impossible for a single individual to achieve. In this *qualitative* concept of truth Steiner differs from Popper who, by truth, understands *quantitatively* "the quantity of all true statements", which naturally needs must remain "an unattainable target quantity". An example for truth as Steiner sees it which Popper also admits is an "undoubted truth" is something like "1+1=2" (Popper, 1972, 57). Truth is therefore seen as the logical coherence of concept-judgements, which Popper – even if rather half-heartedly – refers to as that part of "knowledge in the objective sense" which can definitely be proved: "As soon as we take objective knowledge into account, we must say that at best only a very small part of it can be given anything like sufficient reasons for certain truth: it is that small part (if any) which can be described as *demonstrable knowledge* and which comprises (if anything) the propositions of formal logic and of (finite) arithmetic" (ibid., 75). Incidentally, Popper elsewhere says that "truth is correspondence with the facts (or with reality); or, more precisely, that a theory is true if and only if it corresponds to the facts" (ibid., 44).

29 Popper obviously seems to think that the "objectiveness" of these theories consists of their *publication* (!): "Examples of objective knowledge are theories published in journals and books and stored in libraries". He even has his famous "world 3" consist of the "logical *contents* of books, libraries, computer memories and suchlike" (ibid., 73–74). But anyone who does not *think* when reading written words,

logical content of a theory, however internally consistent, is insufficient for attaining objective knowledge, when this is not seen purely as an objective conceptual understanding such as e.g. in pure mathematics, but an objective knowledge of the world of phenomena accessible to observation e.g. by means of applied mathematics. Naturally, Popper is also aware of this when he says that "truth is correspondence [of theory] with the facts (or with reality)" (ibid., 44).

What is crucial for empirical knowledge is not the thoughts, the theory, but their *agreement with experience*, with the manifest reality. This is why Steiner repeatedly stresses that the judgement of reality must be not merely *logical* but also *in accordance with reality*, for example in a course held in 1920 for students and academics on *"Specialized Fields of Knowledge and Anthroposophy"*: "Two things are necessary nowadays when we wish to make a judgement: first, that the judgement is made on the basis of a correct logical method – the method of calculating is one such logical method –; and, second, that the judgement is also made with a proper insight into the reality. A judgement must be in accordance with reality and logical. The former is usually forgotten nowadays which is why the purely logically correct judgements have such an important part in our scientific life but in certain circumstances have no application to reality" (Steiner, 2005, 42).

In this context the accordance of an idea or concept with reality[30] is tested by the *percept-judgement* in which a concept is related to an actual percept.

---

will only see groups of characters but will not find the *logical content* of the words. This is purely and only to be found inwardly, i.e. intuitively, as a *spiritual content* through intellectual contemplation. This also applies to the spoken word. Popper does not seem to distinguish sufficiently between word and meaning, thought and its verbal expression or between information and the bearer of that information. This is the only explanation for why he places particular weight on *"linguistically formulated* theories" (ibid., 74) and says: "Theories, or propositions, or statements are the most important third-world linguistic entities" (ibid., 157). It is clear that theories can only be *communicated* through linguistic expression; but it is not this but their logical content that is the essential element of the theory. The original Greek meaning of the word *theoria* the display of the godly, in other words spiritual, corresponds rather better with this.

30 With reference to the difference between a concept and an idea, Steiner noted: "Kant pointed already to the difference between intellect and reason. He designated reason [Vernunft] as the ability to perceive ideas [Ideen]; the intellect [Verstand], on the other hand, is limited merely to beholding the world in its dividedness, in its separateness. Now reason is, in fact, the ability to perceive ideas. Here we must determine the difference between concept [Begriff] and idea […]. The concept is the single thought as it is grasped and held by the intellect. If I bring a number of such single thoughts into living flux in such a way that they pass over into one another, connect with one another, then thought-configurations arise that are present only

The effect of this reference is to "specialise" the concept, it ceases to be a pure i.e. general concept. On the other hand, in the epistemological sense, the percept also ceases to be a "pure experience". Because due to the concept related to it, which in its turn is connected to other concepts, it becomes part of a context of ideas. "The judgment under consideration here has a perception as its subject and a concept as its predicate. The particular animal in front of me is a dog. In this kind of judgment, a perception is inserted into my thought-system at a particular place. Let us call such a judgment a *perception-judgment. Through a perception-judgment, one recognizes that a particular sense-perceptible object, in accordance with its being, coincides with a particular concept*" (Steiner, 1988a, 55). For it is the predicate concept which supplies the definition of that which *is* the subject of the judgement, the element before us, i.e. its lawful *essence.*

I see, for example a sunflower and take pleasure in the wonderfully regular arrangement of the seeds on the flower disc. My need for knowledge is stirred and I ask myself what kind of pattern it might be. I therefore start to study geometrical patterns and find out that the seeds of a sunflower are arranged in accordance with a mathematical law which is a Fibonacci series[31]. The visible pattern is *recognised* when I establish that the pattern and law agree completely, that it is therefore its "essence" or, conversely, that the pattern is a manifestation of this law, the appearance of this essence.

This of course means that the Fibonacci series must be "prefigured" in my thinking *beforehand*, i.e. in relation to the moment of the actual act of cognition. In other words a law must be sought and found on account of the need for knowledge, already be intuited and understood, in order to be able to be subsequently linked to the percept in a judgement. "If we therefore wish to grasp what we perceive, the perception must be *prefigured* in us as a definite concept. We would go right by an object for which this is not the case without its being comprehensible to us" (Steiner, 1988a, 56). For instance, thousands

---

for reason, that the intellect cannot attain. For reason, the creations of the intellect give up their separate existences and live on only as part of a totality. These configurations that reason has created shall be called *ideas*" (Steiner, 1988a, 61).

31 A Fibonacci series is a repeating mathematical sequence in which every number is the sum of the two preceding numbers, where the first number = 0 and the second number = 1. The relevant sequence is then: 0, 1, 1, 2, 3, 5, 8, 13, 21, etc. The adjacent sunflower seeds are set at a golden angle to the flower axis, in other words in accordance with the golden ratio. The rational numbers which best approximate this are fractions of successive Fibonacci numbers. The adjacent seeds therefore occupy positions which differ by the Fibonacci number in the denominator. This is what causes their spiral arrangement. The sequence is named after the person who discovered it, Fibonacci (Leonardo Fibonacci da Pisa, ca. 1170 to ca. 1241), one of the most important mathematicians of the Middle Ages.

of people passed the slowly swinging lights in the cathedral in Pisa before Galileo Galilei found the law of this oscillation, the law of the pendulum.

The predicate concept must always be *added* to the subject of the judgement, it is not already present in the percept of the latter. Steiner therefore also describes the act of cognition as *"the synthesis* of percept and concept" (Steiner, 2011, 77). This synthesis is likewise the product of thinking: "To permeate the world, as given, with concepts and ideas, is a *thinking* consideration of things. Therefore, thinking is the act which mediates knowledge" (Steiner, 1981a, 64).

This thinking which penetrates the world of perception with concepts is also prompted in us by a subjective need: "Nowhere are we satisfied with what Nature spreads out before our senses. Everywhere we seek what we call the *explanation* of the facts" (Steiner, 2011, 21–22). We are only satisfied when we *understand* – in the above sense of the word – what we perceive. However, as with the concept-judgement, here again in the percept-judgement the subject, has nothing to say about the objectivity of what is to be cognised.

Looked at like this, we can define *three steps in cognition* through which three things are given to us in the manifest reality: 1. The *observation*; this provides us with the percept; 2. The searching and finding (intuiting), i.e. *thinking* of the law, the idea; and 3. *Thinking* as the actual *act of cognition*, bringing the comprehended law to the percept[32], through which the inner identity of the law and manifestation can be understood. Steiner formulates this as follows (Steiner, 1988a, 55):

> In all cognitive treatment of reality the process is as follows. We approach the concrete perception. It stands before us as a riddle. Within us the urge makes itself felt to investigate the actual [...] essence of the perception, which this perception itself does not express. This urge is nothing else than a concept working its way up out of the darkness of our consciousness. We then hold fast to this concept while sense perception goes along parallel with this thought-process. The mute perception suddenly speaks a language comprehensible to us: we recognize that the concept we have grasped is what we sought as the essential being of the perception.

---

32 As already mentioned above, in terms of *cognitive psychology* there is no percept which is not already permeated with thoughts. However, "perception" here is not meant in the sense of psychological perception, but *epistemologically*. The sensory impression which is already permeated by inherent mental pictures, e.g. the "beautiful pattern" of the "sunflower disc" mentioned above, is epistemologically the percept to which, for the purpose of cognition, thinking must first find and contribute the corresponding law.

The judgement in knowing in fact *formally* performs an identification of the essence of the subject and predicate: This observed animal (subject of the judgement) – *is*[33] (copula) – a dog (predicate).

Whether this judgment is not only a formal but also a *factual* identification does not depend on the knower but on the content to be known. For just as in making the concept-judgement, thinking itself has no part in determining the agreement of its contents[34] but only creates the opportunity for this and experiences the result, the same applies in the percept-judgement: thinking cannot arbitrarily specify which law fits the sunflower's pattern, but in the mutual relationship of the law and the phenomenon, must wait to see if both are congruent. It is only this congruent result which can count as knowledge in the strict sense, i.e. in relation to content as defined by Steiner (Steiner, 1981a, 65):

> When thinking restores a relationship between two separate sections of the world-content, it does not do so arbitrarily. Thinking waits for what comes to light of its own accord as the result of restoring the relationship. And it is this result alone which is knowledge of that particular section of the world content. If the latter were unable to express anything about itself through that particular relationship established by thinking, then this attempt made by thinking would fail, and one would have to try again. All knowledge depends on man's establishing a correct relationship between two or more elements of reality, and comprehending the result of this.

Looked at in this way, everything in knowledge is *experience*: 1. the direct experience which is available to observation, 2. the law which appears through the agency of thinking, and 3. the agreement of law and experience. "In this sense all our knowledge is empirical." (Steiner, 1981a, 68). And in this empirical sense it is also justified in describing knowledge as objective.

This also explains why there is no need for a contradiction to exist between the objective reality of the content of cognition argued by Steiner and the *possibility of error*. Because connecting thoughts with thoughts or thoughts with percepts is an activity of the thinking individual; and for various reasons this activity can be prevented from finding the "correct" thought connection, i.e. those which arise from the thing itself, between the elements of perception.

---

33 The copula "is" can be interpreted as a grammatical and linguistic expression of a semantical identification of the subject and the predicate in the judgement.

34 What was stated above about the concept-judgement (cf. Notes 26 and 27) essentially applies also to the percept-judgement: the need for an agreement between percept and concept would have no content, it would be an empty *desire* for unity, if the given *content* did not itself permit a congruent merging. It could not create this out of itself. The possibility that this view is consistent with the potential for error, emerges from the further discussion.

A simple case is the fact known to everyone that, during observation, you are not always in a position to summon up enough *attention* for the given content of perception, or in thinking cannot find enough *concentration* to work through its lawful connections. Errors may then result from the fact that you assess the percept-content incorrectly and therefore provide it with inappropriate laws (erroneous percept-judgement) or relate thought contents to each other incorrectly (erroneous concept-judgement). In both[35] cases the error is the result of an erroneous judgement, in other words an incorrect thought connection by the subject, which is familiar to everyone e.g. from adding up items in a bill incorrectly or from not paying enough attention to the observation when making a judgement.

The same applies to *pre-judgements and conventions* when the subject fails to wait to see which thoughts turn out to be objectively appropriate for a specific percept situation, but hastily applies judgements to this percept situation. But it is precisely this possibility of ridding oneself of pre-judgements and correcting errors of judgement by improving attention and by keeping up the mental power of concentration that confirms the *factual* character of the content of cognition despite all the subjectivity and susceptibility to error in the activity of cognition.

This is stated as a principle, but is by no means of minor importance in the daily process of science. For instance, it is far from easy to determine the "correct" thought connections between the objects of perception, i.e. those which arise from the thing itself in the case of complex facts in, for example modern biology, as might appear to be the case with the agreement of the Fibonacci law with the observed arrangement of sunflower seeds or the pendulum law with the swinging lights in the church.

However, one major reason for the frequent rejection of the possibility of finding a "correct" thought connection between the elements of perception *in the first place*, may lie in the fact that following Karl Popper and other epistemologists of the 20th century, the objectivity, reality and certainty of natural laws are rejected due to a nominalistic understanding of laws (cf. Section 2.6). This has led to the cognitive thoughts relating to empirical material being at best granted the function of *model representations* which can never or only approximately measure up to reality. And the difference between model representations or hypotheses and cognitive ideas in terms of the objective relationship to empiricism meant here is simply ignored.

---

35 This also applies to the case of inadequate observation. Because the percept (i.e. the object to the extent that it is given by observation) appears in the form of experience, which is not "wrong" *as such*, but appears as a given fact which, for its part, can of course provide the opportunity to assign incorrect thoughts to it in the judgement. The error then lies in this assignment.

A further circumstance which appears to speak against objective knowledge is the *restriction due to perspective* which depends on the subjective limits of experience of cognition (Matthiessen, 2002). Errors can basically be corrected and are therefore no fundamental argument against the objectivity of cognition in the above-mentioned sense. But the restriction to a particular perspective is founded in the nature of the matter. The geographical, temporal, historical, social, economic, physical or otherwise determined viewpoints from which things are observed and judged cannot be avoided *as such* and necessarily lead to limited judgements. This is generally recognised nowadays and is argued by e.g. the genetic epistemology of Jean Piaget, the evolutionary epistemology of Gerhard Vollmer and others, the biologically based constructivism of Humberto Maturana, the consciousness theory of dialectical materialism, the critical social theories of Jürgen Habermas and Pierre Bourdieu, and the feminist epistemology in the USA (Schneider, 1998).

Steiner was also aware of the restricted perspective of cognition such as the external dependence on things like the geographical or temporal position of the subject relative to the object as well as the internal limitation of the percept-picture due to the physical constitution of the subject. "My percept-picture changes when I change the place from which I am looking" (Steiner, 2011, 53); and the conditions of the physiology and pathology of the senses, such as the phenomenon of colour blindness, clearly prove "how our world of percepts is dependent on our bodily and [mental] constitution" (ibid., 53). Steiner calls the dependence of the percept-picture on the place of observation a "mathematical" dependence and that of the subjective constitution a "qualitative" one (ibid., 53). "My percept-pictures, then, are in the first instance subjective" (ibid., 54), and "the percept is partly determined by the [constitution] of myself as subject" (ibid., 55).

However, the restriction of cognition to specific perspectives does not fundamentally exclude the objectivity of knowledge, as understood by Steiner. Because in Steiner's concept of experience or perception, it is not a matter of the things "as such" but the objects "to the extent that [they] are accessible to our observation" (Steiner, 1988a, 20), and "in so far as the conscious subject apprehends them through observation" (Steiner, 2011, 51). This implicitly includes both the external and internal dependence of the percept-picture. *This* percept-picture is the one *given* to us by observation and *to this degree* objective[36]: we do not know any other and could not speculate about it. And

---

36 Due to the physiological *dependency* of the percept-picture it is easy to be tempted to deduce its physiological *causation* and to explain the content of the percept as altogether subjective (cf. Chapter 5). In sensory physiology it has in fact become customary to think like this. However, if the objective nature of the sensory phenomena is rejected and explained by the physiological processes of the sense and

the law of *this* percept must be found through cognition. We can only know the law itself because we can *experience* it in thought activity as a given and objective fact in the above-mentioned sense. We have no other way of approaching reality than through percept and concept.

Objective cognition is therefore always dependent on our perspective and situation i.e. the grasping in thought of the actually observed percept. However, this also implies that objective knowledge is not annulled by objective knowledge from other perspectives because, for the relevant perspective, it is in fact correct: however, it is *expanded by other knowledge to a more comprehensive total knowledge*. Knowledge is not objective primarily because of being *identical* to other knowledge, but by the agreement of percept and concept. Identity, or reproducibility, is indeed *also* a characteristic of objective knowledge, but only assuming an identical perspective.

While one and the same sculpture reveals very differing aspects from different sides, these call for different concepts which correspond to these aspects. But each of these *different* pieces of knowledge has *equal* objective value as long as this correspondence is empirically verifiable. And this applies to everything which is real. It plays a particularly important role in medicine, in which the various sciences and medical disciplines try to describe, understand and treat the human being with his various physical, psychological, mental and social aspects from these different perspectives (Matthiessen, 2008).

This is also connected to the *concept of evidence*. The objective *perception* of the congruence of concept and concept in the concept-judgement and of concept and experience in the percept-judgement is what can justifiably be called evidence in the original sense of the word (evidentia: Latin for being apparent). In *this* sense, the evidence which is sought for in *evidence-based medicine* is actually not real evidence. Because, based on statistical probability calculations, this latter makes outer connections *probable*, but does not reveal the *inner law* of these connections. However justified and necessary this form of statistically based evidence is, from an epistemological point of view it is therefore not only justified but also necessary to try to obtain a clinical knowledge whose evidence is based on insight into inner, lawful connections between phenomena, in other words a *"cognition-based medicine"* (Kiene, 2001, cf. also Section 8.5).

An important consequence of evidence in the sense of the objective knowledge meant here is also *certainty*. This *manifests* in the contemplation of the logical agreement of laws and the agreement of law and phenomenon in accordance with reality in the sense mentioned above. If the objectivity of the

---

nervous system, then the physiological facts as objective reality also have to be rejected, because they also have to be found through sensory perception. However, this would nullify the subjectivity argument itself (Heusser, 2000a).

laws and the possibility of insight into their objective reference to the world of perception is rejected, as for example by Popper, then the possibility of certainty in the sense meant here will also not be accepted.

The view of objective knowledge and certainty held here is therefore contrary to Popper's idea of certainty, at least in its second part. As mentioned in Note 28 Popper at most accepts "for certain truth" the above-mentioned logical agreement of laws (the propositions of formal logic and of arithmetic). "All else – by far the most important part of objective knowledge, and the part that comprises the natural sciences, such as physics and physiology – is essentially conjectural or hypothetical in character; there simply are no sufficient reasons for holding these hypotheses to be true, let alone certainly true" (Popper, 1972, 75). Or: "Apart from *valid* and *simple* proofs in world 3 objective certainty simply does not exist. [...] *From the point of view of objective knowledge, all theories therefore remain conjectural*" (ibid., 79–80). For Popper there is therefore *no scientific verification* but *only falsification*: "The method of science is the method of bold conjectures and ingenious and severe attempts to refute them" (ibid., 81). "All we can do is to search for the falsity content of our best theory" (ibid., 81). This is done by severe testing in which, apart from the logic of the theory to be proved, "the detailed evidence is naturally empirical". However, even when "it passes all these tests", we cannot prove its truth, but only objective reasons for the *"conjecture that the new theory is a better approximation to truth than the old theory"* (ibid., 81).

But this thought process of Popper's is not consistent: how can the falsity of a (logical) theory be tested against the given world of phenomena? Simply by *contemplating the agreement between theory and phenomenon empirically*. The discovery that this agreement is *not* present, i.e. the *falsification*, actually presupposes the potential knowledge of *whether* this agreement is present and therefore the possibility *that* it is present, i.e. the *verification*. This constitutes a crucial contradiction in Popper's thinking. This also becomes apparent by applying his theory to himself. When he writes e.g.: "We can never rationally justify a theory – that is, a claim to know its truth" (ibid., 82), he presents this to us in the form of a truth which he, Popper, believes to have recognised as certain: otherwise this statement would not make sense. So he credits his *practice* of cognition with a capacity which he *theoretically* denies to knowledge, i.e. to be able to know truth. In my opinion the reason for this inconsistency lies in the fact that, in the theory of knowledge, Popper is too one-sidedly rationalistic and pays too little attention to empiricism. In his theory of knowledge he does not truly start from observation of the cognition he himself performs but only "theorises" about it. This is also shown by the fact that, while on the one hand he defines truth empirically as "correspondence with the facts (or with reality)" (ibid., 44), on the other he claims in a one-sided rationalistic way that "increase in truth content" of a scientific theory is "a purely logical affair" (ibid., 81).

For Steiner, certainty is not only a logical but primarily an *empirical* matter, both in purely logical cognition as well as in that related to the senses, in which achieving certainty is not merely a matter of the *formal* agreement of one law with another or of a law and a phenomenon, but the *observation* of the agreement of their *content*. Certainty is the result of observing the objective elements of cognition and their relation to one another (Steiner, 1981a, 68):

> Thinking says nothing a priori about the given; it produces a posteriori, i.e. the thought-form, on the basis of which the conformity to law of the phenomena becomes apparent. Seen in this light, it is obvious that one can say nothing a priori about the degree of certainty of a judgment attained through cognition. For certainty, too, can be derived only from the given.

In connection with the above-mentioned perspective of cognition, the question of certainty plays an important part in the fundamental issues of natural science and psychology which are of importance to medical anthropology. So, for example, the empirically justifiable certainty that I see a red colour is not annulled by the certainty gained from the observation methods of physics that oscillation phenomena of a particular wavelength are present in the same part of space; or the empirically obtained certainty that I am able to act freely out of insight (Steiner, 2011) is not diminished by the certainty that particular neurophysiological conditions are necessary for this (Bieri, 2001); and the same applies to all the emergent levels of the complexly structured reality of human beings and nature (cf. Chapter 5).

2.5 Science and reality

One of the main questions in epistemology concerns the relationship of human ideas to "reality". As with everything else, the nature of reality can only be decided on the basis of perception and thinking. Reality is thus not a question of theory but of empirical knowledge. It goes without saying that acceptance of the outer sensory and inner mental empiricism as a *given* and therefore something objective, as is the case in Steiner's theory of knowledge, must necessarily lead to a totally different concept of reality than, for example, the view of knowledge taken by Kant, Popper and most of the epistemologists of the 20th century who do not accept this objectivity and therefore hold reality in itself to be inaccessible or to be a subjective construct (Schneider, 1998). In terms of Steiner's epistemology it has been shown that all the elements of cognition must be given as experience: the percept, the concept and their congruent agreement (Steiner, 1988a, 36):

In this way experience is deepened with the help of experience itself. Our epistemology imposes the demand for experience in its highest form; it rejects any attempt to bring something into experience from outside it. Our epistemology finds, within experience, even the characterizations that thinking makes.

According to this, the *principle of the scientific process* consists of penetrating the percept given by sensory observation with the lawful ideas brought to light by thinking. It is only through this penetration and not simply by observation that reality is grasped (ibid., 53):

> Science permeates perceived reality with the concepts grasped and worked through by our thinking. Through what our spirit, by its activity, has raised out of the darkness of mere potentiality into the light of reality, science complements and deepens what has been taken up passively. This presupposes that perception needs to be complemented by the spirit, that it is not at all something definitive, ultimate, complete. The fundamental error of modern science is that it regards sense perceptions as something already complete and finished. It therefore sets itself the task of simply photographing this existence complete in itself[37]. To be sure, only positivism, which simply rejects any possibility of going beyond perception, is consistent in this regard. [...] In the true sense of the word this requirement would be satisfied only by a science that simply enumerates and describes things as they exist side by side in space, and events as they succeed each other in time[38]. [Modern natural science] sets up a complete theory of experience in order then to violate it right away when taking the first step in real science.

---

37 In this sense the role of thinking in cognition was in fact seen for a long time as the production of a kind of mental image of reality (Schneider, 1998). With reference to the contemporary Kantian epistemologists Steiner argues that: "As we have already noted above, many people think of the entire system of concepts as in fact only a photograph of the outer world. They do indeed hold onto the fact that our knowing develops in the *form* of thinking, but demand nevertheless that a 'strictly objective science' take its content only from outside. According to them the outer world must provide the substance that flows into our concepts. Without the outer world, they maintain, these concepts are only empty schemata without any content" (Steiner, 1988a, 48). This corresponds to Kant's view, for whom concepts without a sensory perception are "empty concepts", "mere forms of thought, without objective reality since we have no [perception] at hand. [...] Only our sensible and empirical [perception] can give to them body and meaning" (Kant, 1926, 162–163). In Steiner this relationship between percept and concept appears the other way round: it is not the percept which gives the concept sense and meaning, but the concept which performs this for the percept; this is the phenomenon and that is the lawful meaningful idea or essence which brings clarity to the phenomenon.
38 Strictly speaking the objects would not even be allowed to be *described* if science was to limit itself to sensory experience. Because the act of describing also requires the help of concepts which are *added* to the sensory given element of the objects. However, concepts merely for the purpose of description do not have a defining

Because in real science it is not merely a matter of the content of experience as such, but about its *lawful connections* and these have to be found through thinking and *added* to the content of experience.

However, according to what was presented earlier, the matter in question is those lawful connections in accordance with which the objects of perception *actually* prove to be arranged as defined by the above-mentioned empirical concept, e.g. the Fibonacci law in the arrangement of the sunflower seeds. Reality does not therefore reveal its totality as an immediate experience but only the phenomenon accessible to sensory observation. The law according to which this phenomenon is constituted must be found through thinking. Steiner (Steiner, 1988a, 53–54) therefore goes on to say that:

> One disparages thinking if one takes away from it the possibility of perceiving in itself entities [i.e. laws] inaccessible to the senses. In addition to sense qualities there must be yet another factor within reality that is grasped by thinking. Thinking is an organ of the human being that is called upon to observe something higher than what the senses offer. *The side of reality accessible to thinking is one about which a mere sense being would never experience anything.* Thinking is not there to rehash the sense-perceptible but rather to penetrate what is hidden to the senses. Sense perception provides only *one* side of reality. The *other* side is a thinking apprehension of the world.

Cognition would be meaningless if reality did not contain both elements, phenomenon and law (Steiner, 1981a, 69–70):

> We have established that the nature of the activity of cognition is to permeate the given world-picture with concepts and ideas by means of thinking. What follows from this fact? If the directly-given were a totality, complete in itself, then such an elaboration of it by means of cognition would be both impossible and unnecessary. We should then simply accept the given as it is, and would be satisfied with it in that form. The act of cognition is possible only because the given contains something hidden; this hidden does *not* appear as long as we consider only its immediate aspect; the hidden aspect only reveals itself through the order that thinking brings into the given. In other words, what the given appears to be *before* it has been elaborated by thinking, is not its full totality. [...] Therefore, the given world-picture becomes complete only through that other, indirect kind of given which is brought to it by thinking.

---

function but simply a *referring* one: they then serve only *to guide the observing glance to the phenomenon*: "Look at this pattern in the sunflower seeds!" (Admittedly "pattern" is also a characterisation which is attached to the sense impression, but actually in terms of the description, not the cognition.) The concept has a *defining cognitive function* in the following judgement: "The pattern follows the geometrical law of a Fibonacci series".

Thinking actually adds the *essential part* – the one that *constitutes the phenomenon lawfully from within* – of this totality, and it is only the totality, in other words the unity of phenomenon and law, which creates reality (Steiner, 1981a, 71):

> Knowledge therefore rests upon the fact that the world-content is originally given to us in incomplete form; it possesses another essential aspect, apart from what is directly present. This second aspect of the world-content, which is not originally given, is revealed through thinking. [...] *The world-content can be called reality only in the form it attains when the two aspects of it described above have been united through knowledge.*

Like the logical concept of truth discussed above, this concept of reality should be understood *qualitatively* rather than quantitatively. It is not a matter of "reality in general", but of knowledge of that aspect of reality *accessible* to the individual: knowledge of reality exists to the extent that the percept available to observation has been understood according to its law and connection. This also implies that this reality is put into perspective in the framework of a more complete reality in which, however, it retains its quality of reality for a brief moment as long as no errors are present.

In view of the perspectivity of knowledge discussed above, this means that reality, obtained from *one* point of view (e.g. the facts of neurophysiological processes in the brain) cannot be annulled by the reality seen from other points of view, nor does it need to contradict it (e.g. the reality of thinking). This does however mean that – in medicine in particular – there is need of an epistemologically based multi-perspectivity if the "real human being" is to be treated adequately in theory and practice (Matthiessen, 2002). This has become even more evident with the advent of integrative medicine (Matthiessen, 2008). One aim of this book is to show that Steiner's concept of knowledge and reality can make an important contribution to this kind of multi-perspectivity.

The objectivity of the percept, the concept and their agreement shows that it is due to the *nature of our cognition* and not to reality, that the phenomenon and the lawful part of reality are acquired in separate acts and then need to be united. Reality is lawfully organised phenomena. The phenomenon is given directly to observation. But its law must first emerge through the activity of thinking and then be brought into alignment with experience. "It is not due to the objects that they are given us at first without their corresponding concepts, but to our mental organization. Our whole being functions in such a way that from every real thing the relevant elements come to us from two sides, from *perceiving* and from *thinking*" (Steiner, 2011, 74). Or put differently: "If, in the world-content, the thought-content were united with the given from the first, no knowledge would exist" (Steiner, 1981a, 70).

In summary, Steiner's concept of reality in connection with his view of knowledge is as follows: "The percept is thus not something finished and self-contained, but only one side of the total reality. The other side is the concept. The act of knowing is the synthesis of percept and concept. Only percept and concept together constitute the whole thing" (Steiner, 2011, 77).

2.6 Objective ontological idealism

According to Steiner's epistemological view, the reality of a thing therefore lies not only in what we perceive of it, but also in the law which constitutes the phenomenon. The reality is the totality, the unity of phenomenon and law. *The law is part of the reality.* In fact it is its *important, essential* part. It is the task of science or the different sciences to find this. For these sciences are characterised by the fact that they address different areas of manifestation of the manifold branches of reality.

Steiner's view of knowledge and the concept of reality which arises from this provide the epistemological basis of a modern *objective, ontological idealism* which is also known in the history of philosophy and science as *"idea or law realism" or universal realism* (Schneider, 1985), i.e. the view that the laws brought forth by thinking have an objective real nature and are inherent in the corresponding phenomena as their defining essence (Steiner, 1981a, 87):

> Our theory of knowledge supplies the foundation for true idealism in the real sense of the word. It establishes the conviction that in thinking the essence of the world is mediated. Through thinking alone the relationship between the details of the world-content becomes manifest [...].

The objectivity of the content of the laws as such was already mentioned in Section 2.3 in an attempt to provide an argument for objective idealism from the viewpoint of inner mental empiricism. In addition, in Sections 2.4 and 2.5, it was pointed out from the standpoint of outer sensory empiricism that the phenomena follow these laws so that they must actually be manifest in nature itself.

In this sense, three conditions of the laws or universalities must be distinguished which, in the high scholasticism of the Middle Ages, were described as follows (Willmann, 1907, 357): 1. The *Universalia ante rem*, the general laws (universals) as such, quite independently of the particular form of their appearance in individual things or consciousnesses, 2. The *Universalia in re*, the same laws, but as far as they constitute the individual things, and 3. The *Universalia post rem*, once again the same laws but in as far as they are thought as abstractions by human beings[39].

---

39 The three conditions of the general laws or universals were summarised by Albertus Magnus as follows: "Resultant tria formarum genera: unum quidem ante

Objective idealism or universal realism is a well-known position in the history of western philosophy and science which reappears frequently in its various forms. Important proponents of this were Plato (427–347 B.C.) and Aristotle (384–322 B.C.) in ancient times, Anselm of Canterbury (1033–1109), Albertus Magnus (around 1200–1280) and Thomas Aquinas (1225–1274) in the Middle Ages, Nikolaus of Kues (1401–1464) and Baruch Spinoza (1632–1677) at the beginning of the new scientific age, F. W. J. Schelling (1775–1854), G. W. F. Hegel (1770–1831), J. W. Goethe (1749–1832) and F. Schiller (1759–1805) in the period of German Idealism, Rudolf Steiner (1861–1925), Nicolai Hartmann (1882–1950) and Alfred North Whitehead (1861–1947) and the physicists Werner Heisenberg (1901–1976), Walter Heitler (1904–1981) and Carl Friedrich von Weizsäcker (1912–2007) in the 20th century and, currently, the philosophers Dieter Wandschneider (born 1938) in Aachen and Vittorio Hösle (born 1960) in Notre Dame, Indiana. All these thinkers hold the opinion – in various forms and approaches – that laws and ideas are objective real entities according to which – independently of the human subject – the world of phenomena is formed "outside", but which can be made manifest in the human spirit by thinking. What is important here is not the philosophical position of idea or law realism itself, but its epistemological justifiability on the one hand and its *applicability* to the facts of empirical science on the other, for example to molecular biology (Heusser, 1989), evolutionary theory (Heusser, 2001a; Hösle & Illies, 2005; Ziegler et al., 2015) and the understanding of illness (Heusser, 1997).

Historically and principally, law realism has been set in opposition to what is known as *nominalism*, ever since the famous argument about universals in the Middle Ages. According to nominalism, the laws produced by the subject's thinking are not considered as objective real entities, but merely *the human mind's subjective criteria for creating order* which serve the purpose of arranging the sensory phenomena of the world. They have no existential meaning for objective reality, but are only ideational terms or "names" (Latin: nomina) for it. Well-known nominalists of this kind were the stoics in ancient times, Johannes Roscelin (1050–1124) and William of Ockham (1288–1347) with whom the term originated in the Middle Ages, Francis Bacon (1561–1626), René Descartes (1596–1650) and John Locke (1632–1704) at the start of the present scientific age, later David Hume (1711–1776) and Immanuel Kant (1724–1804), Herrmann von Helmholtz (1821–1894), and Karl Popper

---

rem existens, quod est causa formativa; aliud autem est ipsum genus formarum, quae fluctuant in materia; tertium autem est genus formarum, quod abstrahente intellectu, separatur a rebus" (Alb. M. de nat. et orig. an. I, 2. In: Willmann, 1907, 357).

(1902–1994) and constructivism in the 20th century. The nominalist position generally dominates present-day natural science and medical theory.

Nominalism inevitably leads to *one-sided empiricism* or *sensualism* as is well-known in the case of Hume; because if the laws of nature do not belong to the world of phenomena, then the search for reality can only take place in the world of phenomena itself. The result is that this world is then either itself held to be reality in the naive realistic sense, directly as it appears, or the world of phenomena is explained as pure semblance on the basis of physical sensory physiology or other considerations and the "actual" reality sought behind this (Steiner, 2011, Chapters 4 and 5). Sensualism develops further into *materialism* when actual reality is only sought behind or beneath the physical sensory phenomena (and not for example in a similar way behind the psychological phenomena, which could also be possible) and is simply transferred into the molecules, atoms or sub-atomic particles. Materialism finally becomes materialistic *reductionism* when the higher phenomena are explained by the "reality" of the atoms and elementary particles and these actual phenomena are thus made into mere epiphenomena of the "actual" reality.

Since the dawn of the scientific age and in the spirit of Francis Bacon (1561–1626), natural science and in particular medicine and the sciences which serve it have indeed followed this path: the renunciation of the real ideas of Plato and Aristotle and what has remained of this as mere "idols" (Bacon, 1990a, Aph. 63–77), which are "nothing more than arbitrary abstractions" (ibid., Aph. 124), the development of the experimental inductive method in order to move from empirical details to systematic rules and to the discovery of sensory empirical causes (ibid., Aph. 82–120). However, this requires the body to be divided up right to "its really smallest parts", everything must be measured and counted until finally "physics [ends] in mathematics" (Bacon, 1990b, Aph. 8).

The history of medicine offers an accurate picture of this extremely successful dissection of the whole into its parts: from the founding of modern anatomy by Andreas Vesalius (1514–1564), via the discovery of the circulation by William Harvey (1578–1657) as a starting point for physiology, to the discovery of cells by Matthias Schleiden (1804–1881) and Theodor Schwann (1810–1882), the development of pathological anatomy and the first causal tracing of disease back to purely physical structures by Giovanni Battista Morgagni (1682–1771), the replacement of humoral pathology by a cellular pathology in the 19th century by Rudolf Virchow (1821–1902), the rejection of vitalism through biochemistry (urea synthesis 1826) by Friedrich Wöhler (1800–1882), the development of modern physiology in the 19th century by Emil Du Bois-Reymond (1818–1896), Ernst Wilhelm von Brücke (1819–1892) and Herrmann von Helmholtz, the development of modern genetics and molecular biology in the 20th century and the computer-aided mathematical recording of complex molecular processes in systems biology at the beginning of the 21st century to explain life processes (cf. Section 4.8). In a certain sense Francis Bacon's vision can therefore be said to have come to pass.

Symptomatic of the intention which lies at the source of this whole striving – part conscious, part unconscious – is a letter by Du Bois-Reymond in 1842 where he writes about Brücke's first attempt to explain all the vegetative processes in the organic body by physical means and says (Du Bois-Reymond, 1918, 108; quoted in: Varela, 2009, 124):

> Brücke and I pledged a solemn oath to put in power this truth: No other forces than the common physical-chemical ones are active within the organism; in those cases which cannot at the time be explained by these forces one has either to find the specific way or form of their action by means of the physical–mathematical method, or to assume new forces equal in dignity to the chemical-physical forces inherent in matter, reducible to the force of attraction and repulsion.

The completely mechanistic explanation of the world which this entails is the fundamental paradigm in biology to this day and the dominant theories of body, soul and consciousness as represented in typical textbooks and standard scientific works all over the world, for example 1996 in Thomas Metzinger's *"Bewusstsein"* (Metzinger, 1996), and 2003 in Purves' *"Life: The Science of Biology"*[40] (Purves et al., 2003).

It may be presumed that the tremendous 19th century development in modern medicine with its knowledge of the body's material structures and functions going into ever increasing analytical detail is mainly due to this intention of tracing everything back to chemical and physical principles. This is certainly a unique achievement in the cultural development of mankind, one which cannot be valued highly enough. However, the negative aspects of this development mentioned in the introduction also came to light. What was meant by scientific explanation was an explanation from a physical and chemical perspective. The attributes of life, the inner emotional experiences and the spiritual individuality of the human being were viewed from this perspective, prescientific attempts to explain these facts were naturally abandoned, but the inclination towards an adequate scientific development

---

40 "A major discovery of biology is that living things are composed of the same types of chemical elements as the vast non-living portion of the universe. This *mechanistic view* – that life is chemically based and obeys universal physiochemical laws – is a relatively recent one in human history. The concept of a *'vital force'* responsible for life, different from the forces found in physics and chemistry, was common in Western culture until the nineteenth century, and many people still assume such a force exists. However, most scientists adhere to a mechanistic view of life, and it is the cornerstone of medicine and agriculture" (Purves et al., 2003, 15).

from other perspectives – likewise belonging to the nature of the human being – failed to materialise. It has become apparent to many people nowadays that this situation must change. This work attempts to show that the view of science and reality held by Goethe and Steiner can contribute to the development of science from a perspective of this kind.

2.7 Summary

The aim of the science of knowledge is to find the law of knowledge itself in order to be able to use this activity (knowing) – which is practiced in all acts of cognition but seldom reflected on consciously – in a fully conscious way ("critically"). The science of knowledge therefore rests on the knowledge of knowing itself. The reason that the theory of knowledge must necessarily be based on a direct observation of the mental process of cognition and not on an investigation of the physiological processes in the brain is established using the theory of knowledge as defined by Goethe and Steiner. The concept of experience (in terms of percept, phenomenon) is developed and it is shown to what degree experience can claim to be objective. Next, thinking itself is shown to be an empirical fact and it is established that the content of thinking (the lawfulness) which the thinking activity of the subject brings to light can likewise be considered as objective. In cognition the thought content is added to the percept through thinking. It is demonstrated that objective knowledge can be achieved through the objective agreement of the law and the phenomenon. This contrasts with the dominant opinion which follows Popper and others. This concept of cognition gives rise to the corresponding concept of reality: reality is not only to be found in the percept-object but also in the law which constitutes it. The law (something ideal, spiritual) is therefore recognised as part of reality and in fact as its more important i.e. essential part, because it is through the law that the other part, the phenomenon, is constituted. This gives rise to the foundation of an objective ontological idealism which has repeatedly been represented in the history of western thought since Plato and Aristotle via the scholastic universal realists to Schelling, Hegel, Goethe and Steiner and latterly by 20th century physicists, and again in the present day by several philosophers. The fundamental concepts of anthroposophical medicine are based epistemologically on idea or law realism whereas the fundamental concepts of conventional medicine primarily rest on nominalism following Kant and Popper. In this latter, the lawfulness which science brings to light in the process of cognition is seen as something belonging solely to the thinking subject to which no ontological meaning attaches for the manifesting object. The result of this is the conceptual reduction of reality to what can be grasped with the senses or what is revealed behind these, in the last instance to the world of atoms and elementary particles.

3 Science and ontological idealism in physics and chemistry

3.1 Empirical ontological idealism instead of reductionism

For the purpose of scientific explanation, the physical and chemical view of scientific subjects adopted since the middle of the 19th century has led to the reduction of the *actually observed* chemical, biological and psychological phenomena to their underlying smallest physical elements which are no longer perceptible. This is what is known as the *reductionist method* of modern science which Emil Du Bois-Reymond formulated in his famous lecture *"Über die Grenzen des Naturerkennens"* (On the limits of our understanding of nature) in the following words: "Understanding nature [...] means to attribute the changes in the physical world to movements of atoms which are caused by nuclear forces independent of time, or to *break down the processes of nature into the mechanics of atoms*". He follows this with a psychological reason for this process: "It is a well-known psychological fact that, where such a separation is successful, our need for an explanation feels satisfied for the time being. The theorems of mechanics can be described mathematically and contain the same apodictic certainty as the theorems of mathematics" (Du Bois-Reymond, 1882, 10). A similar description of the task of science is given by Helmholtz in his essay *"Über Goethes naturwissenschaftliche Arbeiten"* (On Goethe's Scientific Researches) published in 1853 (von Helmholtz, 1995, 12–13):

> For a natural phenomenon is not considered in physical science to be fully explained until you have traced it back to the ultimate forces which are concerned in its production and its maintenance. Now, as we can never become cognizant of forces *qua* forces, but only of their effects, we are compelled in every explanation of natural phenomena *to leave the sphere of sense*, and to pass to things which are *not objects of sense*, and are defined only by abstract conceptions. [... In the final instance] this is *a world of invisible atoms and movements, of attractive and repulsive forces*.

The consequence of this course of action is that the *sense perceptions are seen as an illusion* and *the world of atoms as the true reality* from which everything else can be derived. Steiner therefore described the scientific view resulting from this as "natural-scientific illusionism" (Steiner, 1988b). Here the "real" atoms, which are of course invisible, are represented by *hypothetical models*. In order to be able to picture them it is clear that there is no other option due to the non-perceptible nature of these entities, and as such this is quite legitimate. Problems only arise if the models are ontologically held to be more real than the actual perceived phenomena and the latter then viewed as an illusion and therefore unreal.

Eugen Kolisko (1893–1939), a close colleague of Rudolf Steiner and Ita Wegman and one of the first and most notable anthroposophical doctors (König, 2000) and scientists working according to Goethe's methodology (Kolisko, 1989), summarised this development as follows (Kolisko, 1922a, 170):

> Science has consequently developed in such a way that *the pure phenomena have moved into the background* and we now tend to have the *mental picture* of the hypothetical mechanical causes in our consciousness. *The mental image of something mathematical and mechanical takes the place of the phenomenon.* [...] In addition, we view all phenomena from the perspective of *cause and effect.* We always imagine a cause which does not lie within the phenomenon but is added to it, as though working from the invisible and creating the elements which are before us.

But what is the nature of these models, the atoms and molecules underlying the phenomena, which are still used nowadays in standard biology textbooks e.g. such as Bohr's model of the atom (Purves et al., 2006, 23)? Basically that they are composed of elements which are *taken or derived from sense perception*. This applies to both the hypothetical atomic spatial configurations (e.g. in the Bohr model) and to the quantitative bonding characteristics of matter. These are in fact initially found macroscopically in the realm of sense perception and not by direct observation of atoms: but they are then transferred to the *imagined* atoms and their relationships within molecules. Thus – as Kolisko stated – a "massive *hypothesis system* was developed in modern science which does not describe how nature *is* but how it *can be imagined* if you start with certain *assumptions*. And the assumptions usually made are mathematical and mechanical ones. The question which the entire classical scientific way of looking finally resulted in was: how can you explain the phenomena [...] when you assume that they have arisen from *mathematical and mechanical causes?*" (Kolisko, 1922a, 168).

The properties derived from sense perception which are attributed to the atoms in the models do not, however, include all sensory characteristics but only selected kinds, as Steiner pointed out in his commentary on Goethe's natural-scientific writings in reference to Descartes[41], these being magnitude,

---

41 "It is not without justification that one has seen in the following words of Descartes the basic formula by which the modern view of nature judges the world of perceptions: 'When I examine corporeal things more closely, I find that very little is contained in them that I can understand *clearly* and *definitely*, except: magnitude, or extension in length, depth and breadth; shape, that results from the limits of this extension; location, that the variously shaped bodies have relative to each other; and motion, or change in this location; to which one may add substance, duration

i.e. spatial size in length, width and depth, shape, location, motion, number, duration, etc., in other words what John Locke (1632–1704) called "primary" sensory qualities which are objective characteristics of the outer things. On the other hand, the atomic models contain neither light or colour, nor sound, taste, heat or cold, smoothness or roughness, because these "secondary" sensory qualities as defined by Locke are, unlike the primary ones, only attached to the feeling subject i.e. interpreted as products of the human organism.

However, in *"Goethean Science"*, Steiner rightly pointed out that *both* kinds of sense qualities, the primary and the secondary, can only be found through sense perception and to this extent are of equal importance in epistemological and ontological terms (Steiner, 1988b, 241–242):

> Magnitude, shape, location, motion, force, etc., are perceptions in exactly the same sense as light, colors, sounds, odors, sensations of taste, warmth, cold, etc. Someone who isolates the magnitude of a thing from its other characteristics and looks at it by itself no longer has to do with a *real* thing, but only with an abstraction of the intellect. It is the most nonsensical thing imaginable to ascribe a different degree of reality to an abstraction drawn from sense perception than to a thing of sense perception itself. Spatial and numerical relationships have no advantage over other sense perceptions save their greater simplicity and surveyability. It is upon this simplicity and surveyability that the certainty of the mathematical sciences rests. When the modern view of nature traces all the processes of the corporeal world back to something that can be expressed mathematically and mechanically, it does so because the mathematical and mechanical are easy and comfortable for our thinking to deal with.

The satisfaction of our need for causes and the "apodictic certainty" of mathematical and mechanical laws are, however – as stated by Du Bois-Reymond himself (cf. above) – *psychological* and therefore *subjective* reasons for preferring quantifiable primary sensory qualities rather than the purely qualitative secondary ones. On the other hand, the secondary sensory qualities cannot be deemed to be subjective because, as Steiner rightfully objects: "Only what is *perceived* as belonging to the subject can be termed subjective" (Steiner, 2011, 83, ePH). But – in terms of the empirical knowledge referred to above – colours, scents, heat, cold, sensations of touch, etc. are perceived on the object in the same way as magnitude, shape, motion, etc., and are

---

and number. As for other things – such as light, colors, sounds, odors, sensations of taste, warmth, cold and the other qualities that the sense of touch experiences (smoothness, roughness) – they arise within my spirit in such *an obscure and confused way* that I do not know whether they are true or false, i.e. whether the ideas that I grasp of these objects are in fact the ideas of some real things or other, or whether they represent only chimerical entities that cannot exist'" (Steiner, 1988b, 239).

therefore just as objective as these. "But with this there disappears any possibility of seeking anything in the motion of atoms that could be held up as something objective in contrast to the subjective qualities of sound, color, etc." (Steiner, 1988b, 197), and from which these could be causally derived. Because the concept of efficient causation can only be applied epistemologically to the phenomena in as far as it arises from the percept. This is not the case in the relationship of the model *imagined* atoms to the *actual* phenomenon. Apart from that, it is difficult to see how the *content* of the secondary qualities is to be derived from the primary qualities which are accorded only to the atoms.

The same applies in principle to the attempt attributed to the chemist Wilhelm Ostwald (1853–1932) to put *energy* in the place of atoms and their movement to explain sensory phenomena, in which recourse is made to a single property of the perceptible, i.e. the emission of energy, and from which sense qualities are then abstracted: "What we perceive are, as already stated, differences in energy states encountered by our sensory organs [...]" (Ostwald, 1895, 29). But what we actually *perceive* is not energy but colours, sounds, scents, etc. whose content cannot simply be reduced to energy. "Energy" is not something that can be perceived, but something which acts and only manifests through its *effects*. For this reason, in classical physics energy was generally described as the ability to do work, with measurable and therefore calculable results in the perceptible realm (Harten, 1974, 32). In this context the energy concept is defined without needing to take account of the *specific* nature of whatever this work performs. "The energy concept suffers primarily from the fact that, of the different qualities in nature such as light, electricity, heat, gas etc. it only takes account of what is common to all, namely the ability to perform mechanical work, and not what is *unique* to them" (Kolisko, 1922a, 221)[42].

However, regard for this "uniqueness" which appears as the specific aspect of the phenomenon, leads to the concept of *specific kinds* of "energy". For example, the concept formulated by the physiologist Johannes Müller (1801–1858) of "specific sensory energy" in which every sense organ only reacts to different stimuli in its own typical specific way is ontologically

---

42 Cf. also Steiner: "The phenomena of the perceptual world – light, warmth, electricity, magnetism, etc. – can be brought under the general concept of force-output, i.e. of energy. When light, warmth, etc., call forth a change in an object, an energy-output has thereby taken place. When one designates light, warmth, etc. as energy, one has disregarded what is specifically characteristic of the individual sense qualities, and is considering one general characteristic that they share in common" (Steiner, 1988b, 255). But this does not invalidate what is specific about this energy.

justified from this point of view. But these specific forms of energy cannot then simply be transferred to the subject's sense organs but for phenomenological reasons must also be recognised in the objective outer world (cf. Section 2.1). Particularising reductionism would then no longer be necessary in energetics[43].

Otherwise, the *consequence* of atomic or energetic reductionism would be the complete elimination of *all* qualities from the scientific world view, including the so-called primary qualities (Steiner, 1988b):

> Someone who makes the human organism into the creator of the happenings of sound, warmth, color, etc., must also make it the producer of extension, magnitude, location, motion, forces, etc. For, these mathematical and mechanistic qualities are, in reality, inseparably united with the rest of the content of the world of experience. The separating out of conditions of space, number and motion, as well as manifestations of force from the qualities of warmth, sound, color and the other sense qualities is only a function of our abstractive thinking. The laws of mathematics and mechanics relate to abstract objects and processes that are drawn from the world of experience and that therefore can find an application only within the world of experience. But if the mathematical and mechanistic forms and relationships are also explained as merely subjective states, then nothing remains that could serve as content for the concept of objective things and events. And no phenomena can be derived from an empty concept.

The basis for *reductionism* is thereby epistemologically and ontologically removed. It cancels itself out.

---

[43] For example, the *photoelectric effect* does not necessarily imply the corpuscular nature of light, an interpretation which has been held since Einstein's work in 1905 (Einstein, 1989). Just because the *constant* irradiation of an unoxidised negatively charged metal plate leads to the *discrete* release of electrons from this metal, *phenomenologically* this result in the first instance only means that the metal *reacts* to exposure to light with electrical "energy quanta" and not that the light is *composed* of quanta, something Steiner correctly pointed out in his first scientific lecture course (Steiner, 2001, Discussion Statement 8.8.1920). Based on the phenomena, it is also possible that, as Einstein himself admitted, light has *the character of a continuum*: "The energy of a ponderable body cannot be broken up into arbitrarily many, arbitrarily small parts, while according to *Maxwell's* theory (or, more generally, according to any wave theory) the energy of a light ray emitted from a point source of light spreads continuously over a steadily increasing volume" (Einstein, 1989, 86). Accordingly, it is basically possible that the effect of the light on the electrons in the metal develops *continuously over time* until they have reached a threshold above which they break away as a *discrete* energy quantum. In contrast to the electrons, light could be regarded in this case as a qualitatively different, non-particulate form of energy.

This final step, i.e. the total elimination of *all* sensory qualities has been accomplished by modern physics (quantum mechanics)[44]. In terms of its important physical and chemical properties, the atom consists primarily of electrons. However, the electrons themselves are no longer entities with a clear spatial existence or sensory properties (such as magnitude, shape, location, motion) independent of context. The physicist's view is that the electron "in and of itself" is best classified by its wave function, likewise the electrons in an atom (Eisberg & Resnick, 1985). This wave function is a mathematical function in an abstract space which has nothing directly to do with normal three dimensional space: the function itself also no longer has any visual meaning (Messiah, 1961). It is only through mathematical operations that this wave function is projected into normal space where it acquires the meaning of a probability of defined events for specific processes, primarily for energy parameters which are then measured indirectly in optical experiments (Bergmann & Schaefer, 1981). In actual fact there is not a single *qualitative* sense-perceptible phenomenon which can be derived from the concepts of quantum mechanics because the electrons do not display any sensory properties which are independent of the context. But what obviously works well are the calculations of quantitative energy relationships (e.g. wavelengths of specific colours in the emission spectra of gases) under the assumption of the manifestation of a non-sensory universalia ante rem (the pure mathematical wave function or the still-pure virtual electrons) into a universalia in re (electromagnetic energy states which are linked to actual observable light emissions of gases at high temperature) under specific sensory conditions (such as temperature, electromagnetic forces or potentials, etc.) (Bergmann & Schaefer, 1981). Interestingly, there are apparently no sense-perceptible conditions which determine the actual phenomenon (in actu) (known in quantum mechanics as principle indeterminism) (Messiah, 1961).

The way out of the dilemma, brought about by the non-recognition of sensory qualities leading to the ultimate loss of all sensory qualities and therefore reality, can only be the acknowledgement of *all* sensory qualities as epistemologically of equal value, and of a phenomenological science arising from this. Steiner comments (Steiner, 1988b, 244):

> Just as little as the processes of the corporeal world can be "broken down" into a mechanics of atoms, so just as little into states of energy. Nothing further is achieved by this approach than that attention is diverted from the content of the real sense world and directed towards an unreal abstraction, whose meager fund of characteristics, after all, is also only drawn from the same sense world. One cannot explain one group of characteristics of the sense world – light, colors,

---

[44] I wish to thank Dr. Stephan Baumgartner for the discussion in this section and the accompanying references.

sounds, odors, tastes, warmth conditions, etc. – by "breaking them down" into another group of characteristics of the same sense world: magnitude, shape, location, number, energy, etc. The task of natural science cannot be to "break down" one kind of characteristics into another kind, but rather to seek out the relationships and connections between the perceptible characteristics of the sense world.

Or: "Theory must encompass what is perceptible [...] and *must seek the interrelationships within this area.*" (ibid., 238). And again here: knowledge of reality as the intellectual quest for the objective lawful connections of the *perceptible*[45] phenomena as defined by Goethe (cf. Chapter 2).

---

45 This view, i.e. that reality or what is real can be found by intellectual penetration of the *percept*, can be countered by the argument that energy or forces are not fundamentally perceptible but must still theoretically be made responsible for the phenomenon produced. This is perfectly correct. Nevertheless, it does not change anything about the *principle* of cognition and the determination of reality. For it is *perceptible phenomena* which are seen to have been "caused", and in fact due to the *perceptible, lawful expression of this action*. Therefore the assumption of something acting in the realm of phenomena is logically justified, in fact Steiner claims that "it becomes necessary to speak not of immediately perceptible elements, but of non-perceptible quantities as in the case of lines of electric or magnetic force. It may *seem* as if the elements of reality of which physicists speak had no connection either with what is perceptible or with the concepts which active thinking has wrought. Yet such a view would be based on self-deception. The main point is that *all* the results of physical research, apart from unjustifiable hypotheses which ought to be excluded, have been obtained through percept and concept. Elements which are seemingly non-perceptible are placed by the physicist's sound instinct for knowledge into the field where percepts lie, and they are thought of in terms of concepts commonly used in this field. The strengths of electric or magnetic fields and such like are arrived at, *in the very nature of things*, by no other process of knowledge than the one which occurs between percept and concept" (Steiner, 2011, 110, ePH). It is precisely for *this* reason that it is not possible to make deductions from the *specific aspect* of any one percept. The *phenomena* of light, colour or heat are *not in the least the same* as those of electricity or magnetism, even though they may be *connected* to them. It is therefore completely inappropriate to describe the entire spectrum of "energy waves" from the radio band via infrared, visible light and ultra-violet all the way to X-rays and γ-rays simply as "electromagnetic" (Harten, 1974, 247). This suggests that light or thermal radiation are "no more than" electromagnetism, which certainly does not fit with the phenomena and is also not viewed in this way by the differential scientific treatment of the qualities in "thermodynamics", "electricity and magnetism", "optics", "atomic and nuclear physics", etc. (Harten, 1974). The task of science working from *empirical* arguments is not to reduce the different physical qualities and effects to *one* type alone, but to find and acknowledge the *specific qualities of natural phenomena* and the *qualitative differences of the physical*

The search for connections and relationships between the perceptible characteristics of the sense world: this is actually a description of Goethe's scientific method, namely phenomenology as "rational empiricism" as Schiller called it in his correspondence on this topic with Goethe (Schiller, 1935, Vol. 2, 237; Heusser, 2000c), or "empirical idealism" as Steiner calls it in his introductions to Goethe's scientific writings (Steiner, 1988b, 232).

This *rational phenomenological method*, as Kolisko also described it (Kolisko, 1922a, 183) is based, as shown in Chapter 2, on the exact observation of the actual phenomena and the determination of *their* lawful connection through thinking, in other words, without this thinking attaching preformed hypothetical models to the phenomena. In this respect the phenomenological science referred to here has to be "hypothesis-free", e.g. as in "hypothesis-free chemistry" whose basic principles were described by Kolisko as far back as 1920 (Kolisko, 1922a).

The starting point for rational phenomenology is therefore the determination of the pure[46] phenomena, *without the presupposition of interpretative concepts*[47]. Of course concepts are also needed to *describe* these phenomena.

---

effects with their own laws and forces *based on the phenomena*, and *only then* to examine their *relationships*. This is what forms the basis of Goethean science.

46 "Pure phenomena" in this context, as explained in Section 2.2 should be taken not in the sense of sensory psychology but in an *epistemological* sense, as *pure experience*. Such phenomena are initially free of all concepts added by thinking, including those which are taken from mechanics. Whether mechanical and mathematical concepts can be applied to them in the act of cognition can only emerge from the phenomena themselves, not from preconceived hypothetical concepts about how the world ought to be.

47 Cf. (Heusser, 2000c, 18ff.): for Goethe the starting point for knowledge of nature is *"the empirical phenomenon* which every person perceives in nature" (Goethe, 1950i, 869–871). The first duty of the scientist is therefore "diligence in observation" and: "The very first duty is the attention through which the phenomenon is secured" (Goethe, 1950f, 857). Goethe calls this "the experience" (Goethe, 1950i, 869) or "pure experience" (Goethe, 1950e, 770). But at this stage of cognition great care is required in establishing the object to be cognised. You need to guard against immediately yielding to the natural human need to judge the phenomena, "in order to be able to come to a conclusion as quickly as possible". For then you are in danger of merely imprinting your "opinion" which is "subjective" onto what is given empirically (Goethe, 1950f, 856–857). This caution, the conscious restraining of these kind of opinions, is what Kolisko calls "a kind of negative activity" of thinking by Goethe. Goethe says: "You cannot exercise enough care in this respect [...] not to jump too quickly to conclusions: because in moving from the experience to the judgement [...] is where, as though on a pass, all a person's inner enemies lie in wait: fancifulness, impatience, overhastiness, self-satisfaction, rigidity, thought-forms, preconceived opinions, indolence, carelessness, fickleness and whatever the whole company with its entourage might be called, all lie here in

But these only have a *referring* or *describing* character for the phenomenon, and do not *define* it by *judgemental* thinking. This is what is meant by the following statement by Kolisko (Kolisko, 1922a, 172–173):

> *Goethe* wanted something totally different from *this* kind of scientific approach. *He wanted to bring out the pure phenomena.* He considered it essential only to approach the objects with thinking after the phenomena had been described in their full purity. He knew that you can never achieve an insight into the being of the things [...] if you do not have the pure phenomena before you. Only the *pure* phenomena can behave in such a way that their relationship suggests itself to thinking as though of its own accord. For *Goethe* thinking is a kind of *negative* activity. He tries to exclude everything which could arise from prejudice in the human subject, including any favourite intrusive mathematical and mechanical or teleological ideas. He tries to become aware of the mental images which fit the phenomena and to throw light on the phenomena by means of these *appropriate* mental images. Goethe viewed the world not through the *lens of teleology* nor through the *lens of causality* but as demanded by *each set of facts*.

---

ambush and unexpectedly overpower both the active man of the world as much as the quiet observer who appears safe from all passions" (Goethe, 1950g, 848–849). It is only this caution which enables the scientist's task to be fulfilled, i.e. "observation of the objects of nature for themselves and in their relationships to one another". The scientist "as both a detached and simultaneously godly being" must "seek and study what actually is and not what feels right". He must examine all the empirically given objects "with an equally calm gaze and take the measure of this knowledge, the facts of the judgement, not from himself, but from the sphere of the things which he is observing" (ibid., 845). The observation should therefore not immediately be converted into the judgement, but be built up as systematically as possible. This is done by "raising the empirical phenomenon which every person perceives in nature [...] to the *scientific phenomenon* through experimentation, by presenting it in a more or less successful sequence under different circumstances and conditions to those in which it was first found" (Goethe, 1950i, 871). By *"experiment"* Goethe means the experimental, reproducible procedure like the one he himself followed rigorously in his *"Theory of Colours"* (Goethe, 1950o). "When we intentionally repeat the experiences which were made previously, which we carry out ourselves or with others, and again describe the phenomena which have arisen partly by chance, partly intentionally, then we call this an experiment. The value of an experiment primarily consists in the fact that it [...] can be reproduced at any time, as long as the determining conditions are the same" (Goethe, 1950g, 848). By *"sequence"*, Goethe means the production of a systematic *series* of individual experiments or observations with successive variations like those he had carried out in an exemplary way in his morphological work on plants (Goethe, 1950n) and on comparative anatomy and zoology (Goethe, 1950m) and through which he was also able to discover the basic principle of metamorphosis and the human intermaxillary bone (Goethe, 1950p).

In a certain sense this approach to cognition corresponds to the avoidance of what Eugen Bleuler (1857–1939) described from his more practical viewpoint as "autistic and undisciplined thinking"[48] and in contrast to which he demanded a "disciplined thinking" in all areas, especially in the field of qualitative, non-quantifiable facts (Bleuler, 1976). For it is possible "to think exactly and inexactly in all fields, with or without numbers and measurements [...]. The type of thinking which is demanded in the sciences is an observant and purely realistic one: we call it *disciplined* thinking and contrast it with the other forms as *undisciplined*" (ibid., 167).

The task of this disciplined thinking is to look for the lawful connections between the actual phenomena, in other words to put into practice the concept of knowledge described above. As we know, this is no simple matter. For, as Goethe remarked, "*the empirical* phenomenon which every person perceives in nature" (Goethe, 1950i, 871) rarely shows its elements arranged in such a way that their *lawful* connection is immediately apparent to thinking. Its *apparent* connection, as it arises for observation, can be accidental and is so in most cases. The empirical phenomenon must therefore be elevated "to a *scientific phenomenon*" (ibid., 871).

---

48 What *Bleuler* seeks for is *"disciplined thinking"*, which he places even higher than "scientific" and "exact" thinking for good reasons: "People like to contrast normal thinking with its sloppy and autistic sideways leaps to '*scientific*' thinking. I do not use this word, because it is subject to too much misuse [...]. [...] The expression not only refers to rigorously realistic and logical thinking freed from all idling and autism, but also thinking in terms of a particular science exactly as it is interpreted at the time, i.e. in terms of specific '*basic laws*' and thought formulas which are like scientific mnemonics and may well be useful for the pupil who needs to navigate a familiar realm, *but whose reliability must always first be tested when you enter a new field*" (Bleuler, 1976, 104). Mnemonics like this are e.g. the above-mentioned atomic mechanical and mathematical models, although Bleuler does not mention these. However, it is clear that he does not limit disciplined thinking to quantifiable properties but wishes to have it applied especially to qualitative phenomena which of course play a large part in medicine: "The demand for *exact thinking* is very popular and there are people who do not want to recognise a science if it is not 'exact'. However, we cannot use the expression here if only because it needlessly tends to recognise only thinking in dimensions and numbers. This only covers the smaller part of the thinking required for science; because in accordance with their nature all qualitative differences are inaccessible to numerical treatment [...]" (ibid., 106). It is important to be aware that the physiological and pathological, the physical, living, psychological and spiritual phenomena in the human being are to a large extent *qualitative* and therefore require the disciplined thinking demanded by Bleuler, without needing to be able to be expressed mathematically.

This is achieved through *experiments*, be this systematic comparisons with other phenomena or setting up artificial conditions with intentional repetition (Goethe, 1950g). It is then possible to analyse the initially incomprehensible complex phenomena in relation to the simple elements which are involved in their production. Coincidental aspects are eliminated and the necessary systematic aspects linked to each other in such a way *that their lawful connection becomes directly evident to thinking*. In this way, complex phenomena are reduced to simple comprehensible basic facts and can then be understood. Goethe called an empirical basic fact of this kind whose explanation *arises directly from the nature of its elements*, a "*pure phenomenon*", "*basic phenomenon*" or "*archetypal phenomenon*" (Goethe, 1950i, 871; Goethe, 1950o; Steiner, 1988b, 244–245), but also "the empirical law" (Goethe, 1950i, 870) which, in the words of Schiller, "is one with the objective natural law" (Schiller, 1935).

Sufficiently well-known examples of objective natural laws or pure phenomena are for example: "When two bodies of differing temperature are touching each other, then warmth flows from the warmer one into the colder one until the temperature is the same in both. When there is a fluid in two containers connected to each other, the water level will be the same in both. When one object is standing between a source of light and another object, it will cast a shadow upon this other object. Whatever is not mere description in mathematics, physics and mechanics must be [an] *archetypal phenomenon*" (Steiner, 1988a, 81). An archetypal phenomenon is therefore a basic fact to be found *in the percept* in which "the character of the process follows directly and in a transparently clear way out of the nature of the pertinent factors"; and it is therefore also "identical with [the] objective natural law. For in it is expressed not only that a process has occurred under certain conditions but also that it *had* to occur. Given the nature of what was under consideration there, one *realizes* that the process had to occur" (ibid., 80). A natural law of this kind therefore has the form: *When* this fact interacts with that one, *then – of necessity* – this phenomenon arises (ibid., 81). A natural law therefore only applies under specific *circumstances*. However, whether in fact all conditions in the presence of which the law is necessarily true are always known is another matter entirely which does not call into question the essential validity of the natural law.

This view of the *understandable, necessary* natural law differs from the inductivism dating back to David Hume (1711–1776) who attributed the laws simply to phenomenological *habituation* in a certain respect. Steiner commented on this view as follows (Steiner, 1988a, 80):

> It sees a phenomenon that occurs in a particular way under the given conditions. A second time it sees the same phenomenon come about under similar conditions. From this it infers that a general law exists according to which this event

must come about, and it expresses this law as such. Such a method remains totally outside the phenomena. It does not penetrate into the depths. Its laws are the generalizations of individual facts. It must always wait for confirmation of the rule by the individual facts.

This also applies to Popper, although in contrast to Hume he rejects a generalisation of the rules gained from repetition of the facts. However, according to Popper every so-called "natural law", such as that the sun rises over London every day, must also hold good through every new perceptible case. So for Popper there is *no verification* but *only a falsification* of laws or theories (Popper, 1972). He is apparently unaware of the fact that the validity of laws is always linked to the relevant *conditions* which apply in the area of perception and that, when these are present, the laws must apply *of necessity* due to observable reasons lying in these conditions, and therefore that this validity is also verifiable – *under the relevant conditions*. This also applies to the daily rising of the sun, for example; and the obvious predictability of the mechanical processes on which its indisputable success rests, is a consequence of this fact (cf. also Popper's rejection of objectivity in cognition, Section 2.4). This also applies to purely mathematical and geometrical laws. The unalterable basic fact that the sum of the angles of a planar triangle always add up to 180° is only valid under the *condition* of a non-curved surface. This is why "that which we express as judgements in mathematical or geometrical laws, [...] also [requires] an empirical verification just as much as what we express in phenomenology" (Steiner, 2005, 126).

This understandable and necessary form of the natural law is nevertheless independent of whether the observed facts themselves are quantitative or qualitative in nature i.e. whether we are dealing with primary or secondary sensory qualities. This is obvious from the examples mentioned above. Because these are of a *qualitative* nature, even before their more precise quantitative examination by means of size, number and weight, their laws can be read purely *from the phenomena* without deriving everything in a reductionist manner from hypostatised causal factors *behind* the phenomena.

The same applies to the basic phenomena established by Goethe in his *"Theory of Colours"*, such as in the formation of what are known as the dioptric colours from the interaction of light with darkness or opaque media (Goethe, 1950o, 63):

> The light with the highest energy, such as the sun, [...] is dazzling and colourless. [...] However, seen through an only slightly opaque medium, this light appears yellow to us. If the opacity of the medium rises or if its depth increases, then we see the light gradually taking on an orange colour which finally intensifies to deep red. In contrast, if darkness is viewed through an opaque medium illuminated by a light falling on it, we see a blue colour which turns increasingly lighter and paler as the opacity of the medium increases but appears increasingly darker

and deeper the more transparent the opacity becomes until, with the minimum degree of purest opacity, the most beautiful violet becomes perceptible to the eye.

Goethe goes on to show how, from these and other simple basic facts, the formation of the prismatic colours and other more complex colour phenomena can be understood, without having to have recourse conceptually to hypothetical assumed realities *behind* the phenomena for their explanation, but by remaining *purely within the realm of the phenomena themselves* and determining *their* lawful connections. Goethe's famous aphorism is also meant in this way: "The highest is to understand that all fact is really theory. The blue of the sky reveals to us the basic law of color. Search nothing beyond the phenomena, they themselves are the theory" (Goethe, 1950c, 723[49]), which Steiner comments on as follows: "Theory is the facts looked at *according to their essential* character, not an abstraction removed from the things" (comment by R. Steiner in: Goethe, 1975a, 376). In this respect Steiner has characterised Goethe's intentions in his Theory of Colours very aptly: "The aim was no other than to trace the rich variety of the world of colour back to a systematic whole so that from this whole each colour phenomenon is as easy for us to understand as any connection of spatial dimension is from the system of mathematics" (Steiner, 1961, 320).

The fact that Goethe did not develop his Theory of Colours mathematically does not detract from it at all. Logical scientific thinking in the realm of qualitative facts and laws basically means working with the same stringency as in the realm of quantities, with stricter adherence to the percept on the one hand and the logic on the other. Goethe therefore consistently distinguishes between mathematics and a *mathematical approach* (Ziegler, 2000; Steiner, 1988a). What is meant by "mathematical approach" is that the methodical strictness with which mathematics and geometry must proceed in their realms should penetrate the whole of science (Goethe, 1950g, 852):

> We must learn from the mathematicians to take care to place next to each other only the elements that are closest to each other, or rather to deduce from each the elements closest to it, and even where we use no calculations, we must always proceed as though obliged to render account to the strictest geometrician.

Goethe's Theory of Colours is itself the best example of a *systematic rational phenomenology*. It shows how qualitative phenomena with what are known as secondary sensory qualities can be studied scientifically with the same methodical consistency as quantities i.e. the so-called primary sensory qualities which are quantifiable through size, number and weight.

---

49 Translation taken from Wikipedia "Theory of Colours": (https://en.wikipedia.org/wiki/The ory_of_Colours. Viewed 16/12/2015).

This has often been misunderstood, and the value of Goethe's Theory of Colours has been called into question[50]. The reductionist approach thinks it necessary to attribute the qualia causally to quanta *behind the qualia*. An example of this is Helmholtz in his essay published in 1853 *"Über Goethes naturwissenschaftliche Arbeiten"* (On Goethe's Scientific Researches) (von Helmholtz, 1995, 12–13, ePH):

> Thus, in the Theory of Colours, Goethe remains faithful to his principle, that Nature must reveal her secrets of her own free will; that she is but the transparent representation of the ideal world. Accordingly, he demands as a preliminary to the investigation of physical phenomena that the observed facts shall be so arranged that one explains to the other, and that thus we may attain an insight into their connection *without ever having to trust to anything but our senses*. This demand of his looks most attractive, but is essentially wrong in principle. For a natural phenomenon is not considered in physical science to be fully explained until you have traced it back to the ultimate forces which are concerned in its production and its maintenance. Now, as we can never become cognizant of forces qua forces, but only of their effects, we are compelled in every explanation of natural phenomena *to leave the sphere of sense*, and to pass to things which are *not objects of sense*, and are defined only by abstract conceptions. In the final instance this is *a world of invisible atoms and movements, of attractive and repulsive forces*.

But there was even a desire to attach hypothetical properties from primary sensory qualities to the models in this world, qualities taken from the sensory world which it was hoped to leave behind. If, therefore, *these* qualities are also disregarded, then there is nothing left in the worlds of matter and energy which form the basis of the sensory phenomena which could be envisaged as a conceivable reality in line with the model of the perceptible world, something which was clearly expressed by Steiner[51] and in particular by the

---

50 A complete misunderstanding of the scientific principle and the scientific value of Goethe's Theory of Colours appears in Albrecht Schöne's book with the revealing title *"Goethes Farbentheologie"* (Goethe's Theology of Colour) (Schöne, 1987). In contrast, Olaf L. Müller, professor for philosophy and epistemology at the Humboldt University of Berlin, recently proved the high scientific value of Goethe's Theory of Colours by precise experiments and epistemological arguments (Müller, 2015).

51 Such as e.g. in 1916 in *"The Riddle of Man"*: "If the world were as [modern, e.g. following on from Max Planck and Ernst Mach] natural science depicts it, no being would ever experience anything about it. To be sure, the world pictured by natural science is there, in a certain way, within the reality from which man perceives his sense world; but lacking in this picture is everything by which it could be perceived by some being. What this way of picturing things must posit as underlying light, sound, warmth does not shine, sound, or warm. [...] What underlies tactile

exponents of modern quantum physics such as Werner Heisenberg[52] at the beginning of the 20th century. What remains is merely *the pure lawfulness, the mathematical structure, without any visible content*, although it has been

---

sensations [which have access to e.g. spatial extension] also cannot be felt by touch. Let it be expressly stated here that we are not merely saying that the world lying behind sense impressions is in fact different from what our senses make out of it; we are emphasizing that the natural-scientific way of picturing things must think of this underlying world in *such* a way that our senses could make nothing out of it if it were in actuality as it [is] thought to be. From observation, natural science draws forth a world picture that through its own nature cannot be observed at all" (Steiner, 1990, Ch. New Perspectives). Goethe's opposition to Newton's colour theory came about because, in contrast to Newton, he was not interested in a hypothetical physical world *behind* the colour phenomena, but wanted to examine the conditions and laws of the *world of colour phenomena itself*. To this Steiner remarks: "Goethe understood that Newton's color theory could provide a picture representing only a world that is not luminous and does not shine forth in colors. Since Goethe did not involve himself in the demands of a purely natural-scientific world picture, his actual opposition to Newton went astray in many places. But the main thing is that he had a correct feeling for the fundamental issue. When a person, by means of light, observes colors, he is confronting a different world from the only one Newton is able to describe. And Goethe does observe the real world of colors" (ibid., Ch. New Perspectives).

52 This was discussed by Heisenberg in his famous book *"Physics and Philosophy"*, based on Democritus but also holding true for the atomic models of the 19th century: "Democritus was well aware of the fact that if the atoms should, by their motion and arrangement, explain the properties of matter – color, smell, taste – they cannot themselves have these properties. Therefore, he has deprived the atom of these qualities and his atom is thus a rather abstract piece of matter. But Democritus has left to the atom the quality of 'being', of extension in space, of shape and motion. He has left these qualities because it would have been difficult to speak about the atom at all if such qualities had been taken away from it. *On the other hand, this implies that his concept of the atom cannot explain geometry, extension in space or existence, because it cannot reduce them to something more fundamental.* The modern view of the elementary particle with regard to this point seems more consistent and more radical. We say, for instance, simply 'a neutron' but we can give no well-defined picture and what we mean by the word. We can use several pictures and describe it once as a particle, once as a wave or as a wave packet. But we know that none of these descriptions is accurate. *Certainly the neutron has no color, no smell, no taste*. In this respect it resembles the atom of Greek philosophy. *But even the other qualities are taken from the elementary particle, at least to some extent.* [...] The only thing which can be written down as description is a probability function. But then one sees that *not even the quality of 'being' [...] belongs to what is described*" (Heisenberg, 1959, 66–67). However, this also means that it is *even less* possible for the characteristics of the sense world to be derived from elementary particles than from Democritus' atoms or those of the 19th century.

derived from this latter. So, according to Heisenberg: "The 'thing-in-itself' is for the atomic physicist, if he uses this concept at all, finally a mathematical structure; but this structure is – contrary to Kant – indirectly deduced from experience." (Heisenberg, 1959, 83). This structure no longer corresponds to Democritus' clearly pictured atoms, but to Plato's *ideas* (ibid., 64–67), or, to use an Aristotelian expression, the *forms* underlying the substance: "The smallest parts of matter are not the fundamental Beings, as in the philosophy of Democritus, but are mathematical forms. Here it is quite evident that the form is more important than the substance of which it is the form" (ibid., 66). Carl Friedrich von Weizsäcker wrote basically the same thing in an article in the Deutsches Ärzteblatt, through which he and his coauthors wished to give doctors the impulse for a complete rethink (Schmal & von Weizsäcker, 2000, ePH):

> Already in the 20th century this (the quantum theory) showed that the atomic structures cannot be viewed as objects, as physically tangible particles corresponding to the clear world of ideas of classical physics in the sense of Descartian res extensa. To date they have been consistently described *only as mathematical structures.*

The consequence of these thoughts expressed by leading physicists would accordingly be to acknowledge *ideas as the basic reality underlying sensory phenomena* and science would then in fact be ontological empirical idealism. The hypothesis based on quantum physics recently advanced by Thomas Görnitz, a theoretical physicist and pupil of Carl Friedrich von Weizsäcker, also concurs with the view that, in the final instance, matter is information and therefore *spirit* (Görnitz & Görnitz, 2008)[53]. Nevertheless, mathematical

---

[53] For Görnitz substance *is* information (i.e "abstract information" or "Protyposis") which stands for itself and does not require a basis, but can become the basis for material properties. This means that information is not a property of matter or energy, but in a certain sense identical with matter. (From the viewpoint presented here you could say that information is the *essence* and a material property is a *manifestation* of this essence.) This thesis is developed using the laws or "information" of quantum phenomena, which are stripped of *all* sensory characteristics such as hardness, tangibility, etc. However, when the authors interpret the emergence of life, goal-oriented behaviour and consciousness purely as "internal existence-preserving quantum information processing", the brain as multi-layered quantum information and thinking as a process which obeys quantum theory, then the old reductionism and materialism reassert themselves, in other words the attribution of all higher levels of being and laws to the *lowest level of the material realm*, that of quanta. And based on this they then believe the separation of body and soul to be overcome, as in their most recent book (Görnitz & Görnitz, 2016). The fact that even the higher levels of substance and above these, life and consciousness, are subject to *different and higher-ranking emergent* laws or information is overlooked

forms, taken on their own, are abstractions of the human spirit in terms of form and certainly not the whole reality in terms of content. Because, in reality, there is firstly never a quantum without quale[54] and secondly there is not only a world *behind* or *beneath* the perceptible phenomena, but also *the world of these phenomena itself*. And this has different classes and types and various levels arranged above, beneath or beside each other. And *all these* have *their* laws. The task of science is not merely to take the perspective of the lowest levels and declare the quantitative laws of these as the basis and cause of the qualitative laws and of all higher-order levels, but to investigate the quantitative and qualitative phenomena of *all* levels and connect these in an overall relationship in which each phenomenon is assigned to the place which objectively befits it. Then reductionism is replaced by empirical idealism and the scientific gaze is free for higher perspectives than those of the smallest physical and chemical entities. This is already clear for the concept of matter, but also applies to the phenomena of living entities, the psychological realm and the spiritual. *It is one of the basic problems in chemistry, biochemistry, biology, psychology and medicine that, due to the reductionist way of thinking, there is always pressure to reduce phenomena of a higher order to lower levels and to explain these causally from them. In so doing, it is easy to lose sight of the fact that these "layers" have their own properties and laws which*

(cf. Section 3.2). Logically, the same must apply to these as to the information in the quanta: they are real and, in addition, non-reducible spirit.

54 This was clearly explained by Steiner particularly in his introduction to *"Goethe's natural scientific writings"*: "Mathematics deals with magnitude, with that which allows of a more or less. Magnitude, however, is not something existing in itself. In the broad scope of human experience there is nothing that is *only* magnitude. Along with its other characteristics, each thing also has some that are determined by numbers. Since mathematics concerns itself with magnitudes, what it studies are not objects of experience complete in themselves, but rather only everything about them that can be measured or counted. It separates off from things everything that can be subjected to this latter operation. It thus acquires a whole world of abstractions within which it then works. It does not have to do with things, but only with things insofar as they are magnitudes. It must admit that here it is dealing only with *one* aspect of what is real, and that reality has yet many other aspects over which mathematics has no power. Mathematical judgements are not judgements that fully encompass real objects, but rather are valid only within the ideal world of abstractions that we ourselves have conceptually separated off from the objects as *one* aspect of reality. [...] It is therefore definitely an error to believe that one could grasp the whole of nature with mathematical judgements. Nature, in fact, is not merely quantity; it is also quality, and mathematics has to do only with the first. The mathematical approach and the approach that deals purely with what is qualitative must work hand in hand; they will meet in the thing, of which they each grasp *one* aspect" (Steiner, 1988b, 185–186).

cannot be explained from the processes and laws of the individual parts. An attempt will therefore be made in what follows to sketch out an understanding of substance, biology and the human being which is epistemologically and ontologically appropriate to these emergent levels.

3.2 Spirit in matter

In the preceding section the attempt was made to show that the observable reality which surrounds us cannot be explained from the tiniest elements or elementary particles which are the ultimate outcome of analysing this reality. From the standpoint of Goethe's and Steiner's understanding of knowledge and the view of reality resulting from this, this explanation *is bound* to fail, even in the realm of the understanding of substance in physics, chemistry and biochemistry. While, in terms of its appearance, matter is "sensory" – inasmuch as this appearance is accessible to sensory perception – in terms of its lawful essence it is *ideal, spiritual*. For to the extent that all substance, even the tiniest speck of dust or the most minute elementary particle, is determined by the lawfulness specific to it and is therefore permeated by this, *matter must be viewed as completely pervaded by spirit*. However – as described above – "spirit" is not to be thought of as something irrational but the most rational comprehensible content of the laws which define matter, which is what is dealt with especially by the exact sciences.

This view is also shared by modern physicists. Walter Heitler, professor of theoretical physics at the ETH in Zurich, summed this up even more precisely than Werner Heisenberg (Heitler, 1972, 14–15):

> They [the physical laws] are strictly formulated in the form of mathematical expressions and the physical processes follow them. A mathematical law is a spiritual entity. We may call it this because it is the human spirit which discerns it. The expression spirit may not be very popular nowadays where an excessive materialism and positivism bear their often very unsavoury fruit. But for this very reason we need to become clear about what knowledge of nature and natural laws is. Nature therefore follows this non-material element, the law. Consequently, spiritual elements are embedded in nature itself. Amongst these are mathematics which is necessary for formulating the laws, in fact higher and even the highest mathematics. On the other hand, the researcher who is lucky enough to make a discovery is in a position to recognise this element which permeates nature. And this is where the connection appears between the human knowing spirit and the transcendental elements which exist in nature. The best way of understanding matter is if we make use of the Platonic way of expressing ourselves, although Plato was not aware of these kind of natural laws. According to this the natural law is an archetype, an 'idea' – in the sense of the Greek word idea – which nature follows and which the human spirit can *perceive*.

In the classical form of expression in universal realism it can be explained as follows: It is *one and the same general* law (universal) which 1. constitutes the structure of all particular crystals of cooking salt (as universalia in re), 2. is thought by individual consciousnesses (as universalia post rem), 3. also exists in a purely spiritual state independent of 1. and 2. (as universalia ante rem, cf. Section 2.6). The realisation of a law from the state of "ante rem" to the state of "in re" is synonymous with the transition of its possible state of being of "potentia" (Greek: dýnamis) into the actual state of its realisation (Latin: actus, Greek: enérgeia) (Schneider, 1985, 44).

A *purely spiritual* form of laws and ideas is rejected by nominalism, does not correspond to habitual ways of thinking and is therefore difficult for many scientists to accept. However, it is the universality of the laws which results in their *independence* from each individual thing and consciousness and, along with the empirically demonstrable objectivity of ideas (cf. Section 2.6), gives rise to the concept that ideas exist in a purely spiritual form, independent of all forms of manifestation. Aside from this, the law as what determines the manifestation is of a higher order than it. Besides the manifestation of the law in our thinking, we empirically know only phenomena of a spatial or temporal kind (psychological phenomena do not take place in space but in time and belong to the latter) and these manifestations are determined by laws which we can find in thinking. The pure law of the snowflake is valid – when the necessary conditions for it are met – for snowflakes of all millennia and regions of the earth, and also survives phases in which the conditions for its manifestation in space and time are not met and in which it is not actually thought. The law is therefore of a purely spiritual nature which is independent of the existence of the manifestation. This can be called "eternal" as long as this is not taken to mean "endless" in the sense of "spatially and temporally without end". In contrast to spatial and temporal manifestations, laws have no characteristics of coming into being and passing away. We are therefore justified in acknowledging the "world" of universalia ante rem as a purely spiritual world, beyond space and time. They do not lose their – eternal – spiritual character, even if they emerge as universalia in re or post rem.

Admittedly, the notion of a physical law existing as a universale ante rem in a purely spiritual state, beyond space and time, might be hard to digest for a world view brought up on reductionism. However, also in this respect newer physics have opened the door to the spiritual. The physical phenomena that initiated this move are summarised as *"nonlocal correlations"* of physical systems such as atoms or photons which are spatially completely separated but correlate lawfully yet without local interactions (Valentini, 1991). An example of this would be two photons emitted from an experimentally activated atom, moving in different directions but still behaving in a correlated fashion: if the state of *one* of these photons is modified experimentally, the state of the other distant photon will be modified too. The photons move away from each other with the speed of light. It is agreed that "the nonlocal correlations of physics are nonsignalling. That is, they do not communicate information. [...] Hence, nonlocal correlations happen without one system influencing the other" (Gisin, 2009). But how is this possible? It is like "Einstein's famous spooky action at a distance. [...] All of today's experimental evidence points to the conclusion that nature is nonlocal. This has implications both for *our world view* and for future technologies" (ibid., ePH). One important implication for our world view is the necessity to accept that the source of these nonlocal correlations must be sought beyond space and time: "In modern quantum physics, entanglement is fundamental; furthermore, space is irrelevant – at least in quantum information science, space plays no central role and time is a mere discrete clock parameter. In relativity, space-time is fundamental and there is no place for nonlocal correlations. To put the tension in other words: No story in space-time can tell us how nonlocal correlations happen; hence, nonlocal quantum correlations seem to emerge, somehow, *from outside space-time*" (ibid., ePH).

This means that nonlocal correlations that *emerge in space-time ("in re")* have their origin *outside space-time ("ante rem")*. Why should this not be possible for other lawful relations that emerge in space-time, for patterns, forms etc.? The *emergence* of these relations and laws can then be characterised as the transition of their states from "ante rem" to "in re", from a purely spiritual form to a material manifestation. Of course, as we have already seen (cf. Section 3.1), the spiritual character of these universalia is not lost when they manifest as universalia in re or post rem. So spirit exists *in* matter, as its lawful essence, but also *outside* manifest matter, in a realm beyond space and time. In its essence the cosmos is spiritual: it only adopts a sense-perceptible form when it becomes manifest. And this is only one part – the material part – of the cosmos.

3.3 Emergence, self-organisation and causality in physics, chemistry and biochemistry

Only through the realisation of a law can its form of manifestation, the phenomenon, emerge in the first place. Prior to this there can be nothing in the world of phenomena which could be characterised by this law and could therefore reveal its existence to cognition. So when a new phenomenon e.g. a new order which was not previously present appears in the world of phenomena, then it cannot be claimed that the law of *this* order or phenomenon was already present *in the matter* in which the new phenomenon arises. However, this is exactly what reductionism does, although it continues to fail to provide the empirical proof of *where and how* the new law is present in the earlier phenomena[55].

Now there is the specific case of *chemical processes* in the course of which *substances as phenomena completely disappear and totally different, new ones appear*. The yellow sulphur and the warm red copper combine into deep blue copper sulphate. The foul-smelling, poisonous, greenish-yellow chlorine gas and the white amorphous sodium powder combine into transparent cubic cooking salt. The phenomena of the initial substances differ completely from those of the final substances and the latter cannot be guessed from the former. This applies to all chemical processes. Rozumek describes this aptly as follows (Rozumek, 2008, 49):

> In each case we observe a more or less abrupt replacement of qualities. One set disappears completely, others arise as though from nothing. There is no bridge between them. Nothing about the first quality reveals anything about the second. Naturally there are also transitions and one quality can persist while the other

---

[55] You could raise the objection that ideal realism also fails to provide empirical proof for the Where and How of the law in the state of "ante rem". But this is a misapprehension of the facts under consideration here. Because, firstly, empirical – in other words non-speculative – natural science can never prove the objectively applicable laws of nature on an object when the law does not show itself as a phenomenon in the object (as universalia in re); and this is only ever possible *after* the emergence of the phenomenon. Secondly, all phenomena which appear in space and time and are accessible to observation by the senses demonstrate the property of coming into being and passing away, which is not the case for the laws and ideas which are only accessible to thinking. Anyone who does not concede this must provide positive proof of the coming into being and passing away of the laws. Anyone who says that it is impossible to know whether the law still exists as universalia ante rem – in other words when it is not actually manifest (as universalia in re) or is not actually being thought (as universalia post rem) – is arguing with the same logic as someone who says that it is impossible to know whether a chair continues to exist while you shut your eyes for a while and cannot convince yourself of its existence during this period by looking. The *content* of a phenomenon must be clearly distinguished from the *occurrence of the same for consciousness* and the former is dependent on the latter. See also Steiner (1988b, 196–200). This applies both to sensory phenomena and to the laws which appear in thinking. However, in the case of the former a *manifestation* (a phenomenon) appears, in the case of the latter the *lawful essence* itself. While you could exchange one *manifestation* for another, identical to it, (e.g. while your eyes are closed one chair for an externally indistinguishable but different one), this is not the case for the *essence*. The essence is the universal which remains identical to itself, irrespective of when or by whom or whether in fact it is thought or at what time and in which spatial phenomenon it manifests. "The essential being of a thing cannot be destroyed; for, it is outside of all time and itself determines time" (ibid., 213), and the same applies to space.

is already emerging. For example, a mixed colour arising in this way can be derived from the initial and final colours. However, there is likewise no bridge of any kind between these.

Reductionism explains this clear *discontinuity of the phenomena* by the thought of the *continuity of the atoms which form the basis of the phenomena*. The same atoms are present before and after, only in a different compound: Na + Cl = NaCl. But this is no explanation for the phenomena. Firstly, as demonstrated above, the observed qualities cannot be derived from those ascribed to the *imagined or inferred* atoms alone. This applies to the substances on both sides of the chemical equation. And secondly, the substances on both sides of the chemical equation must be treated *epistemologically the same way*. Cooking salt has *its* manifestation and this corresponds to a law. The same applies to sodium and chlorine. So for this reaction we are dealing with *three* phenomena and *three* laws: that of the "parts", sodium and chlorine, and that of the "whole", cooking salt[56].

What does this mean? The law and the manifestation of the whole are something *specifically unique* compared to the law and manifestation of its parts. There is no epistemologically justifiable argument for holding the law and manifestation *of the whole* as less objective and real than the laws and manifestations of its parts. Because the law and manifestation present themselves as the complementary factors of reality in the case of a *whole* in the same way as is the case for its *parts*. Looked at in this way, cooking salt or copper sulphate as phenomena do not appear as *a composition* of copper and sulphate but as a *unique* substance whose *reality* is given by its individual, newly arisen manifestation and lawfulness.

This new appearance of phenomena is called *emergence* (*emergere*: Latin for to appear, arise, become visible), i.e. the spontaneous appearance of phenomena or structures at the higher level of a system on the basis of the interaction of its elements where, however, the emerging properties of the higher

---

56 What is interesting about this is that, phenomenologically, pure sodium and chlorine are almost never found *as such*. The original substance, i.e. what is found in nature, is cooking salt (!) and sodium and chlorine gas are mainly known as artificially derived phenomena which are very short-lived in the phenomenal world due to their reactivity. From this viewpoint *the whole is primary and the parts secondary*. It cannot be ruled out in advance that this does not apply to other natural phenomena (a comparison could be made with the development of the organism by consecutive cell division). From the standpoint of actual empiricism the apparently paradoxical – somewhat exaggerated – question can even be asked as to where the characteristics *of the parts* actually come from, and it would likewise be found difficult to derive them from the whole ("holistic reductionism"), as is the case for the derivation of the content of the whole from the parts in the usual particularistic reductionism.

system level *cannot* be attributed to the properties of the elements as such (Stöckler, 1991). This applies to all levels of a system. Hence, the phenomenon of emergence empirically demonstrates that reductionism has failed, as Bernhard Kiefer tried to make clear a few years ago in the periodical of the Swiss National Foundation for Scientific Research (Kiefer, 2007, 33):

> Emergence is one of the most puzzling but fundamental phenomena of the universe: the appearance of new characteristics at each higher level of complexity which cannot be predicted from the previous level. An example: the characteristics of life cannot be deduced from lifeless matter. However far we pursue research in physics and chemistry, this route will never enable us to predict the specific behaviour of living organisms. It appears to be a generally valid principle that the (complex) whole cannot be traced back to its (simple) parts. This includes all stages of increasing complexity. At the level of the atom: observing hydrogen and oxygen atoms in isolation gives no clue to the characteristics of a water molecule. Or, at the end of the scale: the characteristics of consciousness do not result from the extrapolation of behaviour. [...] Emergence gives rise to an important conclusion: reductionism is a false doctrine.

The reductionist attempt to explain emergent properties by a kind of *self-organisation* is well known, but it is envisaged only as *self-organisation "from below"*: atoms or molecules meet randomly under appropriate conditions and arrange or combine "by themselves" into the composite substances. The calotte model, well-known to students of medicine, lends visual plausibility to this mode of explanation: higher-ranking molecules are combined from individual components like a Lego model, their surfaces fit together according to the lock and key principle. The lock and key structures on the corresponding calottes symbolically illustrate specific complementary physical properties belonging to the parts, such as electrical charge relationships etc. According to this model, the whole therefore arises *automatically* from the parts i.e. through a type of atomic mechanism in the random encounter of these parts in the chemical reaction.

However, the problem is that the model only shows additive properties *of the parts*, whereas the resulting *whole* actually demonstrates *new* properties which were not there previously. And even the factual calculation of the properties of the whole from those of the parts appears to be impossible in terms of a real ("strong") reductionism. This was once expressed by John Maddox in an editorial in the scientific journal *Nature* as follows (Maddox, 1988):

> One of the continuing scandals in the physical sciences is that it remains in general impossible to predict the structure of even the simplest crystalline solids from a knowledge of their chemical composition. Who, for example, would guess that graphite, not diamond, is the thermodynamically stable allotrope of carbon at ordinary temperature and pressure? Solids such as crystalline water (ice) are still thought to lie beyond mortals' ken.

Although Maddox expresses his hope that a particular group of Japanese scientists will finally be able to mathematically derive the structures of silicate from silicon and hydrogen, this also requires the introduction of variables which have already been obtained from actual (!) silicates: "On the face of things, the most obvious artificially introduced variables are the known crystal structures of the polymorphs of silica, used as a starting point for the molecular dynamics" (Maddox, 1988). In other words: this does not represent an actual *derivation* of the higher structures from the subordinate components but a calculation of the degree to which the two are *compatible* with one another. And the silicate variables could obviously only be obtained *from silicates*, i.e. from the observed whole itself, not from its parts. From the viewpoint of water or silicates as primary substance, this effort actually appears to be completely absurd, an unnecessary playing about without any real epistemological value.

As explained by Hans Primas, emeritus professor for physical and theoretical chemistry at the ETH Zurich (Primas, 1985a), the reason for this is to be found in the "emergent properties in the molecular hierarchy" of the substances, i.e. "the appearance of qualitatively new properties when we ascend from a hierarchically lower level to a higher level of organisation". What is important here is "that the world possesses a layered structure and that each level has its own properties and laws" (ibid., 117–118).

A reduction of the properties from the higher-order system levels to those of the lower level is therefore not possible in the "strong sense", i.e. by a purely *logical derivation* of the higher order from the lower order, but only in the "weak sense". It is possible "through *asymptotic limit processes* to successively create a whole system of molecular descriptions from the first principles of quantum mechanics" (ibid., 118). As it is possible to control the limit processes between the levels of organisation "through a pedantically careful mathematical analysis", it appears to the non-expert that it is possible to calculate the *content* of the higher levels from the lower ones. However, this is not the case: the limit processes are characterised by a *disappearance* of properties on the other side of the boundary and an *appearance* of new properties on this side of it[57]. "A hierarchically higher-lying description of nature

---

57 An example of this is as follows: "A molecular system consists of electrically charged elementary systems (such as electrons and atomic nuclei) which are always linked to their electromagnetic radiation field. In a fully quantum mechanical description the molecules and radiation field are entangled by EPR correlation so that a molecule no longer exists as an individual object. If the field is made classical by a suitable limit process, then these EPR correlations disappear. The new classic observable to emerge at this limit describes the individuality of the molecule" (Primas, 1985a, 118).

cannot therefore be derived from the first principles of quantum mechanics by purely logical operations. It is a creative construction which is consistent with the first principles of quantum mechanics" (ibid., 118–119). The laws of the higher emergent system levels are therefore *compatible* with the laws of the lower levels but their content does not arise from the latter. The theoretical predictability of crystal structures from their components which has become possible to a certain degree nowadays (Misquitta et al., 2008; Stone, 2008) rests on this mathematically calculable compatibility and therefore on a reduction in the "weak" sense, i.e. the specific nature of the properties and laws of the higher system levels are retained with respect to the subordinate levels. Another reference from Hans Primas (Primas, 1991, 170) on this:

> Higher-level theories do possess a certain autonomy and cannot be deduced from universally valid first principles without taking into consideration the pattern recognition devices necessary for the observation of higher phenomena. Any mathematically formulated reduction of a higher-level description to a fundamental theory is conceivable only if in the basic theory a new contextual topology is introduced. This new topology is never given a priori but depends in a crucial way on the abstractions made by the cognitive apparatus or the pattern recognition devices used by the experimentalist. This program can be realized in terms of modern algebraic quantum mechanics. In this framework it is possible to describe the behavior of matter in many, mathematically precisely characterized, mutually exclusive complementary ways. Each hierarchical level requires an autonomous, non-reducible language which should not be eliminated in favor of an empty 'universal language'. Mutually exclusive complementary descriptions of nature are not only admissible, but they are equally entitled and necessary. That is, *science is necessarily pluralistic.*

In simpler terms: cognition has multiple perspectives even in the realm of physics and chemistry. The parts and the whole each need to be observed *individually*, in separate acts of cognition and be understood according to their lawful properties – two completely equally valid perspectives. Only *then* can the relationships between the hierarchical levels be identified and the compatibility of their laws tested.

But how can the relationship of the parts to the whole be thought about if the former *disappear* as observable phenomena when the latter appear? It is helpful here to remember the ingenious concept of Hegel's *"sublated moment"*[58]. For on the one hand it is obvious that copper sulphate can only

---

58 While the concept of the "sublated moment" was first developed by Hegel purely conceptually in his *Logic* from the concepts of "being" and "nothing" and "becoming" which sublates both of these because it encompasses them, however, as Hegel himself remarked, this concept is also of fundamental importance in our view for the empirical sciences. The appearance and disappearance, i.e. the

form from the meeting of the initially separate components of copper and sulphur and that these must *still be contained* in it as they appear again when the crystal dissolves. But on the other hand the manifestation of these parts as components of the whole is actually *sublated* and in this condition their laws are no longer traceable, but only become so again after the crystal has been chemically dissolved. Instead, in their place, the whole appears with *its* properties; and *their* laws can then be found in the act of cognition. The parts are therefore *"sublated moments"* of the whole in a double sense of this Hegelian concept. The parts as phenomena are "sublated" in the whole in the sense of *"have disappeared"* but at the same time they are still *"contained"* in it and in this – second – sense "sublated". The result, the higher-order substance, is therefore *not the sum* of its parts but *the unity of these sublated parts with its whole*. This result could never manifest without these parts, it requires them as its inner momentum and is a "whole" only in relation to them[59]; however, its lawful manifestation is that of an independent substance.

How is this sublation of the parts in the whole, i.e. the suppression of their lawful properties in favour of those of the whole to be understood? From the viewpoint of ontological empirical idealism it can be said that the law of the whole permeates its parts to such an extent that it imprints them with *its* law and supresses the effects of *their* laws. It prevails against them with the

---

"emergence" and its polar "submergence" of the properties of substances in chemistry can scarcely be better formulated conceptually than by the use of Hegel's concept of the sublated moment for these empirical facts: "Becoming [coming into being and passing away] is the unseparatedness of being and nothing, not the unity that abstracts from being and nothing; as the unity of *being* and *nothing* it is rather this *determinate* unity, or one in which being and nothing equally *are*. [...] In this unity, therefore, they *are*, but as vanishing, only as *sublated*. They sink from their initially represented self-subsistence into *moments* which are *still distinguished* but at the same time sublated" (Hegel, 2010, 80). And: "The German *aufheben* [to sublate in English] has a twofold meaning in the language: it means both to keep, to *preserve*, and to cause to cease, *to put an end to*. Even to preserve already includes a negative note, namely that something, in order to be retained, is removed from its immediacy and hence from an existence which is open to external influences. – That which is sublated is thus something at the same time preserved, something that has lost its immediacy but has not come to nothing for that" (ibid., 81–82). Applied to chemistry, this means that the phenomena i.e. the visible properties of the parts are "sublated" in the whole. The whole is then an immediate autonomous phenomenon, just as the parts were before they were sublated.

59 Both scientifically and philosophically speaking, "part" and "whole" are correlative concepts which determine each other: a part is only a part in relation to the whole and vice versa but they are not identical and cannot be replaced by each other but are entities on their own.

result that the outward manifestation or phenomenon *of this* law takes the place of the other.

This is no longer self-organisation from below due to a simple interaction of the parts, but *"self-organisation from above"*, as Steiner explained in detail in 1911 in relation to the physical laws of form in nature and in the human being using the example of the structure or form-principle of cooking salt (Steiner, 1951a, 173):

> Even though you descend to the crystal, you find that the substances which enter into it, if they are to become what is manifest as the crystal, must be seized as it were by form-principles, which in this case are the principles of crystallisation. Take for example kitchen salt or sodium chloride: here you have, according to our present-day physics, the physical substances chlorine and sodium, a gas and a mineral. You will readily see that these two substances, prior to their entrance into the entity which lays hold upon them in such a way that, in their chemical union, they appear crystallised into a cube, have nothing in them that can indicate to us such a form-principle.[60] Before they enter into this form-principle they possess nothing in common, but they are seized upon and yoked together by this form-principle and there is then produced this physical body, kitchen salt.

Self-organisation in crystals or chemical compounds "from above" therefore means not merely a combination of the parts *into* the whole, but that the whole *takes hold* of its parts, and that their presence under the appropriate conditions actually creates the *circumstances* for the manifestation of the whole. Or, expressed in terms of universal realism: chlorine and sodium are actualised lawful manifestations whose laws (universals) are to be found in the state of universalia *in re*. Under the necessary conditions, the cooking salt law is actualised in them from the state of universalia *ante rem* into the state *in re* and, in this process, the effects of the chlorine and sodium laws are subordinated as the sublated moments of cooking salt. However, this can only last as long as the relevant conditions for it are present. If these change, then the laws of the subordinate parts once again prevail and the cooking salt disappears from manifestation. The reaction equation of a chemical compound can also be understood as the expression of this inter*relationship*: depending on the relative ability of the laws of the parts or the whole "to assert themselves", the chemical equation will tend to be more heavily weighted towards the side of the parts or the whole[61].

---

60 The expression "entity" is consistent here with the expression of the lawful "essence", in other words, with the law in Steiner's epistemological writings (Steiner, 1988a).
61 In many chemical reactions there are several reactants on both sides of the chemical equation (e.g. in organic chemistry with very complex molecules). This makes the relationship between the part and the whole more complex. We have to deal

Looked at in this way, the substances are not conglomerates of subordinate parts, but hierarchically arranged wholes whose parts are "sublated" in the Hegelian sense. As such the substance *cannot* be "put together" from modules, as is still suggested to medical students by the calotte model. This may appear paradoxical, but it is in accordance with a modern understanding of substance arising from quantum physics. Compare what has been said with e.g. the following statements by Hans Primas in his remarkable article *"Umdenken in der Naturwissenschaft"* (Rethinking the natural sciences) (Primas, 1992, 9–11):

> Atomism claims that matter is composed of the tiniest building blocks which cannot be divided any further and that natural processes must be explained from the properties and motion of these atoms [...]. However, molecules, atoms, electrons, quarks or strings are *not the building blocks of matter* [...]. Modern physics has retained nothing of the original concept of matter. [...] If we believe quantum mechanics to be a good theory of matter, then the statement, 'matter is made up of elementary building blocks' is scientifically incorrect. The decisive point is not that the chemists' atoms can be further divided – that would be a trivial matter of nomenclature –, *but that the material reality is a whole that is not built up from parts at all.* [...] The dialectic of whole and part in the quantum world is fundamentally different from the usual descriptions in classical sciences. Due to empirical scientific reasons physics was forced – in the face of stiff resistance from most philosophers – to recognise the holistic nature of the world of matter. [...] In the view of modern quantum physics, material reality is a whole, and what is more, a whole which is not composed of parts. [...] Quantum mechanics is the first logically consistent and mathematically thought-through theory. The concept of wholeness in modern physics is much more comprehensive than the holistic approaches of the other natural sciences.

Matter is not "composed of parts" *because* these parts become sublated, implicit, latent or submerged in the whole, comparable with a nascent state, whereas the whole is explicit and apparent and therefore emergent as a phenomenon. This occurs when its law manifests under specific conditions from the potentiality of the state ante rem into the state in re.

This is also fundamental to Steiner's view of substance when, in a lecture in 1923 on the "composition" of protein from carbon, nitrogen and oxygen, he said (Steiner, 1985a, Lecture V, ePH):

> Today we generally think of things as being formed by combination, but that is nonsense. What we know as certain higher substances are not always composed of what appears when they are analysed, for *these things cease to be present*

---

with *different* emergent wholes of the subordinate components such as carbon, oxygen, nitrogen, etc. Epistemologically and ontologically these are all equal.

in the higher substance. Carbon is not present there as carbon, nor oxygen as oxygen, but there is a *substance of a higher order.*

The emergent whole which is normally called a molecule in chemistry is actually a substance of a higher order when seen in this way, a unified whole in itself with the sublated, "submerged" parts in it. There is also absolutely no indication from observation that carbon is still present as "carbon" in the *intact* protein (Rozumek & Buck, 2008).

In his contribution *"Hypothesenfreie Chemie im Sinne der Geisteswissenschaft"* (hypothesis-free chemistry according to spiritual science) to the academic anthroposophical seminars in Dornach in 1920, Eugen Kolisko stated in a similar vein (Kolisko, 1922a, 206–207, ePH):

> This is the right place to go into the concept of chemical composition. This does not exist *in terms of a phenomenon.* Water has to viewed as just as *elementary* as hydrogen. The compound is only a *dynamic one. Hydrogen is as little contained in water as what is known as 'latent heat' is contained as warmth in the gas or liquid.* Water is one form which hydrogen assumes under specific conditions, i.e. when oxygen and a specific temperature etc. are present. Water is therefore not a compound, it is modified hydrogen, in fact hydrogen modified by oxygen or oxygen modified by hydrogen. Hydrogen is contained in water in the unique way which can only be called a chemical way. There are indeed countless ways of being contained in something. *The purely spatial concept of 'containment' is only the lowest form.* It is not applicable to chemism. [...] Chemical compounds are therefore something quite unique. They cannot be explained mechanically. The latent heat is actually not present in the gas as heat, but that which is released as heat when the gas liquefies appeared earlier in the more expandable, formless properties of the gas[62]. What previously worked spatially in the gas

---

[62] By this Kolisko means the basic property of every gas, depending on its temperature, to expand in all directions in a more or less formless way, something in which it can only be hindered by external confinement. Kolisko also shows in this treatise that the particularistic models of gases as a sum of atoms or molecules moving in space are completely unnecessary and that the gas laws of Boyle, Mariotte and Gay-Lussac can be found entirely without these models (Kolisko, 1922a). These models come about because, when concepts are formed about gases, *the perspective adopted is only of the solid aggregate state* of matter (model of solid little balls), although the actual phenomenon displays a gaseous state. The same applies to the basic models of the liquid state. From the viewpoint of the epistemological method of Goethe and Steiner put forward here, the liquid and gaseous states and even heat must be recognised as emergent properties in accordance with their specific phenomena and laws without having recourse to solid models, i.e. the *perspective of each kind of phenomenon* must be adopted. In agreement with this view, Primas points out that it is simply "claimed" by the more recent "natural philosophy" that gases and liquids are nothing but "swarms of molecules" and temperature "simply another word for the average kinetic energy of

makes an appearance as the release of heat. *Likewise, hydrogen is not somehow contained in water as atoms*, but presents itself in its interaction as water and, when released from this influence of oxygen by other influences, it takes on the form that we call hydrogen.

But it is important to be aware that the chemical compound represents "*something completely new* compared to the so-called components" (Kolisko, 1922a, 212), actually a new emergent phenomenon with corresponding laws which are not contained in the components. This is what is shown by the actual phenomenon whereas the particularistic model suggests that what is qualitatively new in terms of the law and phenomenon "can be traced back to what is purely quantitative, mechanical and mathematical. Following this approach, combustion of a substance with its varied and lively phenomena, for example, will not produce *something new* in the place of oxygen and the other source materials, but the tiniest particles of the oxygen will have moved a little closer to those of the other materials than was previously the case" (ibid., 170).

Kolisko called the latent containment of the components in the compound a *dynamic one*. As expressed by this term, this dynamic state must really be conceived as a truly active one. For in the sublated moment in the second meaning of this Hegelian concept, the components are still "contained" in the whole and still make their properties felt as potential. However, this also means that the components must manifest again from their latent condition as the effect of the law of the whole recedes. Because being sublated merely means that the law of the whole obtains the upper hand in the interaction with the laws of the parts, but loses this again when this effect recedes, e.g. if the necessary conditions for it are no longer present. In other words, the interaction between the whole and its parts is a dynamic one.

The "whole", i.e. the new emergent substance, is indeed *complex* in terms of the "parts" sublated in it, but on the other hand *simple* in itself, as their *higher unified whole*. This arises from the fact that "a so-called chemical compound works as a unity just like a chemical element" (ibid., 229), something which is expressed in such things as isomerism[63] i.e. the functional groups in

---

the molecules". However, the temperature cannot be deduced from mechanical laws, "therefore *the temperature is an irreducible concept*" (Primas, 1985b, 161). This shows the justification for treating heat – in addition to the gaseous, liquid and solid states – as an independent emergent entity, which permeates the others (the gaseous, liquid and solid), but should not to be identified with them. This is also the view of heat taken by anthroposophical medicine and science (Steiner & Wegman, 1996; Husemann & Wolff, 1982, 1987, 1989; Basfeld, 1998).

63 Isomerism: "A compound of methyl, methylene and a primary alcohol group (CH_3, CH_2 and CH_2OH substance) has different properties than a compound of

organic chemistry. The fact that these kind of groups or their subunits appear as single entities under suitable conditions and react with other substances gives rise to the well-known diversity of chemical processes and compounds, especially in biochemistry[64].

The phenomenon of the *crystal lattice* can also be understood in the light of this holistic concept of substance, *without* having recourse to the particularistic mental aids like the calotte model. Kolisko said of this (ibid., 227, ePH):

> It is quite different, if you take something like Laue's crystal lattice for an *illustration of a spatial molecular* structure of the crystal which is invisible under normal circumstances, or for an image which arises through the interaction of the X-ray light with the form which permeates the whole crystal dynamically, which does not need to be material [i.e. does not need to appear in space]. The phenomenon [i.e. the *result* of the X-ray radiation of the crystal made into a picture] is then understood as *arising* from the interaction of the chemical structure with the light. The atomic image is not conceived as a preformed fact but *arising through dynamic interaction* in the experiment.

In other words, the fact of the X-ray crystal lattice cannot be taken as proof of the atomistically conceived self-organisation of the crystal "from below" but can just as well be understood as self-organisation "from above", as the actualisation of the law of the whole in its parts. Rather, the latter can better be reconciled with the fact of emergence and with Primas' holistic interpretation of substance than the former.

An important consequence of this holistic view of substance is the *updating of the Aristotelian concept of substance*: substance, *true* substance, is not simply the world of subatomic particles, "subsensible" force elements which can be inferred experimentally from their effects on what can be observed by the senses and in contrast to which the world of sensory phenomena is a mere illusion. In the modern scientific sense – actually also in agreement with Aristotle and Thomas of Aquinas – substance can be thought of as the entity, the "composite" of "form" and "matter" (Aquin, 1977, 55, Contra Gent. 2, 54) *at each particular system level*. Form and matter are two aspects of a whole which, while not separable, can nevertheless be distinguished. "Matter" is the *particular* which is accessible to sensory perception; and "form" is the *general* lawful What (quidditas), which underlies or "subsists" this particular matter, but is only accessible to the intellect in a spiritual (intelligible) way (Aquin, 1977, 26; 32, Summ. Theol. I, 84,7; 85,1). In this sense the substance

---

two methyl and what is known as a secondary alcohol group (CH3, CH3 and CHOH)" (ibid., 229).

64 Cf. the more detailed explanations by the anthroposophical doctors Eugen Kolisko (Kolisko, 1922a; Kolisko, 1989) and Otto Wolff (Wolff, 2013) and the chemists (Rozumek & Buck, 2008).

sodium is not merely a composition of protons, neutrons, electrons, etc. but, compared to these, actually the emergent, higher-order real "sodium" which sublates these and which appears and is lawfully determined in *itself*. And the substance cooking salt is not simply a compound of chlorine and sodium but likewise, as compared to these, the emergent, higher-order actual "cooking salt" which sublates these and appears and is lawfully determined in *itself*. And the same applies to more complex inorganic and organic substances.

The Aristotelian concept of substance is therefore independent of the hierarchical position of a substance in relation to the substances which are subordinate to it and also sublated in it. At each level of the hierarchy the actual substance is a "simple" phenomenon; the "complexity" of the substances sublated in it has literally disappeared and the lawfulness of this substance is not the law of the elements sublated in it, but that of the *actual phenomenon present*. This is the *"visible form"* or manifestation of its law as it were which *as such* is of course only accessible to thinking. This concept of substance therefore corresponds to the clearly established relationship of percept and concept in Steiner's epistemology (Steiner, 1988b, 220):

> Through our conceptual grasp of something given in the sense world, the *What* of that which is given to our view first comes to manifestation. We cannot express the content of what we look at, because this content consists only in the *How* of what we look at, i.e., in the *form* of its manifestation. Thus, in the *concept*, we find the What, the other content of that which is given in the sense world in an observed form.

It is clear that this holistic or Aristotelian view of substance which arises from ontological empirical idealism also entails a *review of the concept of causality*, in other words, an updating of the Aristotelian understanding of causality.

The causal explanations of reductionism are initially *mechanical* even if they are at the molecular level. However, in mechanics, as Steiner so aptly put it in *"The Science of Knowing"*, "a process results entirely from factors that *confront each other externally*. Here, an event or relationship between two objects [...] is called forth solely by the fact that one thing, in its workings, exercises a certain influence upon another, transferring its own conditions onto others. The conditions of the one thing then manifest as the consequence of those of the other" (Steiner, 1988a, 75). The mechanical "whole", i.e. the overall relationship or the system of coordinated parts is therefore calculable in principle from their states: in other words, the mechanical "whole" is derived from its parts of necessity. This is of course the basis of the calculability of mechanical engineering. In present-day science causality is taken to mean almost exclusively this linear form of external factors which affect each other.

In contrast, Aristotle distinguished between *four kinds of causes* (aitía, causa or arché, principium) which were picked up by Albertus Magnus and

Thomas of Aquinas in particular in the high scholastic era and which defined the entire subsequent period. What is common to them all is "to be the first [the principle], out of which something is or becomes or is known" (Aristotle, Met. Δ 2; 1013a, 18f. In: de Vries, 1983, 98). The four causes are:

1. *Causa materialis* (hýle): the matter e.g. the ore for the statue.
2. *Causa formalis* (eīdos or morphé): the form of the lógos dwelling within the being, i.e. the spiritually graspable content which defines its essence, from which the artist e.g. gives the ore its form.
3. *Causa efficiens* (arché tês kinéseos): the active cause with which the artist e.g. shapes the statue.
4. *Causa finalis* (télos): the aim, for the sake of which the active cause works.

Only the causae materialis and efficiens are left in the linear causality of modern science: for example in a game of billiards the balls as the "causa materialis" and the force of impetus applied externally to a ball as the "causa efficiens" which, when the ball is hit, is transferred to another one. This is how all actions in the mechanical realm are caused *from outside*. The rule according to which this outer causation of the mechanical process occurs is the content of the mechanical law of nature. But this *law* itself is not active in the process. Quite justifiably it is therefore not viewed as a causa formalis as it merely states the law according to which the interaction of the parts occurs. Steiner therefore says: "The [inorganic] natural law [applicable in mechanics] is abstract, standing over the sense-perceptible manifoldness; it governs inorganic natural science" (Steiner, 1988b, 222). In mechanics there is no final aim, no "causa finalis".

However, mechanical action is only the "simplest way of working" (Steiner, 1988a, 75). Everywhere nowadays where self-organisation is seen in physical, chemical, biological or psychological processes, this is a matter of *higher forms of action* which result in *emergence*, in other words, in phenomena which obey higher laws than those in the components contained in them, which cannot be derived from these but which characterise these dynamically "from above". Processes of self-organisation do not simply "result [...] from factors that confront each other externally", but "here, an event or relationship between [...] objects [...] determined by an *entity*[65] expressing

---

65 As described above, "entity" here means the lawful content, the "idea" of a manifestation which, however, cannot be found in the sensory perception of this manifestation, but only in thinking and must be added to the perception in the act of cognition: "We are first able to understand some element or other in the sense world when we have a concept of it. We can always simply point to what sense-perceptible reality offers us, and anyone who has the possibility of perceiving precisely this element to which we are referring knows what it is all about.

itself in outer forms of manifestation, by an individuality that makes its *inner* abilities and character known by working *outward*" (Steiner, 1988a, 75, ePH). However paradoxical it may appear, it is the lawfulness i.e. the "logic" of the emergent higher substance, structure or function *itself* which appears as the cause of the self-organisation[66]. In Aristotelian terminology, this lawfulness is the intelligible "form" of the emergent "matter". This form is the *causa formalis* which penetrates and suppresses the substances and forces subordinate to it – the *causae materialis* and *efficiens* – from inside i.e. "sublates" them as its momentum. The causa formalis can also be called the *causa finalis inasmuch* as *an "aim"* is realised[67] through manifestation of the form which occurs under appropriate conditions, such as in the processes of quantum physics[68] or the teleonomic processes of organisms. And for

---

Through the concept, we are able to say something about the sense world that cannot be perceived. From this, however, the following immediately becomes clear. If the essential being of the sense perception consisted only in its sense-perceptible qualities, then something completely new, in the form of the concept, could not join it. The sense perception is therefore not a totality at all, but rather only one side of a totality. And it is that side, in fact, which can be merely looked upon. Through the concept it first becomes clear to us *what* we are looking at" (Steiner, 1988b, 219–220). Or: "By essential we mean that by which a thing actually is precisely what it presents itself to be. [...] This essential being can only be grasped ideally" (ibid., 212–213).

66 This has already become a topic of discussion in physics, for instance by George Ellis when he argues for multiple causality (top-down, bottom-up and same-level) with reference to the hierarchical and emergent structure of nature as follows: "The point is that higher-level properties themselves [...] are key variables in the causal chain. Paradoxically, although the higher-level properties emerge from the lower-level processes, they have a degree of causal independence from them: Higher level processes operate according to their own higher-level logic. Physics makes possible, but does not causally determine, the higher-order layers" (Ellis, 2005, 53).

67 This *lawful* concept of causa finalis or *teleonomy* should not be confused with the *anthropomorphic* concept of causa finalis, which applies to the human intellectual goal-setting in the same way as an intellectually formulated plan for the manufacture of a machine and which, transferred to nature in a speculative manner, leads to *teleology* or the "intelligent design" of a naive realistically justified theology.

68 This interpretation of self-organising principles in physical processes as "causae formales" or, if realised in time, as "causae finales" arising from epistemologically-based ontological idealism is corroborated by physicists such as Carl Friedrich von Weizsäcker and David Bohm who themselves draw on the four Aristotelian causes to arrive at a full account of causality in the mathematically describable phenomena of quantum physics. For Weizsäcker the "mathematical form" is "indeed a kind of *causa formalis*" (von Weizsäcker, 1986, 260). Likewise, Bohm connected Aristotle's causa formalis and causa finalis with holistically organised phenomena

processes of self-organisation, the causa formalis is also *causa efficiens*. For, what takes place by *external organisation* in the case of machines or in a game of mechanical dominoes, the self-organising system achieves *by itself*: the arrangement (the structuring causa efficiens) of the parts according to a specific design (causa formalis). Self-organisation therefore does not mean self-organisation *of* the parts, but self-realisation of the organising law *in* the parts[69]. Grasped in this way, Hans Primas' dictum mentioned above that *"the material reality of a whole is that it is not constructed of parts at all"* can then be understood without contradiction.

In summary, the epistemological position of Goethe and Steiner and the view of reality resulting from this gives rise to a holistic view of substance and causality whose strength lies in being able to conceptually explain several apparently totally unconnected aspects of the understanding of substance and to link these to one another: 1. In relation to modern natural science, it enables a consistent understanding of the holistic concept of substance which is nowadays gradually coming to the fore due to the phenomena of self-organisation, emergence and a hierarchical system order and which Hans Primas argues on the basis of empirical facts in quantum physics, the science dealing with the smallest particles and which in one respect forms the most successful endpoint to the detailed analysis of nature initiated by Francis Bacon. 2. Epistemologically based lawful realism also leads to an updating of the central principles of Aristotelian natural science in line with modern approaches. These principles are something which Francis Bacon – at the beginning of modern science which he was largely responsible for setting in motion – considered to be mere "idols" (Bacon, 1990b). 3. The holistic view of substance arising from this also enables the study of natural substances or those produced in the laboratory not only from the perspective of their sub-microscopic imagined components but particularly in terms of *their own intrinsic nature* which is revealed macroscopically in its real observable emergent phenomena and also in its effects in the organism. This results in rational options for understanding its use in therapy *without* having recourse to "modes of action" based on

---

in physics and living nature and called attention to the fact that in Greek philosophy the term "form" did not only designate an *external* form or shape as is often the case in modern language, but instead an *"inner forming activity"*, for the designation of which he proposed the term "formative cause" (Bohm, 1986, 276–277).

69 Self-organisation must therefore be distinguished from *self-aggregation or self-assembly* in which the parts cluster together solely due to intrinsic forces. Some authors also interpret crystal formation as self-assembly (Krischer, 2007) in contrast to the interpretation argued and justified here of the self-organisation of crystals.

reductionist models, as applied in anthroposophical medicine and to a certain degree in homoeopathy (cf. Chapter 8).

Epistemology as presented by Goethe and Steiner therefore proves to be a basis for building a bridge in a scientific sense between the new spiritually compatible holistic concept of substance which has developed in physics and chemistry from the quantum physics of the 20th century and the holistic concepts of complementary medicine which otherwise appear difficult to integrate with the basic ideas of conventional medicine. In actual fact Goethe's scientific method has already been recommended as the one epistemologically founded approach which enables the holistic aspects of the various complementary medical approaches – including Chinese and Ayurvedic medicine – to be made scientifically accessible in the modern sense, an intention which is not possible with conventional scientific methods (Whitelegg, 2003). However, it also enables a scientific explanation of the holistic thinking *in conventional* science and medicine (Heusser, 2000b; Heusser, 2006a). This applies to the concept of substance, but beyond this also to modern genetics and biology and likewise to the higher forms of science of relevance for human medicine.

3.4 Summary

It has been demonstrated to what extent objective empirical idealism can lead to a replacement of scientific reductionism in the substance concept in physics, chemistry and biochemistry. Substances are commonly explained by their smallest (elementary) particles and their hierarchical structure viewed ultimately as the consequence of the combination of these particles in atoms, molecules and macromolecules (self-organisation "from below"). However, these system levels are *emergent* compared to what is subordinate to them, i.e. their lawful properties can*not* be explained from what is subordinate. From the viewpoint of empirical objective idealism, the properties of every system level are epistemologically and ontologically of *equal value*, both in terms of their phenomena and their laws. Hierarchically higher-level emergent substances are therefore just as "real" as the subordinate ones which are "sublated" in them (the concept of the "sublated moment" as defined by Hegel). This gives rise to the transition from a particularistic understanding of substance and systems to a holistic one, something which also agrees with more recent scientific facts and viewpoints. The consequence of this, however, is that *"self-organisation"* shows itself to be not merely a process acting "from below" but simultaneously one "from above". A higher-order system law manifests – as it were "from above" – when the lower elements and conditions are present "from below". The lower-order elements are then not the *cause* but the *conditions* or the *material* for the actualisation of the higher-order. And this is causally produced by itself. In other words, substances are causally governed by their inherent laws. But laws as such are "ideal", i.e. *spiritual* in nature. This means that only the *appearance* of substance is sense-perceptible, its lawful *essence* is spiritual. This view is derived from an application of epistemology to physics and chemistry, but it is also substantiated by the views of leading 20th century physicists. This leads to a re-appraisal of the causal concepts as defined by Aristotle (causa materialis, causa efficiens, causa formalis, causa finalis) with consequences for the natural sciences.

4 Ontological idealism in biology

4.1 Chemical explanation of life? Genes, genetic information and proteins

The objective ontological idealism put forward here does not lead to different facts – except when scientific attention is led to other aspects than primarily molecular ones, which is the case for the simplest morphological investigation – but to a *different understanding of the same facts* compared to conventional medical anthropology.

This applies particularly to genetics, which plays a central part in modern biology inasmuch as leading biologists and authors of textbooks still support the view established by Francis Crick (Crick, 1966) that the genes must contain *all* the information necessary for the life of a cell or organism, a view which, with the progressive decoding of genetic dependencies, has been extended to psychological, behavioural and social susceptibilities (Robinson et al., 2008). And because the bearer of this information, the DNA, is a chemical substance, Crick hoped for a "chemical explanation of life" (cited in: Commoner, 1968). But during the 20th century the focus of attention has gradually shifted from the genes to gene *regulation*. For its part, this has revealed that the genes are not simply the central controlling entity which it was hoped at the time, but that they are also subject to control which is dependent on information *from outside the DNA*, some of it from the environment. This has become evident especially through the rapidly evolving field of epigenetics since the 1990s (Holliday et al., 1990). But because any gene regulation demonstrably occurs through the mediation of particular *substances*, at the start of the 21st century hopes of eventually finding a chemical explanation of life grew even stronger (Kirschner et al., 2000). This further consolidated the reductionist causative explanation of the organism "from below", i.e. from molecules and their interactions, advanced since around the time of Du Bois-Reymond.

In terms of the conventional mind-set in genetics, it is only possible to speak of a chemical explanation of life if you do not distinguish clearly between *information* and information *carrier*, a mistake of which even Crick was guilty (Crick, 1966). Because it is only what is known as an information *carrier*, the DNA, RNA, hormonal or other carrier of "information" which is "chemical" or material, the information *content* is immaterial[70]. This was

---

70 Someone who, for example, looks for the information contained in this writing will not find it in the letters however long and thoroughly he looks at them. Anyone who does not *think* while doing this will not find any information at all. This can

also expressed clearly by Norbert Wiener, the founder of cybernetics which is so important for the current concepts in physiology: "What is information? Information is information, not matter or energy. No materialism which does not admit this can survive at the present day" (Wiener, 1948).

The "central dogma of molecular biology" formulated by Crick originally implied that biological information flows from DNA to proteins and thus to the organism, but not vice versa (Crick, 1970). This led to the generally acknowledged notion that *all* the information necessary for the organism lies in the genes, an opinion still widely held to be valid by the public: "In one sense, almost all information about a person's health and physical well-being can be called 'genetic information'. A casual glance reveals information about a person's gender, race, height, weight, and other features that are related, in whole or in part, to that person's genetic inheritance" (Australian Government, 2014). However, this does not hold true either in quantitative or qualitative terms. In terms of quantity, it has been shown for example that the maximum available amount of information in the human being would not be adequate to explain something like the complexity of the neural networks in the central nervous system; this also applies to information networks in other systems such as e.g. the immune system (Stent, 1981; Kauffmann, 1995).

But also in qualitative terms the information coded by the DNA only provides *part* of the total information in the organism. What exactly is the information coded by the DNA in qualitative terms, i.e. what is its content? Simply expressed, it is "the genetic potential of an organism carried in the base sequence of its DNA [...] according to the genetic code" (Oxford University Press, 2014). More precisely it is *the lawfulness of the primary protein structure, i.e. the ordering principle for the beginning, sequence and end of the amino acids* in the amino acid chain of the primary protein strand, corresponding to the base triplets in the DNA associated with these amino acids. What *this* information does not contain is the laws for the structural instructions, i.e. for the sequence of the basic building blocks of other carbohydrate and fatty macromolecules and also the structural instructions for higher-level structures, shapes, arrangements and functions of the cell organelles, cells, organs and organ systems and for their complex relationships and coordination in the whole organism (Stent, 1981; Heusser, 1989; Holdrege, 1999), as will be discussed further.

The normal argument against this is that the realisation of many of these structures and functions is dependent on the catalytic function of the enzymes and therefore on the proteins or their structural information in the DNA. But a catalyst *facilitates* or *enables* the *realisation* of a process, it does not define

---

only be found in the thinking spirit as *spiritual content*. The letters, on the other hand, as the bearer of information, are indeed visible to the senses.

its content or law. This applies in particular to enzymes as catalysts (Ringe & Petsko, 2008). In terms of the genetic information concept as such, the "information" in the structures of substances or organic compounds formed by the mediation of enzymes cannot be attributed to the DNA. This claim in fact fails to answer the question of where the information comes from.

But the *information for the proteins themselves* is also not present in a one-to-one form, but is only as it were *"contained"* in the DNA *in parts* from which it is then *generated* through activity. This applies to what is known as splicing for a start. In this process encoding "exons", which are separated in the DNA string by non-encoding "introns", are removed from the pre-messenger RNA *after transcription* by active enzymatic mediation and combined into messenger RNA (mRNA) which in turn corresponds to the later protein strand. This means that the "genetic information" of a specific protein as defined above is *not actually present in the DNA* but is *only created* by an active operation in the splicing procedure. In the case of immunoglobulins, even the gene segments for the heavy molecule chains are enzymatically recombined at the level of the DNA before transcription (Tonegawa, 1983) and are then also spliced at the level of the mRNA after transcription. In addition, the information elements for the heavy and light molecule chains come from different chromosomes. It is accepted that "genetic information for specific antibodies" *as such* therefore does *not* exist in the DNA, it is always created by the organism. The fact that the different gene segments combine freely to increase the variety of antibodies and are also changed during maturation by mutation, argues – despite all "randomness" – for a process actively controlled by the living organism whose *object* is the DNA.

The protein only becomes functional through its tertiary structure. But this is also not attributable to the genome in terms of its information, not even to the information for the primary protein strand finally contained in the mRNA, although this is what is usually taught. In his standard work on genetics, Lewin writes: "How do proteins fold into the correct conformation? A basic principle is that *structures of a higher order are directly determined by structures of a lower order.* This means that the primary amino acid sequence carries the information for folding into the correct conformation" (Lewin, 1988, 9).

But where exactly is the information for the tertiary structure, in other words the *structural law of the three-dimensional conformation*, actually and verifiably "contained" or encoded in the primary amino acid sequence of the protein strand? Admittedly the primary amino acid sequence contains a series of *necessary material conditions* for the realisation and stability of the conformation (e.g. sulphur-containing amino acids for the later disulphide bridges) and to *this extent* is "determined" by them. Protein refolding experiments even "show clearly that the amino acid sequence alone determines the structure of the native condition and that the folding does not depend on the folding pathway or on the initial conditions" (Fischer, 1993). But as a higher system property, the conformation of a protein is *emergent* and therefore cannot be determined logically or in terms of content from the properties of the primary strand[71], such as must inevitably be the case in a process which is thought in a purely mechanical way.

What is amazing about protein folding is the enormous flexibility of the folded protein chains and partially folded intermediates which could in principle permit a vast range of folding options but nevertheless result in a specific conformation. Fischer says of this (Fischer, 1993, ePH):

> Due to the large number of possible arrangements of the atoms to each other, the time requirement for the transition of a small protein to its native state is already astronomically high, when you also consider that all the theoretically possible positions in the conformation space have to be crossed along the folding path. [...] Nevertheless, some denatured proteins can be refolded to the native structure very quickly, in a matter of seconds. The polypeptide chains can obviously fold along a *preferred reaction pathway* without having to search the entire conformation space available.

Service holds a similar view and illustrates it using the following calculation (Service, 2008, ePH):

> Because two neighboring amino acids can bind to each other at any of three different angles, a simple protein with 100 amino acids can fold in 3^{200} different ways. *Somehow, a folding protein sorts through all those possibilities to find the correct, or 'native', conformation. And it's not trial and error.* Even if a folding protein could try out one different conformation every quadrillionth of a second, it would still take 10^{80} seconds – 60 orders of magnitude longer than the age of the universe – to find the right solution. Because most proteins fold in milliseconds to seconds, something else is clearly going on.

This process is normally explained following Peter Wolynes and Joseph Bryngelson using the picture of an "energy landscape" or according to Ken Dill using an "energy funnel" (Service, 2008): During folding, the protein aims at the same conformation *from every starting point* using the principle of energy minimisation, more or less passively, just as water from every possible starting point runs down the mountain to arrive in the same riverbed, as it were "unintentionally". Of course chlorine and sodium in their chemical

---

71 Cf. also the comments on the emergent nature of substances mentioned above. The *"weak" reducibility* of a superior substance level to an inferior one applies equally to the relationship of the tertiary to the primary structure of a protein. This has to be taken into account for understanding the attempt – which has succeeded to a certain degree – of predicting the conformation of the primary protein strand from the characteristics of the substance (Service, 2008).

combination also aim for energetic equilibrium in terms of the lowest and therefore most stable energy level and therefore neutralise in cooking salt. The cooking salt is nevertheless *emergent*, and brings about a *higher order* in which its starting substances are sublated. The energy conditions of the primary strand likewise "neutralise" in the three-dimensional conformation. But compared to the primary protein strand this conformation is emergent. It is the form in which *its* law is realised, which must be distinguished from the law of the primary strand and is superior to this in its conformation. The higher order must therefore have influenced the lower order.

The realisation of a higher order is thus dependent on the environmental conditions as well as on the lower order. For example, with a few exceptions urea can denature this tertiary structure, and the sequence of amino acids on the protein strand after translation permits only particular conformations. But the number of possible conformations is obviously astronomically high so that, in their famous textbook *Biochemie und Pathobiochemie*, Löffler and Petrides actually talk about a *goal-oriented process* (Löffler & Petrides, 1998, 277):

> If each possible individual conformation of the polypeptide backbone of a protein of only 100 amino acids was tested for its functionality, this would take *at least* 10^{50} (!) *years* until the native biologically active form of this protein was reached. This thought points to the fact that protein folding into the correct spatial structure must be based on a goal-directed process.

"The only explanation" that the authors come up with is "that the primary structure, i.e. the amino acid sequence of a protein, must already contain all the information for producing the correct unique spatial structure for the protein" (ibid., 278). However, as shown above, it is not essential that this information is encoded *in the primary structure* in a comparable manner to the information in the DNA. Further evidence against this is provided by the existence of isomeric antibodies which – with *the same* primary structure – alternately assume e.g. two structurally *different* conformations and therefore can bind two different antigens (Foote, 2003). The same applies to the phenomenon of allostery (Fox Keller, 2000, 64), i.e. the conformation change of an enzyme – mostly induced by the binding of a specific effector – through which its biological *activity* is first *produced*. The – always specific – function of a protein can thus never be reductionistically deduced from the specificity of its primary structure. Therefore: "one protein – many functions", as Fox Keller puts it (ibid., 65). Findings like this nowadays lead increasingly to an "'avant-garde view' in which proteins are conformationally dynamic and exhibit functional promiscuity" (Tokuriki & Tawfik, 2009).

It therefore appears to be increasingly conceivable that the actual tertiary structures produced under the conditions of specific environmental factors and primary structures are a part and expression of conformation laws which can manifest in reality in a similarly "fluid" way, i.e. in different forms, as is the case for the organic archetype as defined by Goethe, in other words for the structural law of the organism as a whole[72]. The archetype can also manifest in the form of variable blueprints, depending on the environment and substrate (cf. Section 4.3).

If the manifestation of the tertiary structure as presented above is considered as a *process of self-organisation*, then this can also explain what is meant in scientific terms by the expression "goal-oriented process". "Goal" is then not a *predefined structure* in the sense of a mentally anticipated goal of human actions or an anthropomorphically imagined "intelligent design" of god, but the manifestation of a structural law, which may *vary* in its expression in real terms, corresponding fully with what the substrate and environmental conditions permit. This manifestation is "goal-orientated" despite its potential variability inasmuch as the law can only realise what is contained within it. Its content, a lawfulness of form (causa formalis) is, as that which is to be manifested, the "goal" (causa finalis). This exactly corresponds to the formative and final cause that David Bohm proposed as causal principles for the goal oriented manifestation of "wholes" in physics as well as in biology (Bohm, 1986). The manifestation process as such appears as "self-organisation". This takes place "from above", to the extent that the form is impressed on the matter present, i.e. the primary protein strands under specific conditions (causa materialis); and at the same time this takes place "from below", to the extent that matter and conditions for this manifestation must be present. This is therefore a mutually determining process similar to the formation of inorganic crystals (cf. Sections 3.2 and 3.3), but in the context of the living organism. It is interesting that Goethe used a similar expression in the organismic context to characterise the nature of the causative agency responsible for manifestation of living organic forms "from above" as did David Bohm in the context of physics: "formative force" ("bildende Kraft") (Goethe, 1950j, 240); "forming activity" (Bohm,

---

72 I am grateful to Dr. Johannes Wirz, Dornach (pers. com. 2009) for these ideas. At the *(macro)-molecular level* his corroboration would provide evidence for the thoughts discussed below on the higher-ranking *morphological structures* and the *blueprint as a whole*, that life is due to laws whose realisation takes place *in becoming*, i.e. in *time*, whereas the non-living is connected to what *has become* which then only has a *spatial* meaning. As something which *has become*, the tertiary structure of the proteins can therefore be compared with an inorganic crystal structure and, in its *development phase*, the inorganic crystal still has a distant similarity to the living. In relation to all its structures, the organism as a whole is only something which *has become* once it is dead. In this context it has then literally become inorganic.

1986, 276). Naturally "from above" and "from below" are not to be taken in an extensive spatial sense, but in an intensive, dynamic one: that which gives the law is always of a "higher" order than that which receives it and this is always of a "lower" one.

In summary it can be said that the genetic explanation of life processes cannot be interpreted as a "chemical explanation of life" because this explanation involves not simply the genes or other substances as the actual "chemical" or material, but also the so-called "information" i.e. something immaterial which is "encoded" or conveyed by these substances. What is generally understood as "genetic information" is strictly speaking only the structural information for the primary structure of the proteins: and these are not encoded *as such* in the genes in a fixed way but only by splicing and other regulatory genetic processes in the mRNA. This information only becomes manifest through the process of translation. It is then present "chemically" in a certain sense, i.e. as the protein strand with its structural inherent law. But what then influences this when it is folded into a tertiary structure is a *further* law which is not to be found encoded in the sense of "genetic information" in the DNA or RNA, i.e. "chemically", but manifests in the material present only under suitable conditions, namely in the manner that these conditions and the material permit. The manifestation of these structural laws takes place as a goal-orientated process of self-organisation. In terms of lawful realism this can be interpreted as the actualisation of the relevant laws (universalia) from the state of "ante rem" to the state of "in re". This involves not only "chemical" or "material" but also "spiritual" – which manifests from the spiritual into the material state – in the creation of organic substances. This is a process analogous to the one described for inorganic crystals or chemical compounds (cf. Section 3.3). The biochemical and biological are based on the spiritual, as is the material itself – as the lawfulness inherent in them. It is not only the creation and composition of inorganic and organic substances which can be understood in this way as a spiritual and material process, but also biological processes, such as gene regulation.

4.2 Gene regulation. From a static to a dynamic concept of the gene

As already mentioned, the "central dogma" first formulated by Francis Crick in 1958 states that the flow of information can only occur from the DNA to the protein but not the other way round. "The central dogma is a negative statement […], a hypothesis which says that you can't translate backwards, […] from protein to DNA or from protein to RNA" (Commoner, 1968). But since then it has become evident that there are also regulatory feedback effects from the RNA to the DNA[73] and also from the proteins to the RNA[74] and from the proteins to the DNA[75] (Wirz, 2008). Further, the DNA expression is influenced by enzymatically mediated epigenetic processes such as methylation and histone modification (Jaenisch & Bird, 2003). The DNA thus proves to be not just a *sender* of information but also a *receiver*. The DNA is therefore also determined by information which does not come from the DNA. The "central dogma" was already disproved years ago as far as this is concerned (Heusser, 1989). "DNA has lost its importance as a control centre and has become part of the molecular regulatory network" (Wirz, 2008, 11).

In terms of the central dogma, the genome has usually been viewed – and often still is – as the permanent element which ensures stability and therefore heredity. As everyone knows, this is correct in the sense that heredity is founded on the stability of the DNA: but in another sense it is wrong. For even gene replication, the prerequisite of every hereditary transmission to daughter cells or organisms, is far from a mechanical and chemical process as suggested by the double-stranded DNA structure. A process of this kind would suffer an error rate of 1:300, something which would be incompatible with life. The enzyme-dependent replication by means of polymerases lowers this error rate to 1:10,000 base pairs, the likewise enzyme-dependent "proof reading", i.e. the correction of incorrectly integrated nucleotides during the replication process, further to $1:10^7$ and the "mismatch repair", the correction after completion of the replication process, to $1:10^{10}$ (Wirz, 2008).

In other words: the DNA replication is an error-prone process and the stability of the DNA is not an inherent property of the DNA itself but is *produced actively* by an enzymatically determined activity of the organism (Wirz, 2008). Fox Keller summarises this as follows (Fox Keller, 2000, 31):

> The stability of gene structure thus appears not as a starting point but as an end-product – as the result of a highly orchestrated dynamic process requiring the participation of a large number of enzymes organized into complex metabolic networks that regulate and ensure both the stability of the DNA molecule and its fidelity in replication.

It goes without saying that the lawfulness of this orchestration i.e. the temporal and spatial organisation is *imprinted* onto the genes. In this respect the *recipients* of the orders do not issue any.

So who is determining whom, the DNA the protein or the protein the DNA? Both mutually determine each other, each in a different way. The DNA

---

73 E.g. inhibition of DNA expression by microRNA.
74 E.g. positive for differential splicing or RNA editing, negative for RNA interference and translational termination.
75 For all gene-regulating proteins.

or RNA on the one hand and the proteins on the other are as dependent on one another as the hen and the egg (Fox Keller, 2000, 97). This is both a complement to and rejection of the central dogma of genetics.

The dynamic functional enzyme activity displayed by proteins is therefore the "conserving element" which produces and maintains the stability of the genome. And the tendency to instability in the DNA, on the other hand, enables innovation, evolution: and the organism can apparently evidently regulate this stability. It is apparent nowadays that the precision of DNA replication in organisms can be actively controlled (Wirz, 2008). Bacteria can lower or raise their mutation rate in response to environmental conditions and also determine the *type* of mutation (Björkman et al., 2000). This can be meaningfully i.e. "selectively" related to the environment, as the phenomenon of *adaptive mutation* first discovered by John Cairns shows (Cairns et al., 1988, overview and discussion in: Wirz, 1996). As plants and animals have similar DNA polymerases and repair and other regulatory enzymes to microbes, it may be concluded that active regulation of mutations also occurs in higher organisms (Wirz, 2008). Indeed, adaptive mutation is now held responsible for the recent acceleration of human adaptive evolution (Hawks et al., 2007). The active regulation of the balance between constancy and environmentally-related variation is therefore a genuine organic capacity. *This* is the primary motor in evolution, selection is secondary.

A striking example of this is the heat shock protein Hsp90 which is responsible for the maintenance of the functional conformation of regulatory enzymes in signal transmission, cell cycles and development, and therefore for the stability of these processes in yeasts, plants and animals. However, if cells are damaged, it is removed from this task and used as a heat shock protein, including for maintaining the structures of other proteins. This results in the destabilisation of those regulatory processes which lead to phenotypical variants which can then be selected and after several generations can be inherited independently of Hsp90. The environmentally-related stress on Hsp90 can therefore lead to the release of phenotypical variants which, without this stress, would not have arisen due to the stabilising effect of Hsp90. Rutherford therefore called Hsp90 a "capacitor for morphological evolution" (Rutherford & Lindquist, 1998). The variants arising in this way only come about through lawful gene regulation and are therefore an active organismal output in which the genes *as objects* of regulation are *recipients* of the law in the sense stated above (cf. also Heusser, 1989).

Like gene replication, gene expression is also a dynamic, active, coordinated enzyme-controlled capacity of the organism, the *objects* of which are nucleic acids. Transcription, for example, is dependent on the *epigenetic* regulation already mentioned above, e.g. via enzymatically controlled methylation or acetylation of the DNA or the histone, around which the DNA is wrapped in the chromosome. By removing methyl groups, the packing density between the histones and DNA is decreased and the latter then becomes accessible to transcription while methylation of DNA or histones blocks transcription. Just think what *complex and precise spatial and temporal coordination* of the acetylases and methylases is necessary in order to merely enable the transcription of specific DNA sequences. The same holds true for promotion. The promoter is the section of the gene where, by linking the RNA polymerase, the DNA double helix starts to uncoil and transcription begins. However, the promoter does not necessarily lie on the DNA strand directly opposite the gene which is to be read but can also be located far away on the same DNA strand (on the same chromosome) or on another chromosome. In this case the two chromosomes have to come closer together in order to initiate transcription (what is known as "gene kissing", cf.: Lanctot et al., 2007).

Transcription itself is not a simple conversion of the DNA code into mRNA code but, as described above, a complex enzymatically controlled process in which the "information" to be converted into protein by splicing *only arises as a correlate of the spliced mRNA*. In addition, the so-called "alternative splicing" means that *different* mRNA transcripts can be produced from *the same* pre-mRNA. And the mRNA transcripts which are produced can be modified by additional enzymatically controlled processes such as insertion or replacement of bases. Evelyn Fox Keller therefore justifiably asks: "Which of these different transcripts corresponds to what we should call the gene?" (Fox Keller, 2000, 62–63). Because these facts mean the end of the traditional concept of the gene as a section of DNA which contains the information for a specific protein (ibid., 62–63):

> We have to give up the notion, even for structural genes, that one gene makes one enzyme (or protein). One gene can be employed to make many proteins, and indeed the expression 'one gene – many proteins' has become fairly common in the literature. The problem with this formulation is that the gene has lost a great deal of both its specificity and its agency. Which protein should a gene make, and under what circumstances? And *how does it choose? In fact it doesn't. Responsibility for this decision lies elsewhere, in the complex regulatory dynamics of the cell as a whole*. It is from these regulatory dynamics, and not from the gene itself, that the signal (or signals) determining the specific pattern in which the final transcript is to be formed actually comes.

Fox Keller therefore makes the justifiable proposal, "to consider *the mature mRNA transcript formed after editing and splicing to be the 'true' gene*" (ibid., 63, ePH). Therefore, in contrast to the traditional view of the gene as a structure which is always present, it is clear that *this* "true gene" is *only produced when required and only for a short time* and therefore really has to be viewed as a *functional gene* (ibid., 71). This *new, dynamic gene concept* is in fact unavoidable if you wish to retain the most important aspects of the old static gene concept but do not want to identify it solely with the DNA: the gene as the structure which encodes the information for a specific protein. The "gene" is then, strictly speaking, the final form of the mRNA which is present on the ribosome as a substrate for translation. All previous forms of the mRNA and especially the encoding parts of the DNA can then be described as "genetic material" in contrast to the specific "gene".

It is therefore dynamic regulation which is responsible both for the gene's creation at the moment it is required and its subsequent decomposition, and also the genetic material's constancy over generations. This regulation takes place due to a complex interaction of numerous factors. These include proteins (enzymes) and also RNA molecules or fractions of these, e.g. microRNA or riboswitches (Hobert, 2008; Breaker, 2008), and even the complex three-dimensional network of chromosomal interactions (Dekker, 2008). It is becoming increasingly clear that gene regulation represents a highly complex, *coordinated* process which organises the genetic material in space and time *actively and flexibly* due to the intervention of regulatory factors (Chakalova et al., 2005), and that this organisation is connected to the development of health and illness[76]. *Organisation* is the one thing which distinguishes the organic from the inorganic, the living from the non-living, as Fox Keller clearly states with reference to the biology since the end of the 18th century (Fox Keller, 2000, 106).

4.3 Self-organisation and causality in biology and Goethe's archetype

But who or what is the cause of this dynamic organisation and regulation? It cannot be the DNA as genetic material, because *its* organisation (the sequence of its nucleotides which are further organised into introns and exons) only produces parts from which the organisation of the "functional genes" (the finished mRNA, cf. Section 4.2) is first composed through enzymatically mediated dynamic regulation and then – again enzymatically mediated – converted into the organisation of the primary protein structure through translation. Conformation, through which the protein becomes functional, manifests itself in this primary protein as an emergent property through self-organisation, in part enzymatically assisted by chaperones. This whole

---

76 Evidence is increasing that the onset, pathogenesis and development of a wide variety of diseases such as viral infections, diabetes, cancer and neurodegenerative, autoimmune and cardiovascular diseases are largely connected to a disruption of the balance in this *epigenetic* regulation, such as from under- or over-expression of microRNA, aberrant methylation or histone modification (Couzin, 2008; Halusková, 2010). This is of course once again the result of dynamic regulation.

enzymatic process-organisation is also not the primary cause of dynamic regulation, because, as a catalyst, an enzyme only produces an energetic facilitation of the *self-organisation* of the catalysed structure or function and, in addition, only specifically for *a single* structure or function. But gene replication, gene expression and morphogenesis are complex events which, beyond enabling *individual* molecules or functions, also have an effect on the enzymatically assisted self-organisation of *higher-order* molecular, organelle, cell and organ structures and functions, in short, on *the organism* and in fact "as the result of a highly orchestrated dynamic process requiring the participation of a large number of enzymes organized into complex metabolic networks" (Fox Keller, 2000, 31, ePH).

Orchestration, however, is synonymous with *active determination* of the orchestrated parts – in this case the regulatory enzymes – by *the whole*, i.e. by the lawful *unity* and *organisation* of these enzymes and their functions respectively. These enzymes are of course themselves the result of the orchestrated dynamic processes, even though at a preceding ontogenetic stage of development. The cause of the orchestration is therefore not the cascades of regulatory proteins but their "conductor", as it were, i.e. that active entity which – in a top-down process – gives all the individual "instruments" their entry in accordance with the "symphony" of the whole, to take the metaphor of orchestration to its conclusion. If we dispense with the anthropomorphism in this comparison and hence with the "conductor" (cf. Section 4.4), then it is the *coordinating or orchestrating law itself, the symphonic structure*, which comes about through *self-organisation* in its instruments and in unity with them.

And the symphony does not reside in the genes. As already described, this organisation involves mutually dependent proteins and nucleic acids (DNA/RNA). The protein is a product of the gene and the gene in its way is a product of the protein. Both are polar elements of a *higher organisation* of which they are themselves incorporated as parts. So, as already explained above, the DNA can never contain all the information in the organism, a view also expounded by Richard Strohman who, when referring to this, speaks of a "coming revolution in biology" (Strohman, 1997, 195):

> There is growing recognition that information for function may not be located solely in genomic databases. That is, it is becoming clear that sequence information in DNA, by itself, contains insufficient information for determining how gene products (proteins) interact to produce a mechanism of any kind. The reason is that the multicomponent complexes constructed from many proteins are themselves machines with rules of their own; rules not written in DNA.

As shown above, the DNA only encodes sections of the primary structure of proteins and they contain their own organisational laws (in the state of universalia in re). All the proteins' higher laws of organisation come about through self-organisation, i.e. by active transition from the potential state of ante rem to the actual state of in re: and the proteins catalyse the self-organisation of the DNA/RNA and of the proteins. The further organisation of the DNA/RNA and proteins in space and time takes place through *further laws*, i.e. through those which determine the spatial and temporal *relationship* of these substances and are therefore *at a higher level* than them. *These organisational laws or information cannot be derived either from the DNA/RNA or from proteins; but they also manifest – by self-organisation – through the catalysing effect of proteins.* The same applies to the enzyme controlled self-organisation of innumerable other substances, functions and structures and their lawful spatial and temporal connections. In other words: the class of proteins which owe their primary structure to genes, catalyses the regulated self-organisation of their own members and their genes plus the spatial and temporal *global order* which makes up the organisms as such, from the lowest level of individual materials up to the highest morphological formation of organs, organ systems and the whole organism. And just as they mediate their own *development*, they also mediate the similarly regulated physiological *breakdown* of this organism or its parts and substances; and the organisational laws governing this are also part of the global organic organisation. From this point of view it is not DNA and RNA which are the *crucial substances of life*, but *proteins*. They transmit all the information about the organism and its complexly controlled realisation in the flexible but always lawful dynamic which constitutes life.

The global information of the organism is therefore far more than that encoded by the DNA, i.e. as the genotype: the global information of the body is the lawfulness of the entire spatial and temporal structure of the body, from the individual organic substances via organelles, cells and organs including their complex functions and relationships all the way to the global organisation of the body which integrates everything. *The global information of the organism therefore includes the lawfulness of the genotype and the phenotype.* The phenotype has its own laws. Although, as has been described, these are realised on the basis of the genotype, they themselves manifest on this basis and through the assistance of enzymes, and their content cannot be deduced from the laws of the genotype. This is also the reason why the relationship between the genotype and phenotype has so far remained unsolved (Wirz, 2008).

The actual cause of organic development is therefore the self-organisation of its global information. This is self-realising and to achieve this it uses its own components, especially the nucleic acids and proteins, but also other substances. However, it is not these substances which "regulate" or "control" the organisation as is often stated by uncritical anthropomorphism, but *this organisation regulates and controls itself through these substances*. Because the substances are a part of this organisation. They all develop through self-organisation but always in alternating dependency on each other: sequentially, hierarchically and in spatial/temporal organisation, i.e. in the orchestrated actualisation of those higher-level relational laws that constitute the organism as a whole in space and time. The unity or system of these higher-level laws is the global information in the organism, in other words the *"blueprint"*, which the central dogma hopes only to find in the DNA. It is the – in itself complex and multi-structured – *organisation* of all organic laws of substance, structure and function, of which the DNA forms only a part. The genotype is also a "phenotype" or a specific part of one, to the extent that it is realised as a phenomenon, as is the protein in its own way. If you consider only the lawful, intrinsic content of this genotype-embracing phenotype, independently of its realisation as a phenomenon or "phene", then it is an *"ideotype"*, i.e. the global organisation of its laws as law, idea or even: as its global information. This contains the *spatial and temporal "global plan"* or *the lawful system* of the organism. It is what Goethe referred to as an *archetype* (Goethe, 1950j). As Goethe rightly put it, it is the *idea* of the organism (Steiner, 1988a, 90).

Of course this idea is not to be seen as a finished fixed thought, a preformed design in terms of the creationist "intelligent design" or, as Steiner expressed it, as a "complete, frozen concept form" like the meaning ascribed by Louis Agassiz (1807–1873). Agassiz assigned every biological species and therefore every specific design the status of a fixed idea based on an abstract understanding of Platonism (Hösle, 2005, 217). Metamorphosis or the transformation of one design to another and therefore evolution as defined by Darwin would therefore be impossible. Goethe's archetype, however, should be viewed as a general *principle* which becomes manifest in the phylogenetic and ontogenetic course of development in variable, metamorphosing forms and therefore represents the ideational, the lawfulness which underlies the manifestation of evolutionary processes observed by Darwin. Steiner expresses this as follows (Steiner, 1988a, 90–91):

> One should not picture this typus as anything rigid. It has nothing at all to do with what Agassiz, Darwin's most significant opponent, called 'an incarnate creative thought of God's'. The typus is something altogether fluid, from which all the particular species and genera, which one can regard as subtypes or specialized types, can be derived. The typus does not preclude the theory of evolution. It does not contradict the *fact* that organic forms evolve out of one another. It is only reason's protest against the view that organic development consists purely in sequential, factual (sense-perceptible) forms. It is what underlies this whole development. It is what establishes the interconnection in all this endless manifoldness. It is the inner aspect of what we experience as the outer forms of living things. *The Darwinian theory presupposes the typus.*

In fact, *comparative* anatomy and typology or the establishment of homologies in morphology as a prerequisite for producing evolutionary phylogenetic trees are only possible because we can define intellectually what is common or typical in the Goethean sense as well as what is different for the individual blueprints (ibid., 91, 92):

> *All* forms result as a *consequence of the typus*; the first as well as the last are manifestations of it. We must take it as the basis of a true organic science and not simply undertake to derive the individual animal and plant species *out of* one another. The typus runs like a red thread through all the developmental stages of the organic world. We must hold onto it and then *with it* travel through this great realm of many forms. Then this realm will become understandable to us. Otherwise it falls apart for us, just as the rest of the world of experience does, into an unconnected mass of particulars. [It is also essential to understand] that deriving something later out of something earlier is no explanation, that what is first in time is not first in principle. All deriving has to do with principles, and at best it could be shown which factors were at work such that one species of beings evolved *before* another one in time.

This idea, Goethe's archetype, is therefore not something hypothetical or speculative as is often believed. It is the lawful, universal of that which can be found as actual organisation in the particular organism being observed. It is nothing other "than what manifests in the particular in the form of the general", as Steiner expressed it in *"The Science of Knowing, Outline of an Epistemology Implicit in the Goethean World View"* in the chapter *Organic nature*. It is "an organism in the form of the general" (ibid., 90), i.e. the idea, the law of the actual manifest organism. This lawfulness must be grasped by thinking – fully in accordance with the general principle of scientific cognition (cf. Chapter 2) – *on the basis of the relevant perception* and must agree with this exactly.

The archetype in this context, as understood by Goethe and Steiner, can be nothing else than the result of empirical cognition (Steiner, 1988b). This is in contrast to the old vitalism and the neo-vitalism of the 19th century and to the teleology which was current before Goethe.

The problem of vitalism is "that, in addition to all the causal factors, it also makes use of the additional causal factor of vitality: the vis vitalis or the nisus formativus or the vis essentialis, the expressions used for it in the 18th and 19th centuries. A causal force was postulated in addition to the spectrum of previous causal forces, this being the life force which brings everything to life" (Schad, 2003, 24).

The "entelechy" of Hans Driesch (1867–1941) is generally added to neo-vitalism, although it is assumed by Driesch to be the unifying principle of the organism on *empirical* grounds[77] and some of its features are also characterised more fully due to their effects (Mocek, 1996). Driesch himself distances himself firmly from vitalism, such as e.g. by the statement that entelechy is not quantifiable and therefore could not be a causally effective energy in the material sense. "Entelechy is limited to an *ordering* capacity and nothing more [...]. I therefore reject any type of 'energetic vitalism' in the strongest possible terms" (Driesch, 1921, 426–427). But *who or what* performs this ordering remains undefined, entelechy remains a hypothesis – in contrast, incidentally, to Goethe's "entelechy" (Goethe, 1950l, 435). This may be the reason why Driesch's entelechy is generally considered to be neo-vitalistic (cf. Section 4.8).

Teleology which was current before Goethe, traced the apparently goal-directed adaptation of the organism to *hypothetically assumed* aims which the creator had imprinted into the creation, which is postulated nowadays in a new form in the creationist theory of "intelligent design" prevalent in the USA (Fuller, 2007). But this is an anthropomorphic idea and assumes a purpose based on human ideas of action, as Steiner described as far back as 1886 (Steiner, 2011 155–157):

> *Purposefulness* is a special kind of sequence of phenomena. [...] One performs an action of which one has *previously* made a mental picture, and one allows this mental picture to determine one's action. [...] Man makes his tools according to his purposes; the naïve realist would have the Creator build organisms on the same formula. Only very gradually is this mistaken concept of purpose disappearing from the sciences. [...] Monism rejects the concept of purpose in every sphere, with the sole exception of human action. It looks for laws of nature, but not for purposes of nature. *Purposes of nature* are arbitrary assumptions no less than are imperceptible forces. [...] Nothing is purposeful except what man has first made so, for purposefulness arises only through the realization of an idea.

This anthropomorphic or "extrinsic" teleology is also almost universally rejected nowadays. It is described as "goal *seeking*" whereas an "intrinsic" teleology or "goal *directedness*" of the organism is recognised by many scientists[78] and described as *teleonomy* to distinguish it from the extrinsic teleology (Toepfer, 2005). Teleonomy refers in particular to the "cycle of mutual independence of its parts" (ibid., 47), in other words, what Rupert Riedl has called "circular causality" (also "recursive", "recurrent" or "feedback causality") of the individual processes in the system of the organic

---

77 "We have tried to formulate the relationship between the entelechy and the elementary inorganic agents in such a way that nothing is demanded which is not justified on the basis of experimental facts [...]" (Driesch, 1921, 441).
78 Goethe also recognised that the organism is purposeful in this *inner* sense, i.e. by the logical inter-relatedness of the parts and functions in the whole. See, for example, the quotation in his *Metamorphosis of animals* in Section 5.4).

whole which act on each other in turn, in contrast to the linear causality of inorganic processes (Riedl, 2005).

However, as regards causality it is not just a matter of the interaction of the parts (because this would result in a linear causality which merely had a circular arrangement), but beyond this of the *arrangement i.e. the system of the parts itself* which as a whole makes this reciprocal dependency possible in the first place: "The organised system as a whole consists in fact of nothing more than this reciprocal relationship of individual processes to each other" (Toepfer, 2005, 48). Because this arrangement is a real one which maintains itself remarkably well in the face of disturbances, it must be the consequence of an action. It is therefore acceptable to imagine an activity linked to the arrangement i.e. the system or the organisation itself which brings the parts into this arrangement and maintains them there. Riedl's "circular" causality therefore presupposes a *systematic causality* or, in other words, *the causality of the system*.

This consideration leads to a new view of Goethe's archetype. Because this is nothing other than this very system of the organism as a whole which can be found by empirical research, in which all the parts are lawfully related to each other in terms of the *intrinsic* goal-directedness mentioned above[79]. Goethe definitely views this archetype, the idea or organising law of the organism, as an *activity*. This activity can be seen as the cause of the organic self-organisation and as the reason for the arrangement of the parts in such a way that they appear in a meaningful, systematic interaction. As the organisational law it is the organic causa formalis; and to the extent that it manifests itself in a goal-directed teleonomic form (compare something like embryology) it is the causa finalis (cf. also Sections 3.3 and 4.1).

The justification on empirical grounds for speaking in this way of a causality of the organic system also arises from the logical comparison of organism and mechanism.

4.4 Organism or mechanism?

It is of course very unusual to explain the order in a living organism by the active *idea* which underlies it and which organises its parts in the process of self-organisation. We are accustomed to think of this organisation as the result of causal interactions in the molecules. The organism is therefore ultimately

---

79 Compare for example *Goethe's "Outline for a General Introduction to Comparative Anatomy, Commencing with Osteology"* written in 1795, in particular Chapters 2 and 3: *"On establishing a typus to aid comparative anatomy"* and *"General description of the typus"* (Goethe, 1950j), and also Goethe's description of this *inner* purposefulness mentioned above.

viewed as a kind of complex biological machinery and it is hoped that a full elucidation of the molecular biological interactions will provide a causal explanation of its order.

But this hope turns out not to accord with reality. Let us for a moment consider the organism *logically* as a mechanism or machine. Can a machine be causally explained *in full* by the interaction of its parts? – No, because *the plan, the design*, according to which the parts of a machine are arranged is never a *result* of their interaction, but the interaction of the machine parts *always requires* that the parts have first been arranged according to this plan. And this takes place as a result of forces which do not arise from the parts, but are imposed on them in order to give them their logical place in the whole of the machine.

The *full* causal explanation of a machine therefore requires *two hierarchically arranged systems of efficient causes*: 1. the *system of forces which makes the plan into a reality*. This is *superior* to the parts and their forces because it compels them to occupy their place; 2. the system of forces which brings about the *function*, the *interaction of these parts*. This can never give rise to the plan and its system of forces which puts the plan into action.

A logical definition of the organism as a mechanism therefore results in the need for two different, hierarchically arranged systems of forces to fully explain the life processes. However, there are two points in which the organism differs significantly from a mechanism. First, the organism is autopoietic, it always creates itself, whereas the machine never creates itself but is built by a design engineer or by other machines which are in turn built by an external something or someone else[80]. Therefore, the organism *itself* must contain[81] what in the case of the machine always has to be brought into it *from outside* by the engineer: the design and the forces which put this into practice. And because in the organism these forces implement *the design directly* without the intervention of an engineer, design (idea, law) and force must be a unity: a lawfully acting force or a force-bearing law.

The definition of the design as an *active idea* in the Goethean sense is therefore simply the consequence of a systematic comparison of organism

---

80 The exponents of *intelligent design* actually imagine that the organism is also created by someone or something else like the machine, but in this case by God. However, this kind of idea has no place in *science*. It is an unjustified hypothesis since it cannot in principle be supported empirically, but the empirically observable circumstances of *human* fabrication are transferred in a naive childlike way to the anthropomorphic picture of God.

81 It is also not possible to argue that the parental DNA is exogenous, because the daughter organism is to a certain extent a *continuation* of the parent organism which makes itself independent. It is therefore primarily not external but internal, as part of the parent organism.

and mechanism. The design itself performs its actions in the organism. Steiner already pointed this out in his *"Goethean Science"* (Steiner, 1988b, comment on p. 49):

> This is precisely the contrast between an organism and a machine. In a machine, everything is the interaction of its parts. Nothing real exists in the machine itself other than this interaction. The unifying principle, which governs the working together of the parts, is lacking in the object itself, and lies outside of it in the head of its builder as a plan. Only the most extreme short-sightedness can deny that the difference between an organism and a mechanism lies precisely in the fact that the principle causing the interrelationship of the parts is, with respect to a mechanism, present only externally (abstractly), whereas with respect to an organism, this principle gains real existence within the thing itself. Thus the sense-perceptible components of an organism also do not then appear out of one another as a mere sequence, but rather as though governed by that inner principle, as though resulting from such a principle that is no longer sense-perceptible. In this respect it is no more sense-perceptible than the plan in the builder's head that is also there only for the mind; this principle is, in fact, essentially that plan, only that plan has now drawn into the inner being of the entity and no longer carries out its activities through the mediation of a third party – the builder – but rather does this directly itself.

And here lies the second difference from a mechanism: in the latter the *structure is always separate from the function and precedes this*; and the engine can only be repaired when it is switched off. While this is apparently trivial, it is generally completely ignored in the mechanical theory of the organism so that the comparison between the organism and mechanism is inconsistent. Structure and function in the organism are always manifested simultaneously – in all the parts and in the whole. The typical properties of life which clearly distinguish the organism from the machine such as growth, morphogenesis, metamorphosis, self-regulation, even the ceaseless active maintenance of the form, are nothing other than an expression of the constant maintenance of the (re)construction of the organism, from the beginning of life until its end. In other words, from this viewpoint the living organism can only exist because it does not merely consist of the system of its interactive components (molecules, organelles, cells, etc.) but, in addition, *contains a second active system* which permeates the first one but is at a higher level to it because it creates the order in the first.

This hierarchical "two-levelled structure" in organisms was put forward in 1968 by Michael Polanyi in his remarkable *Science* article *"Life's irreducible structure"* with reference to modern molecular biology and morphology (Polanyi, 1968):

> So the machine as a whole works under the control of two distinct principles. The higher one is the principle of the machine's design, and this harnesses

the lower one, which consists in the physical-chemical processes on which the machine relies. [...] In this light the organism is shown to be, like a machine, a system which works according to two different principles: its structure serves as a boundary condition harnessing the physical-chemical processes by which the organs perform their functions. Thus, this system may be called a system under dual control. Morphogenesis, the process by which the structure of living beings develops, can then be likened to the shaping of a machine which will act as a boundary for the laws of inanimate nature. [...] Thus the morphology of living things transcends the laws of physics and chemistry. [...] The irreducibility of machines [...] teaches us, that the control of a system by irreducible boundary conditions does not *interfere* with the laws of physics and chemistry. [...] Irreducible higher principles are *additional* to the laws of physics and chemistry. [...] The theory of boundary conditions recognizes the higher levels of life as forming a hierarchy, each level of which relies for its workings on the principles of the levels below it, while it itself is irreducible to these lower principles.

Surprisingly, however, Polanyi does not seem to see what these higher principles of life, this "system of causes not specifiable in terms of physics and chemistry" or this "integrative power" actually is. He considers it to be a "missing principle" and merely refers to the morphogenetic field concept of Hans Spemann and Paul Weiss, the content of which remains hypothetical. He fails to see that the "equivalent biological principles" which correspond exactly to the machine design can be nothing other than the principles of the design of the organism which are manifest in the organism and can therefore be *empirically determined* viz. a system of principles which organises and implements itself, naturally only in the material and under the conditions which these principles require for their implementation, something which is true for all emergent principles. This self-realising organic system is the typus, the "idea" or, as might be said nowadays, the total "information" in the organism (cf. Sections 4.1 and 4.3).

Goethe's particular achievement is to have been the first to recognise this clearly and to have demonstrated the method for the scientific *investigation* of these special biological principles without having recourse to mechanistic or spiritualistic explanatory hypotheses. Because Goethe did not wish to found a science of the organic realm purely by continuing physics and chemistry inside the organism – as did the biology which came after him – but by a systematic discovery of the *emergent, specifically biological* laws of the organism, he can be designated as the true inaugurator of *organic* science. This does not contradict the *facts* of molecular biology which have been brought to light by modern biology since Goethe's time, but only the mechanical interpretation of these facts. Goethe's ideas complement these facts by containing the empirically justifiable explanation for their *system*. Steiner in particular pointed out the fundamental importance of Goethe's achievement for biology and compared it to that of Copernicus and Kepler for astronomy (Steiner, 1988b, 76).

4.5 Typus and morphogenesis

Based on the above, the order of the organism can be explained by the *idea* which is fulfilled in the organism itself. However, is there any need for the "typus", an "idea", if it has been shown that the blueprint of the organism is determined by developmental control genes i.e. by morphogens and signalling molecules?

Let us take the well-researched example of the morphogenesis of the fruit fly, *Drosophila melanogaster*. The maternal ovary cells transcribe what are called maternal effect genes and deposit their mRNA in specific peripheral regions of the developing egg cell, e.g. *bicoid* mRNA at the anterior end and *nanos* mRNA at the posterior end. After fertilisation of the ovum and oviposition, nuclear division begins, initially without cell division, so that up to the 13th nuclear division a syncytium develops whose nuclei are distributed uniformly in the peripheral cortical zone along the entire length of the embryo. The Bicoid protein is formed during this period by translation of the maternal messenger RNA at the anterior end of the embryo. This diffuses towards the posterior end, forming a concentration gradient which decreases distally. The Nanos protein is formed at the posterior end in the same way with a decreasing gradient towards the anterior end. These gradients determine the anterior-posterior axis of the body and predetermine its morphological structure, as each cell nucleus is exposed to a different ratio of Bicoid and Nanos proteins in terms of "position information". This controls the further pattern formation of the insect's body which is also influenced by sequential hierarchical activation of three classes of what are known as segmentation genes. First Bicoid and Nanos regulate the expression of what are referred to as *gap* genes whose products determine a number of broad regions of the body. Bicoid regulates their transcription, Nanos their translation. The gap genes then activate what are known as pair-rule genes which bring about a paired finer segmentation. Finally, the segment polarity genes define the boundaries and the anterior-posterior orientation of each segment whose role (e.g. as the bearer of legs, wings, antennae etc.) is then further controlled by homeotic genes (Nüsslein-Volhard & Wieschaus, 1980; Purves et al., 2006, 484–486).

Christiane Nüsslein-Volhard and Eric F. Wieschaus, who received the Nobel Prize for medicine in 1995 for the discovery of these regulatory process cascades, at first attributed the development of the substance gradients which control the body axis to simple *diffusion*. This implied that the developing pattern had to be a passive i.e. mechanical-chemical result of the physical and chemical properties of the regulatory substances and their interaction with the diffusion medium and not the consequence of an "idea" which manifests itself actively. The modern morphogen concept implies that e.g. Bicoid directly determines the domain boundaries of its target gene (Gurdon &

Bourillot, 2001). The idea here is that the local gap gene expression (e.g. of hunchback) is dependent on a specific threshold concentration in the particular cell nucleus.

However, there are various reasons to question why a *purely passive* diffusion should be the cause of the development of form. Even the basically *teleonomic* character of all development processes of organisms clearly described by Ernst Mayr (Mayr, 1999) suggests that activity may be involved in such diffusion. A *directed*[82] process assumes *active control*. In fact, the scientific literature soon reported on a diffusion control of this kind, for example in *Science*: "Developing embryos may actively ship key signaling molecules from place to place, instead of relying on diffusion to carry the messages. [...] And simple gradients can't explain the physical changes that accompany some crucial developmental events" (Vogel, 1999). It was demonstrated that the transport of maternal mRNA and that of expressed regulatory proteins took place via the relevant motor proteins in a directed manner, by what are known as kinesins towards the posterior end and by dynesins towards the anterior end, at least in part along previously created microtubule structures (Brendza et al., 2000). Further, regulation of the Bicoid gradient takes place by active absorption and subsequent degradation of Bicoid by the local cell nuclei (Gregor et al., 2007) i.e. by a dynamically balanced ratio between dominant morphogen synthesis in the anterior region and degradation in the posterior region (Jaeger et al., 2004). The degree to which the total morphogen distribution is actively and lawfully determined by the organism is shown by the fact that the form and geometrical *proportions* of the gradient pattern always remain the same overall, irrespective of the size of the embryo (Gregor et al., 2007). This would not be possible with passive diffusion as an exponential gradient has its own scale which is not affected by the actual length of the body (Houchmandzadeh et al., 2002). This results in a precision equal to that of the physical measurement instruments used and no longer corresponds to the random distribution of molecules in simple diffusion: "The creation of the Bicoid gradient [...] must involve some sort of facilitated transit from nucleus to nucleus: The gradient is not explicable in terms of simple diffusion of Bicoid through the cytoplasm" (Lewis, 2008). The result of this is a morphogen concentration which is coordinated overall and finely adjusted from one nucleus to the next, which then determines the gene expression in each cell nucleus. Thomas Gregor and his team studied these ratios by direct

---

82 Teleonomic or "directed", as described above, should not be thought of as a predetermined fixed objective in the sense of an external teleology, but rather as an internal teleology or teleonomy arising from self-organisation in which an organic structural order is brought about by activity (cf. the discussion of "causa finalis" in Section 4.1 and "intrinsic teleology" or "teleonomy" in Sections 4.3 and 4.4).

measurement of the concentrations and reached the conclusion that "the embryo is *not* faced with noisy input signals and readout mechanisms; rather, the system exerts precise control over absolute concentration differences and responds reliably to small concentration differences, approaching the limits set by basic physical principles" (Gregor et al., 2007). The resulting segmentation pattern does not arise as a consequence of passive diffusion and from the corresponding chemical and physical interactions of the morphogen with its environment, but as a result of an *active and directed positioning* of the morphogen concentration, naturally always under the precondition of these interactions. The more detailed the present-day findings about morphogenesis become, the clearer it is that morphogenesis only takes place thanks to a precise dynamic *orchestration* of all the process cascades involved. Gene expression, synthesis, transport and degradation of the morphogens plus the specific response of the local interactive cell nuclei[83] are active and therefore *always mutually dependent* processes which are in turn *coordinated*

---

83 For example, not only is the organised spread of the maternal morphogens an active and directed process in comparison with which the local gap genes function as passive information receivers in order to activate the blueprint, but the expression of these *gap* genes in the local cell nuclei is itself an active process which – in a dynamic regulatory interaction with the *gap* gene expression of the cell nuclei of other localisations – responds dynamically to the maternal morphogens, can auto-activate and is thus able to e.g. specify variability at the level of the maternal morphogens and therefore correct any lack of precision in the specified "position information" (Houchmandzadeh et al., 2002). The differentiated, integrative autonomy of the local cell nuclei and subsequent cells also turns out to be a process which is *subject to the intended blueprint*. The former theory of passive morphogenesis put forward by Nüsslein-Volhard and Wieschaus is therefore obsolete and has given way to the current more organismic view which recognises the organised autonomy of the tissue and even of each individual cell nucleus. Jaeger and his colleagues express this as follows: "Our results indicate that maternal Bcd, Hb and Cad alone are not sufficient for positioning of gap gene domains and *hence do not qualify as morphogens in a strict sense*. As has been pointed out, *an active role of target tissue in specifying positional information contradicts the traditional distinction between the instructive role of maternal morphogens and their passive interpretation*. The requirement of specific regulatory interactions in the target tissue for *proper interpretation of positional information* can be interpreted as a requirement for specific tissue competence. In addition, the dynamical nature of the positional information, as encoded by expression boundaries, suggests *that positional information in the blastoderm embryo can no longer be seen as a static coordinate system imposed on the embryo by maternal morphogens*. Rather, it needs to be understood as the *dynamic process underlying the positioning of expression domain boundaries*, which is based on both external inputs by morphogens *and tissue-internal feedback among target genes*" (Jaeger et al., 2004, ePH).

spatially and temporally with other processes, such as e.g. the development and alignment of microtubules. In other words, an *active coordination* of all the sub-processes in the sense of a unity in the whole, i.e. towards *a goal*: the actualisation of the morphogenetic idea.

What then is the defining factor which underlies this coordination? In the light of what has been explained here so far this is no less than the *idea*, the lawfulness of the pattern itself. This idea which, with reference to the morphogenetic gradients ought to become obsolete, reappears as their *prerequisite*. In other words, an idea which is not *derived* from these molecular interactions, but *directs* these to their end result, as it were. Morphogenesis is self-organisation: however, not self-organisation from below but *from above*. It is not the organism's molecular interactions which cause its organised pattern, but the content i.e. the idea of this pattern causes this itself, naturally only under the condition of those interactions and *within* them. Organic knowledge in terms of causality therefore requires not only knowledge of the interactions of the parts as in the inorganic realm, but in addition the morphogenetic idea which manifests in these interactions.

4.6 Organic versus inorganic cognition

Cognition of the organic differs significantly from that of the inorganic through an *understanding of the organisational idea*[84]. Unlike a mechanical process, morphogenesis in the organism is not an activity which can be explained simply from the interaction of its sub-processes which are perceptible to the senses (cf. Section 4.4). It should be noted that what is spoken of here is a *process*, i.e. the *function* of the machine which in a causal analysis is merely a matter of the interaction of the parts according to inorganic laws. In order to explain the *construction of the machine*, the structural *idea* and its actualisation must be included in addition to the parts, in a similar way as is argued here for organic cognition (cf. Section 4.6). In the case of the *function*, consideration of the actualisation of the structural idea of the machine is no longer relevant as a cause because the structure is finished. However, the organisational idea of the organism, something which is no longer perceptible to the senses, always needs to be included in the causal analysis because, as

---

84 Inorganic here means the mechanical, not something like a crystal. Crystals occupy an intermediate position between the inorganic and organic realms. While they are inorganic in terms of "not living", knowledge of them is similar to that of the organic realm inasmuch as it involves an understanding of their "organisational idea", i.e. the principle of the crystal. This can also be seen as an active idea or causa formalis (cf. Sections 3.2 and 4.1) like the archetypal law of the plant except that, in contrast to the latter, it manifests only as a law in space and not in time (in growth and metamorphosis).

shown by the comparison with a mechanism, organic structure and function always appear together[85]. In the mechanism the total process is functionally an inevitable passive-mechanical result of its sub-processes which are all perceptible to the senses. This no longer applies to the organism. In the organism, apart from the sub-processes, it is more important to include its determination by the higher-ranking entity, the law of the whole, when explaining the causality, as Steiner explained in his essay *"The Nature and Significance of Goethe's Writings on Organic Development"* (Steiner, 1988b, 49):

> It must be admitted that all the sense-perceptible factors of a living being do not manifest as a result of other sense-perceptible factors, as is the case with inorganic nature. On the contrary, in an organism, all sense-perceptible qualities manifest as the result of a factor *that is no longer sense-perceptible*. They manifest as the result of a higher unity hovering over the sense-perceptible processes.

As the sub-processes proceed from this unity, in other words from the developmental law of the whole, and as this latter – as an active idea – is not itself perceptible to the senses but can only be grasped by thinking, Steiner rightly says (ibid., 49) that in an organism all sensory processes are determined

> by something standing over them that itself is not again a form observable by the senses; these forms do exist for one another, but not as a result of one another. They do not mutually determine one another, but rather are all determined by something else. Here we cannot trace what we perceive with our senses back to other sense-perceptible factors; we must take up, into the concept of the processes, elements that do not belong to the world of the senses; *we must go out of and beyond the sense world*. *Observation* no longer suffices; we must grasp the *unity* conceptually if we want to explain the phenomena.

This unity is precisely the content of the blueprint which is actually realised, Goethe's typus, *the idea* of the organic whole. It corresponds in the organic realm to the physical natural law in the inorganic realm: the law that governs the mechanical phenomenon which we have to grasp through thinking in the form of an idea in order to understand the phenomenon which is accessible to our observation.

However, because the sub-processes do not simply arise from their interactions but *are also the* result of the governing unity which connects all these sub-processes, this unity itself has *causative* significance. This is the crucial difference between the mechanical natural law and the organic typus. The

---

85 The system of forces which brings about the organic design is a constantly active causa formalis and finalis which influences the (inter)action (causa efficiens) of the matter (causa materialis) and assigns it its direction in the whole. Compare what was said in Section 4.4 on the relationship of the organism and mechanism: the organism possesses *two* active systems in a hierarchical relationship to each other.

idea of the organism is manifested in the phenomena *itself* as the actual form, organisation and structure which appear to the senses. The mechanical law simply specifies the ideal structure according to which the elements which appear to the senses are causatively connected but it is not the causative agent of this connection. Steiner expresses this as follows (ibid., 59):

> This relationship [i.e. the understanding of the conceptual connection, the inorganic natural law], which brings the manifoldness into a unified whole, is founded *within the individual parts* of the given, but as a *whole* (as a unity) it does not come to real, concrete manifestation. Only the *parts* of this relationship come to outer existence in the object. The unity, the concept, first comes to manifestation *as such* within our intellect. The intellect has the task of drawing together the manifoldness of the phenomenon; it relates itself to the manifoldness as its *sum*. We have to do here with a duality: with the manifold thing that we *observe*, and with the unity that we *think*. In organic nature the parts of the manifoldness of an entity do not stand in such an external relationship to each other. The unity comes into reality in the observed entity simultaneously with the manifoldness, as something identical with the manifoldness.

This is so because this unity, the law of the whole, is present *in* the organism as a real, i.e. active factor (cf. Section 4.4). Steiner therefore continues (ibid., 59–60):

> The relationship of the individual parts of a phenomenal whole (an organism) has become a real one. It no longer comes to concrete manifestation merely within our intellect, but rather within the object itself, and in the object it brings forth the manifoldness out of itself. The concept does not have the role merely of summation, of being a combiner that has its object *outside* itself; the concept has become completely *one* with the object. What we observe is no longer different from that by which we think the observed; we are observing the concept as the idea itself. Therefore, Goethe calls the ability by which we comprehend organic nature the *power to judge in beholding (Anschauende Urteilskraft)*. What explains (the formal element of knowledge, the concept) and what is explained (the material, the beheld) are identical[86]. The idea by which we grasp the organic is therefore essentially different from the concept by which we explain the inorganic; the idea does not merely draw together – like a sum – a given manifoldness, but rather sets forth its own content out of itself. The idea is the *result* of the given (of experience), is concrete manifestation.

---

86 Steiner does not mean "identical" here as the formal logical identity (because the explaining and the explained are *distinguished* simultaneously), but in terms of having become "one", as is talked about in the same quotation. This is why it also says: "The unity comes into reality in the observed entity *simultaneously* with the manifoldness, as something identical with the manifoldness". The expression "identical" is justified because the unity mentioned, the law of the whole, is itself the determining *essential element* of the manifoldness produced.

In other words: in *inorganic cognition*[87] the power to judge remains purely "intellectual". Our critical thinking *relates* the general law as a concept to the sense perceptible elements and explains their connection by this law. In *organic cognition* the judgement is also "beholding" in that the *general* law (the concept) is not simply related by critical thinking to the *actual sense perceptible* on the observational side of the cognitive judgement, but with what is in this *actually manifested* organic *idea* observed by our senses in its physical manifestation, i.e. actual form, arrangement, structure, relationship, system or *organisation*. In other words, with that which in fact makes an organism an *organism* and which is emergent compared to everything inorganic. On the conceptual side of the cognitive judgement – to the extent that this can be found – is the explanatory element, the *general* law, what is typical, and on the observational side the thing *to* be explained, in other words the *actual individualised form* of this typical element. Both of these are "identical", inasmuch as the latter is the essence of the former, the appearance. In the organic world it is therefore impossible for the whole, the form etc. to be derived from the parts (as their *sum*, as it were, in the sense of the above quotation from Steiner), but it is always necessary to *wait* and *observe* what nature – or the biological experiment – produces as a *result*, i.e. as the actual physical form created by an organic order.

The science of the organic in terms of Goethe's morphology is always based on what is specifically *organic* in the organism, i.e. on its *organisation* or *development*, whether in a spatial or temporal sense. This does not mean that, from this point of view, we are not justified in making use of the methods of physics and chemistry to investigate what is not organic in the organism and what can only be discovered when you destroy or disregard its form (Heusser, 2000c). In his notes on comparative anatomy Goethe wrote (Goethe, 1950l, 415):

> We observe the organic body to the extent that its parts have form, show a certain defined characteristic and demonstrate a relationship to other parts. We do not make use of anything which destroys the form of the part, separates the muscle into muscle fibres, dissolves the bones into jelly. It is not that we have no wish to know about or do not value such further dissection, but because, by pursuing our expressed goal, we see before us a vast and endless day's work.

---

[87] Note that here the principle of cognition developed in Chapter 2 with its two elements experience (perception, empiricism) and concept (idea, law, theory) apply equally to inorganic and organic cognition. The difference is that on *the side of perception* in *inorganic* cognition, actual sensory elements are present whose relationship is explained by the general law, whereas with *organic* cognition, *an actual formation* of the sense perceptible occurs which proves to be a manifested form of the general law.

Indeed, it is very interesting that, in his preliminary studies for a physiology of plants, Goethe mentioned several auxiliary sciences whose synthesis first determined the *morphology in a comprehensive sense*, i.e. the science of the organic, "a remarkable attempt at a transdisciplinary scientific approach", as Johannes Wirz justifiably remarked with reference to modern biology (Wirz, 2000, 314). These include *morphology in the narrower sense*, in other words "the theory of form, development and transformation of the organic body" (Goethe, 1950q), external comparative morphology ("natural history"), and also anatomy, physiology, physics and chemistry. It could even be said that, without these sciences, morphology in the narrower sense, the study of organismic organisation, could never perform its task well, because the typus requires the relevant materials in order to manifest and this is why the study of the *formative law* is associated so closely with that of *what is formed*. In this respect it is not abiological but rather one-sided if the science developed *after* Goethe has tended to concentrate with a reductionist and mechanistic attitude more on the latter and has neglected the former as an entity in its own right. Because via this route findings about the *material conditions* of the life processes have been brought to light which would otherwise not have been possible, and these have been instrumental in enabling what Goethe called morphology in the general sense. An example of this is the modern findings on morphogenesis discussed above.

Another factor is that the modern methods of analysis which were produced during the remainder of the 19th and especially in the 20th centuries were actually not available to Goethe and his era. Nevertheless, even at that time he pointed out the importance which physiological chemistry would have in future for biology (Goethe, 1950q, 116):

> In this discipline there is also a great debt to the chemist who sets aside form and structure and pays attention purely to the property of the matter and the relationships of its mixtures, and an even greater debt will be owed to him because the latest discoveries enable the most detailed separations and combinations and we can therefore hope to be able to approach the endlessly delicate workings of a living body more closely. Just as we have now obtained an anatomical physiology through exact observation of the structure, so with time we can promise ourselves a physical and chemical one and we can hope that both sciences will continue to progress as though each wished to complete the business alone.

At the end of his life in 1832, ten years before the statement by Du Bois-Reymond cited above (cf. Section 2.6), Goethe wrote in connection with the isolation of caffeine and carotin by Runge and Wackenroder which was partly prompted by Goethe himself (Goethe, 1988, 467–468):

> I am very interested to what degree it might be possible to get to grips with the organic and chemical operation of life through which the metamorphosis of plants is brought about in the most diverse ways, according to one and the same law.

This goal has been achieved in a certain sense with the molecular biological methods of the 20th and start of the 21st centuries and Goethe's statements clearly show that he would have been delighted with the above-mentioned molecular biological decoding of the morphogenesis of Drosophila. But Goethe would never have explained these research results *mechanically*: he was "convinced of the eminence of life which works above and often contrary to mechanical laws" (Goethe, 1950l, 417). He therefore clearly opposed a mechanical interpretation of his *Metamorphosis of Plants* (Goethe, 1975b, 206), and the use of atomistic concepts "in organic cases" he attributed to the fact that "the atomistic concept is easy and convenient for us" and we tend to push the problem of the dynamic element to the side: "Because when you put problems which can only be explained dynamically to one side, then mechanical kinds of explanations once again appear as the order of the day" (Goethe, 1975a, 413). Steiner's comment on these points (ibid., 413, Note by R. Steiner) was:

> *Atomistic* is an explanation of nature which starts from the final conceivable details of a thing and explains the phenomena through an interaction of these details. *Dynamic*, on the other hand, is an explanation which tries to make a concept of a *unity* and lets the details arise from this unity. Goethe professed to the latter form of explanation. [...] Goethe is an opponent of the *exclusive* use *of one* way of viewing all natural processes. He wishes to adapt the method of explanation to the objects. In the physics lectures which he held in 1805 he noted [...]: *Dynamic* type of view: becoming, initiating, acting, causing. *Atomistic* type of view: completed, enduring, excitable, static, created.

So here again it is a case of the perspective of investigation (cf. Sections 2.4, 2.5 and 3.1). The science of the *parts* which make up the organism is atomistic by nature. The science of the spatial and temporal *organisation* of these parts, in other words what is specifically organic, which is characterised by its unceasing organised development from a single unity – even after completion of its growth, in regeneration, reproduction and also simply in the preservation of its structures by dynamic equilibrium – can only be a dynamic one in this sense, exactly as was first developed by Goethe in his morphological writings. The *principle* of Goethe's methods remains valid even if the facts have progressed far beyond that of his time. This is why even nowadays it is possible to follow on from Goethe in this respect (Heusser, 2000b). Steiner therefore attached particular importance to Goethe's morphological work in the history of scientific methodology[88] and gave a comprehensive epistemological foundation of this method in his own writings.

Goethe's organic scientific method in fact provides a solution to a difficulty which biology has faced repeatedly throughout the history of science, that of either falling into the *Scylla of an unscientific spiritualism* or the *Charybdis of a mechanical materialism*.

Goethe's method can avoid mechanistic materialism because, while recognising the action of material substances and forces on the one hand, it also acknowledges the lawfulness inherent in these as a real spiritual entity on the other, and because on empirical grounds it not only admits the mechanical but also higher forms of organised effects and can thus do without reductionism. This type of cognition can avoid unscientific spiritualism because it seeks the spiritual not in a speculative way but an empirical one, initially in the form of a law.

The unscientific spiritualism referred to here is that discussed earlier, usually religiously-based external teleology which lives on in the aforementioned theory of intelligent design or which is also held in Europe by theologically motivated philosophers and biologists (Spaemann & Löw, 1991). These forms of teleology are unscientific in that they are founded on hypothetical assumptions about the creator which are not accessible to observation, whereas science can only rest on observation and thinking.

Hypothetical assumptions as such are obviously not a priori unscientific: science very often justifiably makes use of hypotheses. Hypotheses are ideas whose perceptible correlate has not (yet) been found. As soon as the relevant percepts can be brought forward as confirmation, then the hypotheses cease to be hypotheses and acquire the status of cognition. Unscientific hypotheses are those which Steiner calls *unjustified hypotheses* because the perceptible correlate demanded of them by science *basically cannot* be found: "Every kind of existence that is assumed outside the realm of percept and concept must be relegated to the sphere of unjustified hypotheses. To this category belongs the 'thing-in-itself'" (Steiner, 2011) as defined by Kant, but also the ideas of purpose in the mind of god assumed by religious teleology[89].

---

88 "The great significance of Goethe's morphological works is to be sought in the fact that in them the theoretical basis and method for studying organic entities are established, and this is a *scientific deed of the first order*" (Steiner, 1988b, 47).

89 This statement is not the same as a rejection of the divine-spiritual as the basis of creation of the world: it only objects to an anthropomorphic view of this divine which is incompatible with a scientific attitude. It should also be remembered that the lawfulness in the world is itself of a spiritual nature and is therefore justifiably referred to as the "world spirit" by Hegel, or, in religious terms, to the "logos" in the gospel of St. John. The relation of this spirit or logos to the creative "God" expounded by religious denominations is a very pertinent question whose quick

Inasmuch as vitalistic traditions from historical forms of medicine attribute the properties of life and the specific structure and function of organisms to specific "life forces", mostly thought of as immaterial or supramaterial which are *fundamentally* inaccessible to scientific empiricism, they are also unscientific and spiritualistic for the same reason (Lehmann, 1991). It is not the assertion of specific immaterial life forces which is unscientific – because for the reasons presented in this book it is always possible that such forces might exist just as much as radio waves or magnetic fields – but their fundamental empirical inaccessibility. The question of the existence of specific life forces or other immaterial forces must be solved empirically in the same way as the question about physical forces. This is also applicable to the question about the existence of morphogenetic fields which have been used to explain morphogenesis since the time of neovitalism.

4.7 Morphogenetic fields or morphogenetic substances?

The morphogenetic field concepts of neovitalism have also remained hypothetical to a large extent (Lehmann, 1991). However, there is a crucial difference from the older vitalistic traditions. The neovitalists were biologists working in experimental science in the modern sense at the end of the 19th and beginning of the 20th centuries who, for the most part, arrived at the postulate of their force or morphogenetic fields as a result of development and regeneration experiments with embryos of lower animals. In contrast to the old vitalism, it was scientific *empirical* reasons which led the likes of Hans Driesch (1867–1941) to his concept "of relatedness of the whole and causality of wholeness" (Driesch, 1929, 3), and which he called entelechy following Aristotle due to the goal directedness of its effects (entelécheia: Greek for 'that which carries its goal within itself') (Driesch, 1931). Alexander Gurwitsch (1874–1954) viewed Driesch's wholeness factor "as a reality" due to his own especially mathematical and statistical studies on cell and nuclear division patterns (Gurwitsch, 1922). Taking his ideas from electromagnetic fields, Gurwitsch talked about *force fields* or *embryonal fields* i.e. a *vectoral biological field* which, as a supracellular ordering principle determines the fate of the individual cell out of the whole (Beloussov et al., 1997). Paul Weiss (1898–1989) derived his morphogenetic field concept from that of Gurwitsch and his own experiments, and in 1939 summarised what had until then been empirically determined field properties as follows (Beloussov et al., 1997):

---

and easy answer from a creationist viewpoint appears not only scientifically but religiously naive.

1. The field activity is always connected to a material substrate.
2. The field is a unity, not a mosaic.
3. The structure of the field varies in three dimensions and usually has one axis.
4. Like the poles of a magnet, no constituent of the material substrate can be identified with one of the different components of the field.
5. If the dimensions of the field are reduced, the structure of the field is still retained.
6. If the material substrate ("field district") is divided, then each half contains a complete proportioned field which corresponds to the structure of the single original field.
7. The fusion of field districts produces morphological results which depend on the orientation of their axes.

These field properties can readily be verified empirically using the well-known examples of amphibian embryos (Hadorn, 1981) and regeneration in Planaria (Kühn, 1965). In the words of Bernhard Dürken, they demonstrate the "controlling and overriding position of this bearer of wholeness". This "cannot be a separate mechanism, in other words no special system in the gametes" (Dürken, 1936, 116) rather, it is a consequence of the fact "that the *whole* of the embryo is involved as a *causative agent* in development" (ibid., 109). A causative agent which, in addition, is of a *higher order* than the organic matter as is demonstrated by Oberheim and Luther's experiments on the reversal in polarity in the regeneration of salamander extremities (Oberheim & Luther, 1958). The forelimb of a salamander larva was implanted in the dorsal region of another larva. After it had healed in there (which only happens if there is an adequate supply of nerves to the wound area and accompanying vascularisation), the limbs were amputated on both sides. The original animal regenerated the lost limb normally while the second animal regenerated its mirror image. This is enabled by a dedifferentiation[90] of the

---

90 According to recent findings, no complete dedifferentiation takes place: as blastema cells, muscle, cartilage and skin cells retain something of their genetic identity and can only regenerate into other cell types to a limited degree. Cartilage cells even retain their "position information" in the form of a cell regulator protein (MEIS) and specific Hox messenger RNA. This means that cartilage cells which were transplanted into the upper forelimb before amputation, moved distally into the developing hand as blastema cells after amputation, whereas Schwann cells which did not express these substances, were distributed uniformly over the whole length of the extremity (Kragl et al., 2009). This fact does not, however, permit the conclusion that the position of the cells in the whole i.e. the form of the hand or extremity can be attributed *causally* to this or similar substances, however they are the material *condition* for this position or form. First, the expression of these substances requires gene regulation of this position and form and, second, the

previously differentiated cells of the "reversed" implanted limb stump with subsequent redifferentiation of these cells (Hay, 1968), but NB in the sense of a "normalisation" of the limb axis, i.e. by inverting the polarity of the limb "design" by 180°. This in fact indicates *a relative independence of (part of) the design from what was planned*: it can "disconnect" from it and "reconnect" in a new orientation. Paul Weiss therefore distinguished with a certain justification "*differentiation potential* which belongs to each material part, let us say to each cell and includes its entire range of responses *on the one hand*; and *organisation potential* (or *development potential*) which only belongs to a material *whole* as such and which corresponds to the *determining ability* for the lower order material part or that which will be subordinated, on the other. The dynamic bearer of an organisation potential is an organisation field" (Weiss, 1927, 335).

As already mentioned, what is common to the neovitalist force and morphogenetic fields is that they were postulated on the basis of empirical experimental facts (Beloussov et al., 1997), in fact following the same principle according to which science accepts the existence of electromagnetic force fields: conclusions are drawn about the existence of the effector and about its properties on the basis of observable effects, although the effector as such cannot be observed. The postulate of morphogenetic force fields is scientifically justified to this extent.

But what is the effector, the creator of the whole? As long as the lawfulness of this creator of the whole cannot be specified like the lawfulness of electricity, magnetism or gravitation, his postulate remains too indeterminate, no full epistemological value in scientific terms can be attached to it. Concepts such as entelechy, wholeness, force field etc. remain abstract if their content cannot actually be defined: they thus have little explanatory value. By "floating" so to speak above the organic substrate they remain "spiritualistic" hypotheses. For this reason but especially because, in contrast to electromagnetic force fields, *no material field source can be specified* for morphogenetic force fields, the concept of morphogenetic fields has failed to become established. It was finally assigned a *descriptive* function but not an *explanatory* one (Saunders, 1968, 63): it merely offers "a useful circumscription for certain system properties of pattern forming groups of cells", but is otherwise a black box (Sander, 1990, 136).

---

migration of suitably predestined cells and their positioning within the meaningful whole of an extremity is a coordinated event corresponding to the "design", with regard to which the individual cell including all its material is a *receiver* of the command. This is even clearer in the case of the polarisation reversal of this design mentioned in the text.

It is therefore natural that the explanatory pendulum has once again gradually swung back from an unsatisfactory holistic spiritualistic interpretation of morphogenesis to a particularistic mechanistic one, with the starting point for this being the "organiser effect" which was discovered by Hans Spemann (1869–1941) and for which he was awarded the Nobel Prize in 1935. This refers to the *induction* of morphogenetic processes by what are known as "organisers", i.e. particular areas of tissue or *substances* which can be isolated from these areas. According to Spemann's description, an inductor or organiser of this kind is characterised by the fact that it "creates an *organisational field* with a particular orientation and dimensions in the undifferentiated material where it occurs or into which it is artificially transplanted" (Spemann, 1921, 568). A well-known example of this is the induction of a secondary embryonic anlage by transplantation of presumptive chordal material from the lip of the blastopore of one amphibian gastrula to another[91]. Spemann believed in the existence of the organisation field and was of the opinion that "regulation [...] is determined by the system of the whole" (Spemann, 1968, 273).

After the Second World War the morphogenetic field concept was largely abandoned. Better biochemical and later genetic methods of analysis led to the identification of a series of form-inducing substances and to a purely molecular biological and mechanistic interpretation of the creation of form. And

---

[91] An example of the organiser effect of chordal material discovered by Hans Spemann: If tissue containing presumptive chordal material is excised from the dorsal lip of the blastopore of a young amphibian gastrula and transplanted into the abdominal wall of another amphibian gastrula, this induces a secondary gastrulation there, the implant invaginates, expands and duly divides into secondary chorda and somites. This not only uses cells from the implant, but – like a chimeric creation – also mesenchyme cells from the host which have ended up in the area of the implant as a result of the gastrulation. "The process is very strange: the tissue in the implant is not first differentiated according to its origin and then completed by laying down material at the body's 'break points' [...]; but the implant incorporates material which is not yet finally determined into its *field area* within a specific radius before differentiation and then the *field organisation* takes place across the whole area regardless of the material boundaries of the different sources" (Kühn, 1965, 268). A neural plate with bulges at the edge is then induced above this secondary chordal and somite anlage which can also be chimerical and which closes to form the neural tube. Towards the ventral surface the secondary axis anlage has an inductive effect on the endoderm of the host so that a secondary gut lumen is produced and thus gradually a second embryonic anlage which is joined to the primary one abdomen-to-abdomen. "The *secondary embryonal anlage* can thus continue to develop and attract so much nuclear material that it actually forms a duplication" (Kühn, 1965, 269). Spemann therefore referred to the corresponding material from the lip of the blastopore as a primary "organiser".

the fact that morphogenetic fields could not actually be reduced to material factors due to the properties assigned to them was regarded as evidence for the unscientific nature of the field concept (Beloussov et al., 1997).

Nowadays the textbooks attribute this regeneration and likewise morphogenesis to mechanistically interpreted genetic activation and gradient formation of morphogens (cf. Section 4.5). The distribution of the transcription factor β-Catenin in the blastula stage, for example, correlates with the localisation of the Spemann organiser of the dorsal lip of the blastopore in the gastrula stage, induces gastrulation and in experiments proves to be an initiator of the organiser activity, although in interplay with other elements of a complex organised signal process (Purves et al., 2006, 501–502). Orientation of the head-tail polarity in the regeneration and homeostasis of Planaria is also controlled by β-Catenin (Gurley et al., 2008); and RNA interference experiments with Planaria have shown clearly that the formation of a posterior oriented blastema on any cut surface is determined by activation of the *Smedβ-catenin-1* gene (Petersen & Reddien, 2008).

The molecular biological basis for Spemann's famous "abdominal segment" experiment[92] has also been largely explained in recent years and can be summarised in a very simplified way as follows. In the ventral pole of the embryo, gradients of morphogenetic proteins called *bone morphogenetic proteins (BMP 2, 4 and 7)* are expressed which induce a ventralisation by blocking (at the transcription level) dorsalisation morphogens localised in the dorsal pole (e.g. *Chordin, Noggin,* etc.) which, in turn, antagonise the effect of BMP in the upper pole (at receptor level). The typical dorso-ventral blueprint is brought about in the dynamic equilibrium between these and other antagonists (De Robertis & Kuroda, 2004). After dividing the nucleus into dorsal and ventral halves, the dorsal half is able to make a compensatory upwards adjustment of a previously expressed protein called *anti-dorsalising protein (ADMP),* which is related to the ventral MBPs and is also a *Chordin* antagonist. The loss of BMP due to the cut is therefore dorsally compensated and – from half of the original material – a complete if smaller embryo is regenerated. In contrast, in the ventral half there are apparently no or not

---

92 Spemann halved amphibian embryos at the two-cell stage using a loop of hair. When the ligature was placed between the two cells, both halves (cells) produced a whole animal with a fully expressed body plan under appropriate conditions. The same happened when the division into a left and right half of the embryo was carried out a little later, in the blastula stage. However, if the ligature was placed at right angles to the blastula so that the embryo was divided into a dorsal and ventral half, then a whole animal again developed from the dorsal part, but only an incomplete "abdominal segment" from the ventral part (Spemann, 1968).

enough dorsalising morphogens present after the division and a ventral torso remains (Kimelman & Pyati, 2005; Reversade & De Robertis, 2005).

At the beginning of the 21st century the hope has been rekindled of providing a *complete chemical and physical explanation* of morphogenesis on the basis of a genetic programme: transcription of the genetic sequence into proteins, self-aggregation and self-organisation of the proteins into higher structures and systems and selection of these through environmental conditions. According to this view, the induction of developmental and organisation processes described by Spemann does not come about through field-based instruction of previously naive cells, but through chemical, i.e. morphogen-based selection of alternative signal and expression paths. These are already *prefigured* as competences i.e. as specific possibilities in the cells and cell systems and come about by self-organisation, after the morphogen – partly also due to mechanical or chemical triggers from outside – has provided the material stimulus (but not the specific instruction). The driving force for self-organisation is considered to be the achievement of thermodynamic energy minima; and the stability of the structural and functional equilibria and the robustness of the resulting order, i.e. its ability to recover after perturbation is also attributed to this. This ability was previously accorded to vitalistic principles, which is why, in their millennium article in *Cell*, Kirschner et al. wrote: "It is this robustness that suggested 'vital forces', and it is this robustness that we wish ultimately to understand in terms of chemistry. We will have such an opportunity in this new century" (Kirschner et al., 2000, 87).

But – as described above (cf. Sections 3.3 and 4.1) – even in chemistry and the inorganic formation of crystals, self-organisation is not a process to be understood as happening merely "from below" in which the parts combine into a whole simply through the properties of these parts, but an organisation of these parts "from above" through the law of the whole which makes its appearance in them. This appearance is emergent, it is no longer the appearance of the parts – added together – and no longer follows their laws, it is the manifestation of the *whole* and follows *its* law.

This determination of the parts by the whole is even clearer in the organic realm. Because, in contrast to the self-organisation processes in chemistry and in crystals, those in the organism are *flexible*: its order is not fixed but *metamorphoses*, the spatial structure of the organism comes about gradually and flexibly *over time* (growth, growth changes, cell migration, differentiation). Even after development and differentiation are fully complete, the structural development or morphogenesis remains active in that, although the form persists, the material of which this form is composed is constantly changing (dynamic equilibrium). For this reason it is only in the case of a chemical compound or a crystal that homeo*stasis* is really a true "stasis", i.e. a standstill[93] of the self-organisation which has been produced. In an organism the incessant self-organisation, i.e. the process of forcing the parts to become a whole, only stops with its death. Before that it is a constant flux of substances in a dynamic equilibrium, even in seemingly stable structures.

The phenomena of spatial and temporal *coordination and synchronisation*, essential characteristics of organismic self-organisation, are further evidence that the parts of the organism are always actively subject to the whole or subsystems of the whole and to this extent are determined by it by systemic or formative causation (cf. Sections 4.1–4.3). For example, the thousands of cells of the future mesoderm in the zebra fish embryo internalise themselves "in a single synchronized internalization wave around the entire circumference" (Keller et al., 2008, 1086), to mention only one of many examples. Synchronisation and coordination always indicate the subjection of the parts in question to a common order, i.e. the *"syn"* and *"co"*. Riedl talks about *synorganisation* (Riedl, 2005, 124). However, who or what *brings about* this order? Vitalism laid the responsibility on its hypothetical principles of undefined content. Modern molecular biology places responsibility with actual chemically definable substances which have been discovered empirically through experiments[94]. But, as repeatedly established above, these substances

---

93 This statement is qualified by the well-known chemical Belousov-Zhabotinsky reaction in that, in the sense of a chemical oscillator, it does not stand still but always starts from the beginning when the end product is reached, this again providing the starting material. It involves a complex oscillating reaction between repeated reduction and oxidation which appears as an oscillating colour change in time or spatially as an alternately coloured expanding wave of the corresponding substances (self-organisation) (Belousov, 1981). But in terms of inorganic chemistry this self-organisation also oscillates towards its final state (depletion of the reduction-oxidation system) i.e. it does not achieve homeostasis or dynamic equilibrium in the *organic* sense in which this depletion could be constantly compensated by the relevant system organisation. In addition, it lacks the overall organisation which is decisive for the organic realm i.e. the active coordination of its sub-processes as a system.

94 So, for example, the coordination of the migration of the endoderm and mesoderm cells during gastrulation in the zebra fish is induced by specific chemokines and their receptor, resulting in the modulation of extracellular matrix proteins which are necessary for gastrulation (Nair & Schilling, 2008). In gastrulation in Drosophila, fibroblast growth factor (FGF) has been identified as one of the factors which is responsible for the migration of the mesoderm cells towards the ectoderm. The corresponding groups of cells show "a high level of organization" (McMahon et al., 2008, 1550). In the embryonal development of the amphibian limb "a concerted colinear regulation" of Hox genes is responsible for the development of the typical structure (Deschamps, 2004). In migrating future heart cells in ascarid embryos, responsibility for the concerted activation of a network

are not the cause but the *initiator or mediator* of the coordination and already presuppose this. Because these substances always act *individually* on the actual receptors or genes of *individual* cells, their action must already be coordinated in order to produce coordination between the cells. Identification of the inductors does not therefore solve the coordination problem but merely moves it one stage backwards. In the organic realm substances are not simply subjected causally to other substances as in mechanical systems, but to an *order*. *This* itself is therefore – initially at least according to the law – what creates the whole, the coordinating and synchronising element at a higher level than the material which e.g. Driesch tried to find in his unspecified entelechy. However, as this law is to be *manifested*, there must be a *system of forces* connected to it. For this reason the order of the whole is a decisive causal agency that has to be included in the causal analysis of morphogenesis and organismic structure and function (cf. Section 4.7).

The current molecular biologically-based versions of the regeneration experiments like those of Spemann mentioned earlier also show that organismic coordination is brought about *actively* from a *whole* which is at a *higher level* than the material to be coordinated. The body plan of the whole amphibian embryo comes about in a dynamic equilibrium between ventralising and dorsalising morphogens which are expressed at the ventral (e.g. *BMP*) or dorsal (e.g. *Chondrin*) poles. However, the production of an equilibrium between mutually antagonistic *active* processes requires an *activity*, but one that is at a *higher level* than the antagonists. Kimelman's comments on this finding – "Chordin and ADMP exist in a complex dynamic equilibrium, where each factor regulates the other" (Kimelman & Pyati, 2005, 984) – is therefore only correct under the assumption that the equilibrium is actually a dynamic one, i.e. is actively controlled. Furthermore, the example shows that the activity to create this equilibrium encompasses the whole, in that the regulation processes in the dorsal and ventral poles are coordinated to produce the harmonic overall form. Lastly, the example shows that this wholeness-creating activity is *at a higher level than the organic material as a whole*: after dividing the ventral and dorsal poles, what previously occurred in the *whole* is repeated with the help of the ADMP Chordin antagonism in *half* the material i.e. the actualisation of *one and the same* body plan (Reversade & De Robertis, 2005). *The localised morphogen expression occurs in accordance with* this.

---

of genes whose products determine the migration, adhesion, cell polarity and membrane polarity of these heart cells has been attributed to coordination of the transcription (Christiaen et. al., 2008). This is only to name a few of the countless examples of the more or less *chemically* induced synchronisation and coordination of individual processes in the development of organismic form.

It is also apparent from the regeneration experiments with Planaria mentioned earlier when e.g. a segment excised from almost any place in the Planaria body expresses *Smed-βcatenin-1* at its posterior cut margin and therefore produces *β-Catenin* as an inductor for regeneration of the missing portion of the tail on this segment (Petersen & Reddien, 2008). Who tells the cut edge or the cells there and their genome that they are located at the *posterior* end of the new (partial) whole obtained by excising the segment? If *the same* cells end up at the *anterior* cut margin of the distal portion of the body (tail portion), then they do not express *Smed-βcatenin-1* but create an *anterior* blastema with photoreceptors, head ganglia and head structure. This is only possible if the genome of the local individual cell is informed or instructed *in relation to the spatial whole* or part of the whole in which it ends up (position information). This applies not only during regeneration but throughout life even in uninjured animals, when e.g. young cells need to replace old differentiated cells for maintaining cell homeostasis. *Smed-βcatenin-1* prevents cells in the relevant localisation from being able to form a head and enables them to become trunk or tail cells. *Smed-βcatenin-1* itself turns out to be determined by what are known as *Wnt* signals[95] (Petersen & Reddien, 2008), i.e. by a conservative genetic programme which also acts through antagonism and whose proteins are responsible for the regulation of numerous pattern formations in organisms from all the main branches of animal evolution up to human beings. The regulation of the *wnt* gene is thus also dependent on position, i.e. it results from the whole which is at a higher level than all the parts.

An informative example of the position-dependent instruction of cells is the Spemann organiser on the dorsal lip of the blastopore, from where the organisation of the future gastrula is induced. After excising this group of cells at the beginning of gastrulation, other cells from the vicinity become organiser cells, and basically every cell of the animal blastula pole can become an organiser cell if it receives the correct combination of signals from its environment, e.g. by being transplanted to the right place. But as soon as

---

[95] *"Wnt"* (pronounced "wint") is a combination of the abbreviation for the segment polarity gene *wingless (wg)* in Drosophila and the homologous gene in the mouse *int*. *Wnt* proteins are found throughout the animal kingdom and also in human beings and play a decisive role in pattern formation and also in some cases in cell migration and gene mutations during the development of cancer. *Wnt* proteins are secreted outside the cell and are gradient-forming signal molecules which bind to the receptors on the cell surface and upregulate *β-Catenin* via a signal cascade in the cytoplasm. This β-Catenin then acts intracellularly as a transcription factor in gene regulation.

the first organiser cells have been produced, the formation of other centres is blocked (Spratt & Hass, 1960).

The fact that the formation of the organiser is dependent on these chemical signals appears to contradict the theory of position information or the thesis that gene regulation takes place from a higher i.e. *immaterial* level than the individual cells and their genome. However, the production of these signals and their gene expression in the relevant surrounding cells is also position-dependent: it takes place in a spatial arrangement which corresponds to the targeted *body plan*. The same applies to the above-mentioned regulation of *Smed-βcatenin-1* by *Wnt* signals. If the *Wnt* signals are examined further, it transpires that the *Wnt* genes are also regulated by actual transcription factors and a series of enzymes which are all dependent on further enzyme-regulated genes, etc.: The chemical and molecular biological factors are always dependent on more of the same.

Is this a "chemical explanation of life" after all? Because there is no sign of the action of an "idea" *in place* of chemical factors. There is therefore no basis whatsoever for the intervention of special immaterial forces as postulated by vitalism. According to molecular biological findings, organic self-organisation always takes place through sequential, modular and hierarchically organised cascades of molecular interactions (Riedl, 2005).

But this is not just a mechanical process and it does not happen *without an "idea" or "spirit"*. Because not only is every substance, however subordinate, spirit in accordance with its lawful nature (cf. Sections 3.1–3.2) but the same applies to all higher-level emergent substances, structures, processes and arrangements which arise from these subordinate substances through their interactions (cf. Section 3.3). New "spirit" is manifested as it were "from above" if the material conditions for it are present "from below". The *lawfulness* of the sequential, modular and hierarchically organised cascades of the molecular interactions, the spatial and temporal order which manifests as organic self-organisation is in this sense "spirit", spirit which manifests in the material occurrence and which the latter follows. There is nothing mystical, irrational or speculative about this spirit. Its discovery is the result of scientific empiricism, it is identical with the empirically established laws of organisation; and its recognition as spirit is the result of epistemological empiricism (cf. Chapter 2 and Section 3.1).

This applies particularly to the field of knowledge applicable to the development of organic form. Both the spatial arrangement and the temporal sequence of the relevant processes are highly coordinated, synchronised or orchestrated. They are therefore subject to a spatial and temporal organisation which does not come from them but is either *imposed* on them (e.g. the *Wnt*-dependent expression of *Smed-βcatenin-1* on the cut edges of Planaria positioned at the posterior end) or which are *induced* by them (e.g. the *Smed-βcatenin-1* dependent production of missing parts of the tail distally from the posterior positioned cut edges). The induced process is itself highly coordinated and thus subject to an organising function at a higher level.

It is often believed that this "higher level" is simply due to a complex hierarchical organisation of gene expression cascades where activated "master genes" trigger further genes in downstream regulatory networks until the level of specific genes that constitute the phenotype of individual tissues is reached. Indeed, quite a number of such gene regulatory networks have been identified. For example, in ocular lens morphogenesis Pax6, a DNA-binding transcription factor has been identified as a key inducer for lens placode formation and subsequent stages of lens morphogenesis. Pax6 induces the expression of c-Maf which then induces terminal differentiation of lens fibres, including regulation of crystallins, key structural lens proteins required for its transparency and refraction (Xie & Cvekl, 2011). On the basis of mechanistic conceptions such sequential processes in gene expression, morphogenesis and organismic self-organisation are usually envisaged as complex mechanistic processes that are *caused* by the master gene in the way that the first domino is the cause of the row of other dominoes behind it falling over, each causing the fall of the next. In this mechanistic explanation, as in the functioning of a machine, there are no other causal factors than the parts and their interactions. "Self-organisation" is then the automatic downplay of a complex organic "domino" or biological machine. However, as in all such mechanistic explanations, the causal origin of the sequential *order* of the biological "dominoes" is forgotten, i.e. the prerequisite for the correct downplay. In a machine or in the domino game this kind of order can only be caused by an *external* agency. But there are no machines that arrange their functional parts themselves, from *within* (cf. Section 4.4).

In contrast, the organism is autopoietic, it arranges its "dominoes" by itself: the cause of a process lies *within* the process (cf. Section 4.5). Regulatory genes and molecules do not mechanically *cause* but organically *induce* the respective downstream processes which are all autopoietic in nature. Even individual elements of these cascades such as proteins are self-organisers. As we have seen, the higher level structure which is of relevance for their function is *emergent* (cf. Sections 4.1–4.3), i.e. *it follows its own law* and cannot be deduced from the elements which have induced it. It *actualises itself*, but *always under the pre-condition of* the previously actualised elements, processes and structures which precede it. Once actualised, e.g. as a functional enzyme, it induces other self-organising processes, e.g. biochemical reactions, biological factors or even complex biological functions. For this reason morphogenetic substances can only be ascribed the function of *inducing* morphogenetic self-organisation, but not of *causing* it.

The more we know about the molecular biological processes of morphogenesis, the more it becomes clear that it is due to a sophisticated orchestration of complex self-organising subsystems in space and time (Lecuit & Le Goff, 2007): genetic programmes expressing signalling molecules and growth factors, processes of cell division and rearrangement, apoptosis, etc., and in all this the creation and maintenance of regular tissue size and shape and a unitary organisation in which a myriad of such singular processes are part of a lawful but flexible living whole.

In the current literature we still find the attempt to "understand within a single mechanistic framework" how these complex morphogenetic processes can come about (Lecuit & Le Goff, 2007). But gradually even the advocates of a mechanistic and chemical understanding of life start to consider that mechanical explanations are not sufficient to explain organisms. Even Kirschner et al. who seek for a chemical explanation of life, admit that mechanical causality cannot explain organic self-organisation and refer in particular to:

> [...] two of the most archetypal and unusual biological properties: (1) the capacity for unitary organization, also called polarization; and (2) the capacity to generate nearly regular biological structure when size and composition of components are altered, also called regulation. These properties are not what we would expect from mechanical processes, and no machines of human design evince such properties (Kirschner et al., 2000, 80).

One reason why it seems so difficult to overcome reductionism in science may be because mechanical causality and chemical processes always rely on *local interactions of parts and their forces* and that thinking is accustomed so much to mechanics that it can see causality only in such local interactions. However, as we have seen (cf. Section 3.2), even in physics causal processes are known that do not come about by local interactions of parts but by their *nonlocal correlation*. But in that correlation they can no longer be regarded as separated interacting parts but as a *unity*, the causality of which is *holistic* in nature. Hans Primas comments on experiments with photons by Aspect et al. (1982) that convincingly demonstrate this as follows (Primas, 1992, 10):

> You can look at the matter in any way you like, but as soon the discussion of this experiment speaks of two particles you become entangled in logical contradictions. The system, which appears to consist of *two* particles from a traditional viewpoint, is in truth *one* undivided whole.

Nonlocal actions are known not only in physics but *also in biology*. The reason why the organism's "capacity for unitary organization" and its ability "to generate nearly regular biological structure when size and composition of components are altered [...] are not what we would expect from mechanical processes" (Kirschner et al., 2000, 80) lies in the fact that such processes cannot be brought about causally by local interactions alone, but by the coercion of local processes in *nonlocal, i.e. holistic organisations*. The experimentally-based insight that morphogenesis is founded on such nonlocal processes has recently led to a reappraisal of Paul Weiss' *morphogenetic field* concept by Michael Levin[96] in an article with the indicative title: "Morphogenetic fields in embryogenesis, regeneration, and cancer: Non-local control of complex patterning" where he emphasises (Levin, 2012):

> The quintessential property of a field is nonlocality – the idea that the influences coming to bear on any point in the system are not localized to that point and that an understanding of those forces must include information existing at other, distant regions in the system. [...] In this review, I focus on the spatially distributed nature of instructive patterning signals, discussing the evidence from developmental, regenerative, and cancer biology for *non-local control of pattern formation*. Specifically, these data suggest the hypothesis that many diverse examples of pattern formation are best understood not as cell-level behaviors around any one locale but rather at higher levels of organization.

According to the interpretation set forth in this treatise, the agency causally responsible for this nonlocal control of pattern formation is the patterning principle itself, Goethe's typus (or archetype), the acting law of the organisation as a whole. This is the actual and logically intelligible content of what has been repeatedly called the "morphogenetic field" for over a hundred years. The *cause* of morphogenesis is this archetype or "morphogenetic field" and *not* the morphogenetic substances. These only serve to induce the self-organisation of the morphic pattern principles.

4.8 Causality and systems biology

The previous sections have shown that neither interacting substances per se nor purely hypothetical morphogenetic fields can be accepted as causes for the development of form, but the "nonlocal" action or "organisation" that structures the parts into a coherent "whole", the "typus" as defined by Goethe, the "form" or organisational law itself, an acting "idea". Paradoxical as it may seem, from the perspective of epistemologically justified ontological idealism developed in this treatise, the "idea" or "plan" can be recognised as the actual cause of the self-organisation of the developing form. It is of course not just any idea, but the idea or *system* of lawful principles that constitutes the organic organisation as such, with its highly complex functions and structures, in a meaningful temporal and spatial order, sequentially and

---

96 Working in a similar direction and based on a large body of evidence, Ana Soto and Carlos Sonnenschein have recently proposed the "tissue field theory" as an alternative to the somatic mutation theory to explain carcinogenesis (Soto & Sonnenschein, 2011).

in hierarchically organised modules from within itself and maintains them until the organism dies. However, due to reductionist thinking habits, it is not a trivial matter to maintain this view consistently. For example, despite the fact that Kirschner et al. acknowledge the non-mechanical nature of organic self-organisation, they still view it as "a manifestation of complex yet robust chemical processes" (Kirschner et al., 2000, 80), as this self-organisation only comes about – according to everything that can be observed empirically – by chemical and molecular biological interactions and biochemical processes. But what comes about through these molecular interactions and biochemical processes is more than these alone, it is their *consistent systematic organisation*. As Steiner pointed out in his commentary on Goethe's fundamental biological concepts, the parts of the organism *"do exist for one another, but not as a result of one another. They do not mutually determine one another, but rather are all determined by something else"* (Steiner, 1988b, 49, ePH), this being the systematic organisation, the "idea" of the organism (Goethe, 1950j). The fact that all the interactive parts are organised or determined by the systemic whole is also the reason for the point mentioned by Steiner that everything in the organism is functionally related to everything else, as also noted by e.g. Riedl: "Almost all features of an organism – that is, all that are functionally adjoining – are interdependent" (Riedl, 2005, 125).

In his essay *"Biologisches Denken" (biological thinking)*, Wolfgang Schad rightly points out that linear causality is a feature of the *inorganic* realm and is causally related to the past: the *earlier* cause determines the *later* effect. In contrast, teleology is a feature of the thinking human mind and is causally related to the future: the idea of a *later thing* is the reason for carrying out a *prior* action aimed at the imagined goal. In contrast, the organic realm is mainly connected to the present in that the parts of an organism are always related to each other *simultaneously*: "It is only through this *simultaneous connection* that every organism always appears to us as a whole. […] The central method of biology is therefore to find correlative connections" (Schad, 1982a, 19). The causative agent which *creates* the connection is actually the self-realising connection itself, the typus in the Goethean sense, because this "is in fact actually the connection of the simultaneously interdependent parts" (ibid., 20).

Inasmuch as this correlative organisation takes place teleonomically, Goethe was justified in calling the typus "entelechy" (Goethe, 1950l, 435), i.e. that which carries its aim (Greek: télos) within itself. The difference from Driesch's "entelechy" is that, although Driesch inferred the necessity of the existence of this type of self-actualising wholeness-creating entity from empiricism, he did not define the content of this and thus exposed it to the accusation of vitalism (Küppers & Paslack, 1996). Goethe, on the other hand, defines the content of this entelechy, that is to say the "typus", i.e.

the general lawful "organisation" of the organism which has to be *defined purely empirically* (Goethe, 1950l, 421). It is not necessary to have recourse either to supernatural principles which remain hypothetical nor to merely mechanically-interpreted molecular biological interactions in order to *grasp* the organic process of self-organisation, but only to study the real natural[97] *organisation* which manifests through these interactions. This manifestation takes place successively and, in the spatial and temporal coordination of molecular biological processes, brings about the complex, hierarchically arranged *system* which is always organised as a whole, a system which became known in increasingly minute detail during the 19th and 20th centuries. The lawfulness of this system can be found in every part and at every level of organic development – from the biochemical and molecular microbiological level via organelles, cells, organs and organ systems up to the complete form of the organism – because it is *this* lawfulness which is the crucial basis of self-organisation. *Self-organisation means: the lawfulness of the system brings itself into being* (cf. Sections 4.3–4.4). And the information about the parts of this system which is obtained from the relevant scientific sub-disciplines then needs to be combined in a coherent synthesis, exactly as intended by Goethe's concept of a comprehensive morphology with its auxiliary sciences.

This is in fact the approach taken by modern *systems biology*, which does not only study single components but tries to understand cells, organs, etc. as systems[98]. It examines both the whole and the parts, as far as possible under

---

[97] "Natural" is meant here in contrast to the "supernatural" principles of vitalism and not in the sense of "normal" or "healthy" or similar. It is often in fact the abnormalities or other pathologies which provide very good opportunities to get behind the principles of the normal organisation, something demonstrated very instructively by Goethe himself in *"The Metamorphosis of Plants"* (Goethe, 2009) and is particularly obvious in modern times in what are known as "knock-out" experiments in genetics.

[98] Description of modern *systems biology* (official publication of the German Federal Ministry of Education and Research, 25 June 2007, www.bmbf.de/): "A new research approach has been established in recent years, based on the findings of research in molecular biology. This research approach – systems biology – aims at a thorough quantitative understanding of the dynamic interactions between the elements and components of a biological system in order to understand the behaviour of this system as a whole and to enable predictions to be made. Mathematical concepts are applied to biological systems in order to achieve this goal. A key element of this is an iterative process between laboratory experiments and computer modelling. The biological systems research approach is based on the interdisciplinary cooperation of biologists and doctors in their own fields with mathematicians, information technologists, physicists, chemists and engineers". Modern systems biology mainly pursues practically-oriented industrial objectives with applications in medicine, the chemicals industry and environmental sciences.

natural conditions, but always with the aim of explaining the whole from the interaction of its parts in a mechanistic way using mathematical models and computer simulation[99]. However, this leaves out the system law according to which the parts are kept in their places in a meaningful whole and therefore the causality of the system. This is why it was stated above that the causal explanation of the organic realm must include the connecting law as well as the parts (cf. Section 4.6).

The task of systems biology in *this* sense is not to derive the system from its parts and their interactions but to discover the *system law itself* and its connection to the parts at each level of the system. Every level of organisation of the system is equally valid epistemologically and ontologically: the whole and the parts obtained by analysis should be studied equally by observation and thinking. There is no epistemological reason to hold the manifestation and law of the whole or system as less objective or real than the manifestation and law of the parts. In his time Goethe devoted himself primarily to studying the higher level whole and its law, the typus. After him came the ever more detailed analysis of the parts which, in line with Du Bois-Reymond, were the only things which were accorded the status of reality. However, what Goethe wrote in his essay *"Analyse und Synthese"* (analysis and synthesis) applies to this day (Goethe, 1950b, 889–890): "The main thing which would appear not to be thought about when using only analysis is that every analysis presupposes a synthesis. [...] Modern chemistry rests mainly on dividing that which nature has combined: we cancel what nature has united in order to get to know it in separate elements." But the analyst "actually carries out his work in order to reach a synthesis again in the end". And: "What is a higher synthesis than a living being; [...] a complex [...] which constantly establishes itself, irrespective of how many parts we have torn it into".

So, in order to understand the *system* in the organic realm, it is necessary to remember the "spiritual link" which has been lost in the one-sided analysis of the parts, as *Goethe* in his *"Faust"* (Goethe, 2005, lines 1936–1939) has Mephistopheles in the role of the professor say to his student:

He who would study organic existence,
First drives out the soul with rigid persistence;
Then the parts in his hand he may hold and class,
But the spiritual link is lost, alas!

---

99 "To explain behaviour, the parts have to be considered in their natural context, i.e. as components in the functioning system, and at the same time, the system has to be studied as a whole. [...] In the end, a mechanistic explanation of a cellular behaviour amounts to a quantitative, mechanistic model that captures those molecular phenomena in the cell responsible for the cellular behaviour that is to be explained" (Boogerd et al., 2007, 14).

This synthetical link is the idea, the organising law which brings all the parts of the organism into the system of the whole: therefore in fact *spirit*[100], but neither more nor less spirit than that lawfulness which is dealt with by natural science or mathematics. Every bit of matter, even the smallest molecule or elementary particle, is determined out of the spirit. For all material is lawful, and there is absolutely no other way for the scientist to know matter or organisms than by grasping their spiritual essence in their own mind through thinking (cf. Section 3.1–3.2).

The organismic whole – but at their own levels also those of the chemical and crystalloid (cf. Section 3.3) – is therefore a *system* as described by Paul Weiss (Weiss, 1970) and Ludwig von Bertalanffy (von Bertalanffy, 1970) and aptly formulated by Wuketits (Wuketits, 1981, 88–89):

1. The whole (= the system) is always "more" than the sum of its parts (= system elements). It is not the mere addition of the elements but only the specific interconnection of these (*integration*) which produces a system. [...]
2. Systems are organised in a *hierarchy*. Every system consists of sub-systems which are subordinate to it. [...]
3. Systems show particular properties or laws which are lacking in their parts (sub-systems, elements) when looked at alone. These are what are known as *system properties* (system laws).
4. The number of parts (elements, sub-systems) which create a particular system varies. [...]
5. From the perspective of systems theory, the hierarchical organisation of systems [...] shows a hierarchical structure which is in agreement with the philosophical *theory of categories* (e.g. of N. Hartmann 1964). [...]
6. There are interactions, *between* the elements *of a* particular stage (system level) on the one hand and *between the stages* themselves on the other.
7. According to the model of functional causality, a particular element never merely affects another one, but this latter in turn affects also the first via the network of the whole system.

---

100 The expression "spirit" may appear rather unusual in the context of biology or even lead to misunderstanding, because the word "spirit" can have a wide range of connotations. However, as already described above (cf. Chapter 3), the expression here is chosen intentionally in order to draw attention to what already proves to be spirit in scientific epistemology. This is not something irrational but, on the contrary, represents the epitome of rationality and in normal consciousness is only accessible to clear thinking. It is this spirit which provides the bridge between science and the spiritual science developed by Steiner (cf. Chapter 7). And it is initially only *this* spirit which is referred to here, the law.

8. The description and explanation of a system remain incomplete as long as only individual stages (sub-system), or even one single stage, is analysed (reductionism). The study of the whole system without taking account of its parts (sub-systems) is equally incomplete. Consequently neither the pure *analysis* nor the pure *synthesis* does the matter justice.

But although Wuketits' specific system laws assert what the parts lack (i.e. emergent laws) and although he lets the parts act via a network of the whole system (therefore via the system laws), he nevertheless attributes the system to "molecular causes" (ibid., 89). He is obviously unable to let go of the mechanism which is also shown by the fact that his functional causality concept only recognises *interactions*, be they between the elements within a system level or between system levels.

The present-day *complexity theory*, an updated form of the system theory advocated by Wuketits and others, thinks in a similar way, for instance when e.g. one of its main exponents, Sandra Mitchell, views emergence as arising as follows: "In this new sense, just as in the ideas from the 19th century, emergence means that the interactions between the individual parts can lead to new properties not possessed by any of the individual components and that these properties of a higher order can in turn become causally active" (Mitchell, 2008, 47). However, this involves a contradiction, because if "these properties of a higher order can in turn become causally active", then the higher order as such, *the emergent or whole* is accorded *its own efficient causation* and not just the details. And because, through emergence, "new structures arise which *obey new laws*" (ibid., 27, ePH) the efficient causes (causa efficiens) of the whole must be linked with *its* law (causa formalis) in a unity. For it is impossible for the whole to produce an effect without this effect obeying *the law* of the whole. *The law of the whole must therefore be an active one.*

From the viewpoint of systems and complexity theory, the force-bearing law or lawfully acting force of the system can itself be considered as the cause of the higher order of the emergent properties. The "system" itself imposes order on the parts, just as the earth's magnetic field aligns all compass needles. This explains why, in an analogy to physical force fields, Driesch, Gurwitsch, Weiss, Spemann and, more recently, Rupert Sheldrake (Sheldrake, 2009) have made morphogenetic fields responsible for the development of form in organisms or crystals.

However, there is a crucial difference between physical fields and the morphogenesis of a crystal or organism: the iron filings are of course compelled into the force lines of the magnetic field, but their particulate appearance and characteristic features remain intact. The whole remains *external* to them, their relationship to it is one of a physical *interaction*. Interactions in the *real* sense only exist between *externally opposed* causal factors, as is the case in mechanics. But in chemical, crystalloid or organic systems, the whole permeates its parts *internally* and therefore brings about its emergent properties in their place (cf. Chapter 3 and Sections 4.4. and 4.6). The emergent system therefore comes about because the higher (system) law does not confront the subordinate elements externally but *permeates them internally* and lends them *its* appearance and *its* lawfulness, so that a higher whole arises in which the properties of the parts are "sublated" in a certain sense. Accordingly, the system law has to be thought of as an *active* one.

4.9 Biology beyond vitalism and mechanism

The empirical ontological idealism as set forth by Goethe and Steiner enables the founding of a modern biology beyond vitalism and mechanistic materialism. This type of science views knowledge as the agreement of the given percepts with the corresponding laws, and is characterised by the empirically based recognition of objectness[101] both of the percept and the law, on account of the epistemologically based demand to treat every realm of phenomena scientifically in this way and, finally, thanks to a concept of reality that grasps not only the percept but also the law as a real constituent factor of reality (cf. Chapter 2). The hierarchically ordered emergent system levels of the organism must therefore be studied according to their *own* laws and recognised in accordance with their particular reality. As a result, the lawfulness of the higher emergent properties of the organism must be viewed as an active entity which assumes a causative role for morphogenesis and the systemic order of the organism. This enables us to do without the reductionist interpretation of the organism and life, without underestimating the role of the different substances for the creation of the activities and forms of the living realm and without needing to have recourse to speculative principles of life.

Recognition of the real and active nature of the law in fact enables the riddle of emergence to be solved, something which otherwise seems difficult to achieve. "It is clearly a riddle: how can a property of a higher order have its roots in the physical components on the one hand (water is nothing other than H_2O) and on the other hand be something new which cannot be predicted from the properties of the components alone?" (Mitchell, 2008, 38). The reason for this is that the new entity has its own law which *penetrates* its components and thus produces a new – in other words *its own* – manifestation, which obeys this law. This solves yet another aspect of this riddle, as discussed

---

101 The term "objectness" (objecthood) is chosen in order to indicate even more strongly than through the concept "objectivity" (which is what is meant by it) the actual real character or the ontic autonomy of what is given in the form of a percept or law.

by Jaegwon Kim: "But how is it possible for the whole to causally affect its constituent parts on which its very existence and nature depend?" (Kim, 1999, 28). This is possible because, although the parts are the *prerequisite* of the actualisation of the emergent law, they are not its *cause* and because, as something *active*, this law is superior to the parts.

In other words: the phenomenon of emergence can be understood without contradiction when the lawfulness of each emergent level is accepted first in its content and agency (as causa formalis). We have already seen that the holistic concept of matter currently put forward in elementary particle physics on empirical grounds can be understood on this basis (cf. Sections 3.2–3.3). This also applies to a holistic understanding of the organism (cf. Section 4.3).

This enables its content to be defined, something which remained undefined in the hypothetical concepts of morphogenetic fields and the wholeness-creating entelechy of Driesch, that is, the nature of the *cause* of wholeness in organic development. The creator of wholeness lies in the *system law itself*. This has to be found empirically, based on the actual organisation which appears as emergence, be that in the macroscopic morphology as practised by Goethe or in the molecular biological investigation of cell organisation in modern biology.

The speculative element can be avoided by ensuring that thinking holds strictly to the perception and only takes an organisational idea as far as this can be justified by the actually observed organisation. The ideas must "really agree with the objects and coincide with them", as Goethe demanded of the hypothetical ideas of Sömmering's anatomy (Goethe, 1958b, 349). This is indeed how morphology and palaeontology proceed, inasmuch as they deal with the production and comparison of blueprints or study pattern formation in organisms by other means. Due to the basic reductionist attitude, we only rarely make ourselves aware that the *organisational principle itself* should be ascribed the character of reality, in fact that it should be viewed as the determining cause for the pattern formation.

What is required when determining the morphic cause of organic pattern formation is to find the lawfulness of the actualised pattern in the phenomenon itself, because it is this lawfulness which manifests through self-organisation in the material from which it is formed.

Someone who was probably the last representative of morphogenetic field theory as defined by Spemann, Curt Stern, comes close to this view. In 1954 he compared the morphogenetic field to the embryonal prepattern: "The prepatterns of the embryonic tissue in *Drosophila*, which call forth a response of genes involving the differentiation of bristles, are embryonic fields of larger dimensions than the limited points of normal location of bristles" (Stern, 1954; cited in Beloussov et al., 1997, 778). However, apart from the fact that Stern then attributed this prepattern to genes, thus promoting the subsequent causal primacy of the genes (Beloussov et al., 1997), it is not *this* particular prepattern or any other pattern *as such* that is the morphogenetic field, but something much more all-encompassing: the lawfulness of *every* self-organising pattern or structure of a spatial or temporal nature that is or can potentially be enacted in the course of development of an actual organism is *part* of the morphic cause or entelechy meant here. The entelechy is the *system* of this lawfulness. The development of an organism at any of its stages can be interpreted in such a way that parts of the whole system which have already been produced provide the means and the medium for inducing the self-organisation of further parts, such as e.g. the enzymes for the functions or structures catalysed by them (cf. Section 4.7). The entelechy as such is, as already described, the lawfulness of the complex, structured, hierarchically organised whole of all these parts, i.e. the system, the typus or the idea or total information of the organism in its full potentiality.

This idea or total information is the *real* blueprint of the organism of which the blueprint coded in the functional genes, the information for the primary structure of the proteins, only forms a small part (cf. Section 4.1). And it is only of this comprehensive *systematic* blueprint that it is justified to state that which – expressed in the words of Ernst Mayr – scientists in the 20th century wished to ascribe to genetic information: "And the blueprint makes everything!" (cited in Riedl, 2005, 128). For the organic system brings about its conditions itself, in its phylogenetic and ontogenetic *development*[102]. This

---

[102] The application of this basic thought to ontogenetic and particularly to phylogenetic development throws an interesting light on the nature of inheritance and the basic biogenetic law, as Steiner explained in his *"Goethean Science"*. "In Goethe's concepts we also gain an ideal explanation for the fact, discovered by Darwin and Haeckel, that the developmental history of the individual represents a repetition of the history of the race. For, what Haeckel puts forward here cannot after all be taken for anything more than an unexplained fact. It is the fact that every individual entity passes, in a shortened form, through all those stages of development that paleontology also shows us as separate organic forms. Haeckel and his followers explain this by the law of heredity. But heredity is itself nothing other than an abbreviated expression for the fact just mentioned. The explanation for it is that those forms, as well as those of the individual, are the manifest forms of one and the same archetypal image that, in successive epochs, brings to unfoldment the formative forces lying within this image as potentiality. Every higher entity is indeed more perfect through the fact that, through the favorable influences of its environment, it is not hindered in the completely free unfolding of itself in accordance with its inner nature. If, on the other hand, because of certain influences, the individual is compelled to remain at a lower stage, then only some of its inner forces come to manifestation, and then that which is only a part of a whole in a more highly developed individual is this individual's whole" (Steiner, 1988b, 73–74). We can also compare the application of this

is why the content of the law in this system can only be found by following the actual course of development of the organism, as Steiner commented with reference to Goethe's way of studying organic nature (Steiner, 1988b, 61–62):

> The entelechy, built upon itself, comprises a number of sense-perceptible developmental forms of which one must be the first and another the last; in which one form can always only follow the other in an altogether definite way. The ideal unity puts forth out of itself a series of sense-perceptible organs in a certain sequence in time and in a particular spatial relationship, and closes itself off in an altogether definite way from the rest of nature. It puts forth its various states out of itself. These can therefore also be grasped only when one studies the development of successive states as they emerge from an ideal unity; *i.e., an organic entity can be understood only in its becoming, in its developing.*

Goethe pursued this becoming macroscopically in his morphological work. The biology which followed him developed the means to follow this development microscopically and at a molecular level. But both can combine in a modern "biology beyond mechanism and vitalism" (Wuketits, 1981, 88), when cognition operates strictly in the objective world dialectic between percept and concept, when the same cognitive principle applies to every level of manifestation and when not only the sensory phenomena but also the lawfulness prevailing within these is accepted as an objective fact of reality. The epistemology developed by Goethe and Steiner can make an important contribution to this (cf. Chapter 2).

4.10 Reflections on the thermodynamics of organic processes

Ontological idealism i.e. the view of biological system laws as *real, active* self-organising entities can also be justified from an energetic perspective in terms of thermodynamics, despite the fact that, from a reductionist perspective, the driving force of self-organisation is seen as the spontaneous tendency of the system towards a *thermodynamic energy minimum*, in other words a *passive* process. The stability of the structural and functional equilibrium and the robustness of the resulting organic order is also usually attributed to this, i.e. its ability to recover after perturbation (Kirschner et al., 2000, 81):

---

basic idea to the problem of the concept of higher development in the theory of evolution (Rosslenbroich, 2006), to the problem of the heterochronism of form characteristics in the evolution of vertebrates (sometimes misconstrued as mosaicism) (Schad, 1992), to the question of the special morphological position of the human being in relation to the higher mammals and primates on this basis (Verhulst, 1999; Schad, 1977), and to the increasing autonomy associated with higher evolutionary development up to the human being (Rosslenbroich, 2007; 2014).

> The concept of the steady state as a thermodynamic minimum helps us understand the ultimate driving force behind self-organization, and may account in energetic terms for the robustness and pathway-independence of a process like spindle assembly. [...] The steady state resembles equilibrium in the sense that the system moves spontaneously downhill over some energy landscape to reach the steady state, where it comes to rest. If a steady state is perturbed, it will return to its preferred parameters.

Ilya Prigogine, to whom Kirschner refers here, in fact discovered that in open systems which are *not* in thermodynamic equilibrium due to constant energy flow, stable orders can develop in the form of a dynamic equilibrium. However, the achievement of the dynamic equilibrium is thermodynamically linked with energy minimisation and an increase in entropy which is why the system spontaneously seeks this orderly state (Prigogine, 1977). A state of this kind is therefore also known as an *attractor*. Flow equilibria are dynamic systems which are also interpreted as *dissipative* structures because they arise through irreversible processes associated with energy dissipation, e.g. the form of a flame; the self-organisation of honeycomb structures in viscous fluids in shallow containers due to convection which is produced by gentle heating from below (Rayleigh-Bénard convection) (Küppers, 1996); the growth of crystals; biochemical flow equilibria with an inflow of substrate and removal of products; and the flow equilibria in living systems in general.

However, it must not be forgotten that these kind of self-organisation processes which lead *automatically* to dissipative structures or stable flow equilibria can only be induced and maintained by *organised activity*, including particularly the *production of a thermodynamic disequilibrium* from the supply of energy or material and – e.g. in the case of biochemical and structural organic flow equilibria – organised product removal. Küppers, for example, describes the "elementary basis of self-organisation" as follows: "Self-organisation is the formation of dynamic structures which are self-reproducing. In physical systems it is linked to a dynamic which is driven into imbalance by compensation processes. *Compensation of the imbalance has to be prevented by maintaining this imbalance so that it does not grind to a halt after a short time.* Self-organisation therefore relies on an environment which provides the resources for maintaining this kind of imbalance", such as e.g. by a constant precisely regulated supply of heat in the Rayleigh-Bénard convection (Küppers, 1996, 134, ePH).

So what must be artificially supplied *externally* – or to an extremely limited degree also through a favourable constellation of environmental conditions – to a physical or chemical self-organisation system to maintain a stable state of self-organisation, is achieved by the organism *internally* itself: the active organisation of the supply and removal of the relevant energy carriers and substrates. In contrast to purely physical and chemical self-organisation, the organism therefore organises itself in a *double* sense: first through the self-organisation of physical structures which are analogous to those of the physical, chemical or crystalloid self-organisation (e.g. self-organisation of protein or organelle structures). These are the only ones taken into consideration in almost all discussions on self-organisation. Secondly, the organisation of the conditions which make the above-mentioned self-organisation possible in the first place: creation and maintenance of the thermodynamic *imbalance* as the starting and maintenance conditions for the attractor or for the steady state i.e. self-organisation which aims towards homeostasis. This latter self-organisation which determines the first-mentioned self-organisation is generally ignored by self-organisation theories just as much as the engineer with his design and construction process is forgotten in the case of mechanical theories of the organism (c.f. Section 4.4). This is why e.g. Kirschner believes that organic self-organisation, in fact even the large-scale coordination of functional units or wholes, can simply be explained by "collapse" in the sense of the above-mentioned "descent" to the thermodynamic energy minimum. "Induced collapse embodies the idea that self-organizing systems are able to generate single and appropriate functional outcomes by coordinating the activities of many identical components over large distances. [...] Such systems [...] always collapse to a singular state [...]!" (Kirschner et al., 2000, 87). He anticipates that, based on this, the 21st century will produce a purely chemical explanation of life.

But this model forgets the ceaseless creation of the thermodynamic non-equilibrium, the permanent maintenance of energy and material gradients which precede the countless and incessant "collapses" and provide *their energetic opposite*. This thermodynamic non-equilibrium is *hierarchically superordinate* to the automatic tendency towards equilibrium and has to be *organised* in space and time. The organism produces this organisation itself through the activity of enzymes and their coordination. The enzymes themselves are in turn attributable to enzymatically controlled, spatially and temporally coordinated dynamic regulation of gene activity (cf. Section 4.2). It is therefore impossible to avoid contrasting the more or less passive energy minimisation with an *active energy maximisation* and of thinking of both as being united in a complex, meaningful overall organisation which only then constitutes the living organism.

This is the case with thermodynamically dissipative systems as defined by Prigogine and also in the conventional view of thermodynamics as is generally still presented in the text books (Purves et al., 2006, 128–148). After all, the energy-consuming creation of complex highly-ordered structures in organisms and the corresponding decrease in entropy is at first sight in contradiction to the 2nd law of thermodynamics. However, the predominant catabolism is compensated by anabolism: 10 kg of biological material are broken down for 1 kg of body mass produced, leading to much more "disorder" (entropy) than is contained in 1 kg of body mass. This is the reason for the belief that the increase in complexity in an organism is only "*apparently* at odds with the second law" (ibid., 131, ePH). However, the contradiction is not merely apparent but real in the anabolic system but is *compensated* in the overall context by the catabolic system i.e. through the energy supplied to the open system from outside, so that the organism can exist "as an extremely organised, low-entropy system" (ibid., 131). The anabolic system therefore *serves* the energy supplying catabolic system and both are organisationally interconnected, i.e. are parts of the overall organisation which integrates the build-up and breakdown in a meaningful way. Energy carriers have to be available in the right quantities at the right time and in the right place; breakdown products likewise have to be removed in an equally carefully organised way; the functional and histological anatomical structures have to be organised for these purposes both inside the organism and at its regulated open boundaries to the outside world. It is only through this *organisation* that the above-mentioned thermodynamic compensation of the energy-consuming creation of order is possible. The order of the organism does not therefore *arise from* its energy budget, but *controls* this i.e. the order manifests itself *in* the energy budget. The organisation as such is an ideal but at the same time real principle, which brings itself into being in the process of self-organisation.

4.11 Life versus death: physical and etheric organisation

The fact that the lawfulness of the organic system cannot be explained by its components but ontologically requires its own "reality", can be recognised because the system law actually brings its specific characteristics and independence to bear *in contrast to* the material components which are integrated in it, i.e. the specifically organic versus the inorganic, life versus what is dead, mineral.

Thus after death the substances in the animal or human corpse or the dead plant body inevitably decay according to the 2nd law of thermodynamics into the surrounding lifeless realm of nature, the mineral kingdom, as do the substances excreted from the organism during life or body parts separated by trauma. However, this dissolution is constantly prevented in the *living* organism: in fact the exact opposite of this disintegration takes place. In the living organism the inorganic substances are built up into higher-ordered organic substances by the constant well-organised creation of a state *far from a thermodynamic equilibrium* and arranged in a highly complex order which actively and by means of self-organisation (autopoietically) maintains itself and re-establishes itself if perturbed. "Maintenance of the stationary state far from the thermodynamic equilibrium is a *conditio sine qua non* for the existence of living systems" (Penzlin, 2016, 182). In this respect the property of "life" *as life* does not actually lie in the production of complex substances, although life can only develop based on them and as a rule organic substances can only be produced in living organisms, with the exception of the small molecules such as urea. But, as Ernst Mayr says: "These organic molecules are not principally different from other molecules" (Mayr, 1984, 45). For even the higher-order substances such as nucleic acids, proteins and lipids are physical substances and obey physical laws, albeit higher emergent laws as described above (cf. Section 3.2–3.3).

Considered in isolation these higher-order substances also behave like "organic crystals", and in physics are regarded as this in every respect (Head et al., 2000). As these substances are self-organising under suitable conditions – in fact even enzymes (e.g. ribonuclease) can regain their native three-dimensional conformation in the test tube by self-organisation after this has been destroyed by careful denaturation, leaving only the primary protein strand (Lewin, 1988, 11) – there is no reason in a *material* sense to see the organism as anything other than a purely physical body, although a body which behaves like a crystal, in other words *lifeless*, if nothing else were to be added which is *not physical*. Steiner therefore once commented: "But you will recognise: if man merely had his physical body, he would be a crystal – even if perhaps a very complicated one" (Steiner, 1972, Lecture I). This "crystal" is of course complicated because the self-organisation – always under the appropriate conditions – of atoms and molecules does not produce merely macromolecules but these form the basis for even higher structures such as organelles and cells and beyond these cell communities, organs, organ systems and even the physical body as the overarching organisation. This in its complex totality would then need to be thought of as a "crystal".

And as self-organisation is always involved in building up the structure, then only physical laws and forces can be made responsible *causally* for the development of this physical body. But because every emergent structural level has its own laws to which the elements of the lower levels are subject, the different higher level laws can each be regarded as a causa formalis[103] which

---

103 This view of a real active causa formalis as the cause of the self-organisation of crystals and the related physical creation of form in the organism is compatible with modern chaos theory. For example, in his book on chaos theory in the chapter with the revealing title "The ontic conception: causal versus geometrical mechanisms", Stephen H. Kellert compares the mechanical causation, which he calls "causal" with the *geometrical* one which he makes responsible for the "common *patterns* or regularities at a more *macroscopic level* of analysis" and in view of which it is rather pointless "to try to understand the behavior of the *whole* system by tracing each individual process. Instead, one needs to find a way of representing what the system does on the *whole* or on the average,

subjugates the causa materialis of the subordinate levels, which of course means they themselves must be causal agencies (cf. Sections 3.2–3.3). This can explain such things as the formation of the three-dimensional conformation of proteins which, for their part, are integrated in even higher structural laws. Viewed in this way the system of structural laws, which in their overall context comprise the blueprint for the physical organism, can be referred to in the Aristotelian sense as the active complexly ordered *"form"*. The "material" for this form would then be the relevant substances which have to end up in this form e.g. through eating or from intermediary metabolism in order to be incorporated into higher substance and structure. *In itself* this "form" or the system of spatial formative laws naturally has an ideal, non-sensory nature, as is normal for laws (cf. Chapter 2 and Section 3.1); so we can understand why Steiner describes this system of laws, the "physical organisation" or "human form" as a "supersensible system of forces" which as it were "confronts" the nutrition when this enters the organism and why he compares this process with the production of cooking salt from chlorine and sodium (Steiner, 1951a, 173–174):

> Only by reason of the fact that this super-sensible form incorporates the nutritive matter does the human organism become a physical-sensible organism, something that our eyes can behold and our hands can grasp. That which thus confronts the external nutritive substances is called "form" in accordance with the law that is operative throughout the whole of nature, an identical law termed the *"principle of form"*. Even though you descend to the crystal, you find that the substances which enter into it, if they are to become what is manifest as the crystal, must be seized as it were by form-principles, which in this case are the principles of crystallisation. [...] And so everything which enters into the human organism as nutritive substance presupposes the [...] super-sensible *form*.

Seen in this way the physical body can be viewed as the unity of this physical "form" with its "matter".

But this physical body would not be alive if only the stated physical substance and structural laws were active in the various levels of the system. It would still not be alive if, instead of solid crystals, it were to produce processes similar to metabolism with dissipative structures such as the above-mentioned Belousov-Zhabotinsky reaction. Because these are merely oscillating chemical processes that die away until the point of exhaustion which, although they form typical patterns, still lack the *specific organisation of the living*, that being *"a self-enclosed whole"* (Steiner, 1988a, 100) which the present-day neo-Aristotelian

---

which abstracts from such specific causal detail" (Kellert, 1993, 104–105, ePH). However, a "geometrical" causation is of necessity one of spatial form. In this sense the manifestation of the lawful physical forms and shapes of crystals and organisms can be viewed as "geometrical" causation.

ontology also views as a characteristic of living beings (Schark, 2005). The elementary units of life (cells) exist only as organisms, they are "integrated individuals" (Laubichler, 2005) which, as units organised within themselves *actively* hold their own against the outside world and are simultaneously open to it. Belousov-Zhabotinsky processes do not display this basic characteristic of the living world, an organismic organisation. The additional characteristics which are specific to the living world such as growth, organised metabolism[104], defence, reproduction etc. only apply to these kind of differentiated units or entities. Crystals or the cyclical expanding fronts of the Belousov-Zhabotinsky reaction do not grow in the organic sense but acquire substance by apposition. Biological growth, on the other hand, is an active and flexible increase in size organised in metamorphoses towards a developmental aim, in which the matter is absorbed in a sophisticated way into the organismic system. "The difference between inorganic matter and living organisms lies not in the matter of which they are composed, but in the *organisation* of the biological systems" (Mayr, 1984, 50, ePH).

In contrast to the non-living, the important aspect of living self-organisation is therefore this spatial whole organised within itself as a unit and, in addition, that this whole comes into being teleonomically over time. It is not only physical/spatial or morphological/geometric laws which are at work, which would ultimately result in a "crystal", but *temporal laws* are also involved. The organism *actualises* its form not simply like a crystal but it *develops* during this process. Organisms are "developing systems" (Stotz, 2005). The organic form changes in the process of coming into being: morphogenesis through *metamorphosis* as described so well for plants by Goethe and recently by Andreas Suchantke (Goethe, 2009; Suchantke, 2009) and by many other authors in the 19th and 20th centuries for the development of animals and human beings with their unbelievable changes in form and movement, from the spherical egg cell through the stages of morula, blastula and gastrula, via fish-like body plans all the way to the typical mammal and human form (Hinrichsen, 1990). At the molecular level this development is enabled by a well-ordered, accurately-timed sequential application of whole cascades of developmental control genes (cf. Section 4.5). The individual substances and structural elements are supplied in a temporally ordered sequence in such a way as to be appropriate for the overall spatial structure (cf. Section 4.7). The *spatial design* is therefore put into practice according to a *schedule*: in living organisms the *spatial organisation* is subject to a *temporal organisation*.

---

104 This includes the acquisition of raw materials at the correct time and place for what in the organism can be compared to the Belousov-Zhabotinsky reaction, and the analogous disposal of the corresponding end products.

This temporal organisation is responsible for the *rhythmical, cyclical and temporal* structure of all living things (Endres & Schad, 2002; Foster & Kreitzman, 2004). The study of these temporal laws has become established as an independent branch of science in modern chronobiology (Touitou & Haus, 1992; Rosslenbroich, 1994), and these cyclical processes show the unbelievable complexity of the temporal organisation of living things (Hildebrandt et al., 1998). This is not a mechanical oscillating system but a flexible, responsive – in fact living – temporal organism which is organised as a unit and integrates in itself countless rhythms of milliseconds, seconds, minutes, hours, days, weeks, months, years, seven-year and other periods in order to regulate the juvenile emergence, the maintenance and the senescence of the organism. It is comparable to an inner symphony of time which includes the entire lifespan of a living creature and which clearly shows that the *spatial organism* is determined by a *temporal organism*, i.e. from the *organisation of active temporal laws* which are responsible for its development, its healthy functioning and its maintenance. These viewpoints have become very important in applied medicine (cf. Section 8.4).

What makes the organismic body *different* from a crystal is that, *in addition* to the spatial laws, there are these temporal laws governing the becoming, the maintenance, and even the decay (e.g. in apoptosis) of the body, so that a flexible dynamic, i.e. *living* organisation comes into being. This applies to the whole, but also to the individual parts of the organism. Although under special physical conditions proteins behave like static crystals, in the living organism they are dynamic, flexible structures. The interpretation put forward by Karplus as far back as 1986 has now been fully confirmed (Tokuriki & Tawfik, 2009): "Protein molecules are elastically deformable and never at rest. If they were static structures they would often be unable to carry out their important tasks for life. [...] Proteins are therefore dynamic entities which constantly change their conformation" (Karplus & McCammon, 1986, 108). In fact, recent discoveries have shown that enzymes and their functions are much more dynamic than was ever previously imagined. For example, enzyme functions depend on subtle conformational changes that include active-site conformational motion in time scales of nano- or even picoseconds (Lu, 2012).

Just as the spatial structural laws are realised by self-organisation when the appropriate pre-conditions are present, so are the temporal ones. However, the temporal structures appear spontaneously as emergent properties compared to the spatial structures: they can only be observed *in* these spatial structures, but are not caused by them. And just as the spatial laws of substance and structure can be viewed as active realities, so too can the temporal laws. For if the spatial order is *actually* arranged according to temporal structures, these must produce an *effect*. In this respect the self-organised temporal form can be considered as the causal basis for the teleonomic character of the organism. It creates the *temporal form* of the organism, i.e. it is the organism's causa formalis. As this is brought about in time, it is also a causa finalis and, because it is active, a causal agency (causa efficiens). But the *spatial* causa formalis always arises as a unity with the temporal; and for this reason the Goethean typus or the organic "entelechy" contains both as an ideal unity: the spatial and temporal order. This also explains why the organic organisational laws can only be discovered by means of the actually observed *development*, as shown by Goethe in his *Metamorphosis of Plants* (Goethe, 2009) and Steiner in his *"Goethean Science"* (Steiner, 1988b, 61–62):

> The entelechy, built upon itself, comprises a number of sense-perceptible developmental forms of which one must be the first and another the last; in which one form can always only follow the other in an altogether definite way. The ideal unity puts forth out of itself a series of sense-perceptible organs in a certain sequence in time and in a particular spatial relationship, and closes itself off in an altogether definite way from the rest of nature. It puts forth its various states out of itself. These can therefore also be grasped only when one studies the development of successive states as they emerge from an ideal unity; *i.e., an organic entity can be understood only in its becoming, in its developing.* An inorganic body is closed off, rigid, can only be moved from outside, is inwardly immobile. An organism is restlessness within itself, ever transforming itself from within, changing, producing metamorphoses.

But this "restlessness", the actual *living aspect* of the organism, arises from the *temporal* element which the inorganic crystal lacks and it is therefore *the temporal organism that is the cause of life*. Temporal laws are not laws of physical spatial structures and therefore *not physical* laws: they only form the physical over time. Time is therefore represented as the accomplishment of this (physical) process, or expressed diagrammatically in a spatial form as the time axis or something similar. "But we must always be clear that we are then not dealing with time but with its spatial depiction" (Schad, 1992, 117; Schad, 1993). Steiner therefore describes rhythmic phenomena as *"half spiritual"*: "Rhythmical processes are nothing physical, either in Nature or in man. They might be called half spiritual. The physical as object vanishes in the rhythmic process" (Steiner, 1973b, 187). And the laws of these "half spiritual" phenomena are by nature of course just as spiritual as the laws of physical organisation (cf. Section 3.2). With reference to the expression "super-sensible system of forces" cited above, we can also say that the matter which enters the human being is not absorbed only by the supersensible force system of *physical* laws – because this is only "the very first super-sensible element, the human form" (Steiner, 1951a), which makes the absorbed material into a higher substance and a spatial physical form. The substance is actually absorbed by a *second super-sensible system of forces* through the *organisation of active temporal laws* which grants the otherwise more-or-less "crystalline" body the functions of life.

In contrast to the physical organisation, Rudolf Steiner calls *this* one *"etheric"*, thus drawing not only on an older word usage but also on a Greek expression, just as modern science does with "physical" or "psychological"[105].

Unlike the mineral realm, living creatures such as plants, animals and human beings not only have a physical body but also an etheric organisation, an *"ether-body"* or *"life-body"*, as Steiner puts it (Steiner, 1970a, 27). This etheric organisation and its forces are of course as little perceptible to the senses as are the forces of physical organisation. Nevertheless, both are accessible to scientific knowledge, precisely due to their lawful *effects*, the physical forces through spatial structures and laws, the etheric forces through temporal ones. The observed emergent temporal laws of the organism are laws of the etheric organisation. Wolfgang Schad therefore justifiably claimed that: "Chronobiology is etheric research" (Schad, 2003). And Steiner stated: "In order to understand the etheric body and the physical body, it is well to conceive that the physical body is pre-eminently a space-body, and the etheric body pre-eminently a time-being" (Steiner, 1945). "Pre-eminently", because on the one hand the etheric organisation acts on the spatial organisation by directing the operations of the *physical* laws "as a kind of architect" (Steiner, 1969b, 43) of the physical body and because, on the other hand, the spatial organisation then receives a *temporal* character.

The etheric and physical organisations should be thought of as a *unity*: the etheric body does *not remain external* to the physical body as was imagined for the morphogenetic fields in neovitalism[106], but permeates and determines

---

105 The expressions used by Steiner which are otherwise not common in science such as "etheric", "astral" (cf. Sections 4.11 and 5.4), etc. often take people aback (Burkhard, 2000), and some believe Steiner to have borrowed his concepts from older philosophical, esoteric, folk or other historical medicinal traditions (Zander, 2007). But just as modern science and medicine do not mean the Greek expressions "physical" or "psychological" to denote an anthropological or medical meaning which they might have acquired by tradition from the time in which these expressions were coined, so in Steiner's case there is no such intention with expressions such as "etheric" or "astral" which were coined and used in the same ancient times. Steiner wished to use these expressions in order to have names for what can be established by observation and thinking *using the present-day methods of gaining knowledge*. "The name is used because it relates to earlier, instinctive ideas of this world. Compared to the clear perceptions now possible these ideas no longer have validity; but we have to give names to things if we wish to refer to them" (Steiner & Wegman, 1996, 5).

106 In view of a potential confusion of the etheric body with neovitalist ideas of life forces, Steiner explains in his *"Theosophy"*: "The author of this book [...] applied to what is here called etheric or life-body, the name 'formative-force-body'

it *from inside* as its own higher emergent nature. This is comparable to the cooking salt law which, in the chemical process, takes the elements of chlorine and sodium and makes the emergent substance *cooking salt*: as the lawful properties of sodium and chlorine are "sublated" in the double sense meant by Hegel (cf. Section 3.3), so the substances of the physical body "divest" themselves of the action of their physical laws when they are taken hold of by the etheric organisation and thus become emergent *living substance*. This is why Steiner and Wegman wrote in their first joint book on the foundation of anthroposophical medicine: "Because of it [the etheric body] something happens in the human being that is not a continuation of forces of the physical body acting according to their laws, but happens because physical substances rid themselves of their physical forces, as soon as they stream into the etheric" (Steiner & Wegman, 1996, 5–6). But it does not remain like this, just as in the chemical reaction the reaction equilibrium does not remain exclusively determined by the compelling action of the emergent law of the whole over the laws of the parts. But just as the parts there emerge again from the chemical compound due to their dynamic relationship to the whole, so here the physical substances again emerge from the living entity. "Thus we have one stream of substance. Lifeless substance transforms itself into living; living transforms itself into lifeless" (ibid., 25). This has to take place in a manner controlled by the organism if the organism is to be healthy.

An example of this is the formation of physical structures which are required for physical and mechanical purposes of the organism, e.g. when apatite crystals crystallise out in the dental enamel. This is a process which is

---

> […]. He felt moved to give it this name, because he believes that one cannot do enough to prevent the misunderstanding due to confusing what is here meant by etheric body with the 'vital force' of older natural science. In what concerns the rejection of *this* older concept of a vital force in the sense of modern natural science, the author shares in a certain respect the standpoint of those who are opposed to assuming *such* a vital force. For the purpose of assuming such a vital force was to explain the special mode of working in the organism of the inorganic forces. But that which works inorganically in the organism, does not work there in any other way than it does in the inorganic world. The laws of inorganic nature are in the organism no other than they are in the crystal, and so forth. But in the organism there is present something which is *not* inorganic: the formative life. The etheric body or formative-force-body lies at the base of this formative life. By assuming its existence, the rightful task of natural science is not interfered with: viz., to observe the workings of forces in inorganic nature and to follow the workings into the organic world: and further, to refuse to think of *these* operations within the organism as being modified by a special vital force. The spiritual investigator speaks of the etheric body in so far as there manifests in the organism something *other* than what shows itself in the lifeless" (Steiner, 1970a, 27).

carefully regulated by life: the inorganic crystallisation processes must take place at the right time, in the right place and in the right form. Morphologically the tooth still has an *organic* form, even if its substance becomes practically inorganic in the appropriate places. If a substance like this is produced displaced in space and time, then illness may occur. A well-known example of this is what is known in rheumatology as "microcrystalline attacks" e.g. in periarthritis humeroscapularis. Microscopically small crystals of apatite crystallise in the connective tissue of the shoulder joint and, in an attempt to heal itself, the body reacts to its own substance which has become foreign with an acute painful inflammation. Illness in this case is not so much a matter of the substance but of the *displacement of the process*: a process which is quite normal in itself, an inorganic crystallisation, takes place at the wrong time and in the wrong place. Life, health and illness in this sense are not simply a matter of physical and chemical interactions but a question of the relationship of the *spatial and temporal organisation* or, in the terms commonly used in anthroposophical medicine, an expression of the relationship of the physical and etheric bodies in an organ or organ system (Steiner & Wegman, 1996; Selg, 2004).

The phenomenon of death can also be understood from this point of view. If the temporal organisation which is responsible for the life processes is removed completely, then death takes place and the physical body must inevitably decay according to the second law of thermodynamics, unless it is made into a kind of "crystal" by being preserved. After the etheric forces have ceased, then – apart from bacterial decay – only the physical and chemical forces and laws act in the body and the corpse gradually decomposes into the mineral realm. The etheric body as the bearer of life and the basis of organic development prevents this decay throughout life.

4.12 Summary

Objective empirical idealism leads to a replacement of scientific reductionism, first in the substance concept in physics, chemistry and biochemistry and then in molecular biology, morphology and the concept of the organism. It is not only atoms, molecules and macromolecules which are hierarchically organised in their compositions and structures, but also higher organic structures such as organelles, cells, organs, organ systems and finally the organism as a whole. Each hierarchical level has its own emergent laws and properties. These are realised "from above" when the necessary substances and conditions "from below" are present. This leads to a holistic view of substances, systems and causes in molecular biology and the understanding of the organism in biology. The organism is viewed as an emergent *system* which comes into being "from above" in complex and well-coordinated process cascades (organic self-organisation) which, while relying on the appropriate materials and conditions "from below", cannot be explained by these. The total "information" of the organism is seen as the lawful content of this system, in fact as a whole system, i.e. as the unity of the higher and lower aspects of the system which ultimately appear as the phenotype. The *genetic information* (genotype) is only a part of this, specifically the sequence of amino acids in the primary protein strand. It is only *this* information which is coded in the nucleic acids *as such* and, strictly speaking, not in the DNA (distributed there in pieces as "introns" between the "exons") but actually only in the messenger RNA after the splicing process (dynamic gene concept). The tertiary structure of the proteins then manifests under *the condition* of and *in* an appropriate primary structure as an emergent higher structure with its own phenomena and laws. These are not coded in the genome, and neither are the structural laws of other macromolecules, organelles etc. nor the laws of *spatial and temporal organisation* through which all these elements are related to each other as though in a whole. What is more, the temporal organisation is at an even higher level than the spatial organisation. If we designate the spatial organisation as the physical, then the temporal organisation which is emergent compared to the spatial organisation and from that point of view independent can be given its own name. Steiner calls it the "etheric".

On the basis of morphogenetic processes, the attempt was made to show that this overall spatial and temporal organisation corresponds to what Goethe called the typus, in other words the self-*actualising idea* of the organism, which manifests "from above" – under the necessary conditions and in the relevant materials provided "from below". In accordance with its content this idea or law can, in terms of Goethe's epistemological principle, only be what is found *empirically* by observation and thinking as the real actual *organisation*. Accordingly, Goethe's concept of the typus enables a scientific explanation of organic self-organisation, without having recourse to purely hypothetical vitalist principles or morphogenetic fields on the one hand or only to subordinate molecular elements and mechanisms on the other. This perspective of a *biology beyond vitalism and mechanism* seems to be supported by considerations from thermodynamics, and the recognition of the emergent organisation principle of life, the etheric body, as something *active* and additional to the physical body, and appears to be compatible with the unavoidable decay of the body after death in accordance with the 2nd law of thermodynamics.

5 Neurobiology, psychology and philosophy of mind: the reality of the soul and spirit

5.1 Neurobiology and the emergence of consciousness. Is the soul real?

Animals and human beings differ from plants and microorganisms not only on account of their typically different physical structure and their polar respiratory physiology (consumption instead of production of oxygen) but particularly due to the possession of a *psychological inner life and consciousness*. And the bodily functions and structures centred on the nervous system are organised specifically towards the exercise and service of this inner life. There is therefore a fundamental distinction between the life of animals and humans and that of plants and minerals, so that it is justified in viewing the world of *animate* beings as a kingdom of nature in its own right.

Consciousness is an emergent phenomenon distinct from the living body, just as life is in comparison to the purely material realm. This means that, contrary to widespread opinion, the organism and the nerve-sense system is not the *cause* of consciousness, but merely the necessary *condition* for its appearance. However, according to the findings of modern neurobiology and psychology, this condition is such that specific states of consciousness can be assigned to specific processes in the brain[107] (Cleeremans, 2005; Roth, 2007) so that, up to a point, conclusions can be drawn about the presence and nature of the contents of consciousness based on defined neural processes (Haynes & Rees, 2006).

However, it is incorrect on epistemological grounds to deduce that the brain is the *cause* of consciousness because of the *correlation* between the two ("Mental phenomena are caused by neurophysiological processes in the brain and are themselves features of the brain", Searle, 1992, p. 1) or even, like Gerhard Roth, to claim "that spirit is a *physical* state", only because its occurrence "is necessarily connected to physical laws and, in the narrower sense, to chemical and physiological ones" (Roth, 2007, 309). For

---

[107] However, there appears to be a type of bio-psychological "uncertainty relation" about the mutual connection of the conditions in the brain and consciousness. The smaller the measurement site in the brain (micro level) the less precisely the behavioural function (macro level) can be determined (e.g. psycho-pathological multifunctionality of the putamen); and the more precisely the behavioural function needs to be described, the less it is possible to locate a specific site for this in the brain (e.g. "working memory": prefrontal and parietal cortex, cerebellum, etc.) (Tretter, 2007, 502).

the fact that perceived consciousness is *tied* to physical material processes is not the same as being *identical* to them. In reality it is simply "*covariance relationships* [...] which enable mental phenomena to be assigned to their neural correlates" (Schumacher, 2007, 321).

This means that – as already described for other emergent phenomena – both areas, the consciousness phenomena and the neurophysiological phenomena, have to be observed *for themselves* and cognised according to their specific laws. Their legitimate correlation or covariance can only be established as a third step. The result is then merely what was previously called psychophysical *parallelism. As a result*, this correlation cannot be identified as a cause because *there is no empirical observation of this causation*. Between the subjective experience of consciousness and the objective perception of the parallel physiological process in the brain, observation encounters a chasm and consequently an *explanatory gap* as Joseph Levine called it (Levine, 1983). According to David Chalmers this chasm, "the hard problem of consciousness", cannot be bridged as a matter of principle (Chalmers, 1996).

This is due to the fact that these two correlated realms are by nature very different and can never be reduced to or even identified with each other. The phenomena of consciousness are *psychological* phenomena and can only be observed by *psychological introspection* (Brentano, 1911), while those of their neural correlate are *physical* phenomena and are only observable by *extraspection* by means of the sense organs. "Physical and psychological phenomena" are, as Franz Brentano (1838–1917) already demonstrated, clearly distinguishable "as two classes of phenomena" (Brentano, 1907, 96). The psychological experiences such as feelings, intentions and sensations have neither spatial extension nor weight in themselves, nor other properties of physical objects, they are *qualia* of a psychological nature.

This is also mentioned by Gerhard Roth when he says (Roth, 2007, 308–309): "Spirit and consciousness, at least as subjectively experienced phenomena, are not material in the normal sense i.e. they are not composed of elementary particles, atoms and molecules, [...] spirit somehow appears to be non-spatial." And: "At the same time there is obviously considerable inherent order in the spiritual realm which cannot be explained by physics – but this is the case for many of the properties of biological systems" (ibid.). This is because both biological systems and laws and, beyond them, the psychological systems and laws are *emergent* with respect to the corresponding basis in which they realise themselves through self-organisation: the biological system from its physical and chemical basis and the psychological state from its purely biological basis. To claim "that spirit is a physical state of a particular kind with many specific laws" (ibid., 309) is no less than an act of helplessness in face of the actual phenomena, an escape from having to recognise the new, different and in fact particular nature of these psychological phenomena and "spiritual" laws (cf. Chapter 2) on empirical grounds. Things are therefore declared to be physical[108] which do not prove empirically to be physical at all.

Paradoxically, John Searle makes the same mistake. In *The Rediscovery of the Mind* he refers explicitly to the irreducibility of consciousness (Searle, 1992) and distinguishes very clearly between physical and psychological *phenomena*, for example when speaking about pain: "If I now say, 'I am in pain', what are the facts? Well, first there is a set of 'physical' facts involving my thalamus and other regions of the brain, and second there is a set of 'mental' facts involving my subjective experience of pain." And in the case of a reductionist explanation of the psychological: "But of course, the reduction of pain to its physical reality still leaves the subjective experience of pain unreduced [...]" (ibid., 120–121). However he explains "inner qualitative states of consciousness and intrinsic intentionality" as purely "physical" (ibid., xii). He therefore nullifies the distinction which he *had* to make between "physical" and "mental" on empirical grounds by *conceptually* identifying both classes of phenomena as "physical". This, however, is a category error. It is the only means by which he can refer to consciousness as "an irreducible feature of physical reality" (ibid., 116).

Thomas Fuchs (Fuchs, 2005, 1–10) in his "ecological" view of the relationship between the body and soul also recognises the "irreducibility of the subjective experience" on empirical grounds and the "epistemological or methodological dualism" which "allows both subjective and objective access to reality their own validity". Since: "We will not find conscious experiences in the brain – only their neural correlates", and so both the psychological and physical must be perceived and formulated into concepts in *their own* way. "The concepts from the personal level where we are dealing with *my* experiences, feelings and memories cannot in principle be translated into the concepts of action potentials or stimulation patterns". But Fuchs nevertheless ontologically assumes a "naturalistic i.e. non-dualistic position" and rejects the ontological dualism which accepts the psyche and physical nature *existentially* as two independent entities. This takes place not by declaring the psychological as something *physical*, as done by Searle and Roth, but by designating both as a *"psychophysical unity"*: "Thus neurobiological processes on the one hand and psychological experiences and motives on the other would be viewed as different aspects of an ultimately *unified, though multi-level psychophysical event*. It is certainly not two different entities, 'brain and consciousness' or 'body and spirit' which mutually influence each

---

108 Roth himself obviously does not know what is meant to be physical about the spirit: "What the 'physics of spirit' will look like is not clear" (Roth, 2007, 309). It is still astounding that a whole spirit/body theory has been built on a foundation of this kind.

other here. The two ways of describing can equally refer to the system status of the organism in its environment (*ontological monism*). They nevertheless describe very different levels of complexity which cannot therefore be seen as identical (*epistemological dualism*)" (ibid.).

But this is also a merely *declared* monism: for what does "unity" mean here? Certainly not an "identity" or "uniformity", since the dissimilarity of physical and psychological phenomena has just been admitted, in fact it was stated that: "In line with epistemological dualism, no one description has more 'reality' than the other" (Fuchs, 2005, 7). With reference to the comparison of the "objective physical reality" of the bodily processes and the "subjective appearance" of the psychological phenomena, Searle also commented: "But in the case of consciousness, its reality is the appearance" (Searle, 1992, 122).

What does this mean? It means that the psychological element must logically be placed alongside the physical one, not only epistemologically but also *ontologically* as *of equal value*, even though *of a different nature*. Physical reality is complemented by psychological reality. Incidentally, this corresponds exactly to the epistemological and ontological position held by Goethe and Steiner and supported in this work: reality as the phenomenon permeated by its laws, independently of whether the phenomenon is perceived at the lower or upper system level or in the subject or object (cf. Section 2.5). This is in accord with Searle, Roth and Fuchs who concede that the physical and psychological elements have their own phenomena and laws or concepts. If the physical law is therefore declared to be real, this must also be allowed to apply to the psychological element. (NB This also applies to the case where no "causal interaction" between physical and psychological phenomena is postulated due to the above-mentioned observational chasm, a view also held by the author of this work). However, this epistemologically obvious conclusion is not drawn. "Ontological dualism" is avoided at all costs. Why?

There seem to be several reasons for this. The first and possibly most important reason is in fact not a scientific one but a profoundly psychological one and many people are perhaps not even aware of it: *fear* of the emotional or spiritual realms as such. Searle openly admits this: "If one had to describe the deepest motivation for materialism, one might say that it is simply a terror of consciousness. [...] The deepest reason for the fear of consciousness is that consciousness has the essentially terrifying feature of subjectivity. Materialists are reluctant to accept that feature because they believe that to accept the existence of subjective consciousness would be inconsistent with their conception of what the world must be like" (Searle, 1992, 55). "But if it does behave like this," John Eccles responds, "it makes their [the materialists'] philosophy irrational. It is irrational" (Eccles, 1994, 263). He goes on to point out the *nature of the assertion* contained in the theories held by Changeux, Crick and

Koch, Dennett, Edelmann, Sperry and Searle of the identity of self and brain or of spirit and matter: "This strange postulate of identity is never substantiated, but it is believed that it will be explained once we have a fuller scientific understanding – perhaps in 100 years. This is why we have ironically named this belief 'IOU materialism'" (ibid., 243).

A second reason for the rejection of ontological dualism is the *insufficient epistemological basis* of many neurophilosophical considerations, which Gerhard Roth himself diagnoses as a fundamental lack in his own discipline: "Brain research has [...] until now not carried out any fundamental methodological or conceptual self-criticism" (Roth, 2004, 67; cited in Tretter, 2007, 501). This fundamental methodological or conceptual criticism would have to consist primarily in an empirical analysis of the process of knowledge itself with an explanation of the principle relationships of phenomenon and law and with a differentiated application of this principle to the different emergent phenomena of hierarchically ordered nature. However, this produces not simply an ontological dualism but actually an *ontological emergentism*: *every* emergent layer of hierarchically ordered nature has *its own* reality which reveals itself through its phenomenon and its law and cannot be reduced to the reality of the subordinate layers. This applies even at the level of the physical and chemical realm of inorganic and organic substances and structures (cf. Section 3.3). But it also applies to the further emergent realm of life (cf. Section 4.8) and beyond this to the realm of consciousness.

A third reason for the predominant resistance to an ontological twofold or even multi-part picture of the human being in medicine is the understandable dislike of the ontological dualism *as defined by Descartes*. This is usually what is under attack (Searle, 1992, 2). According to Descartes the spatial body (res extensa) is a machine which is controlled by the non-spatial mind (res cogitans) via the pineal gland (Streubel, 2008, 31). Compared to the body, the mind is something completely different and *external*, so that on the one hand the interaction between mind and body has a *seemingly linear causal mechanistic character* but on the other it remains simply incomprehensible or in a physicalistic perspective breaks through the closed physical causal nexus, thus throwing doubt on the unity of nature (ibid., 36). The "mind" in Descartes' system is therefore the same for the mind/brain relationship as is the hypothetical morphogenetic field or the speculative "life force" of vitalism for the life/body problem. Ontological dualism is justifiably rejected in *this – external – sense* because it breaches the laws of nature.

John Eccles (1903–1997) with his "dualistic interactionism" held a view of this kind and is therefore dismissed by most people. According to his model the mind *interacts* with the sensory and motor areas of the brain responsible for conscious experiences from which the normal neural circuits are then activated[109]. This is still a mechanistic view, even though it accepts the mind as an independent entity. The mind is presented like a piano player who remains totally outside his instrument, hammering externally (mechanically) up and down on its keys, after which the mechanical transfer to the strings releases the sound. "The most important statement by dualistic interactionism is that mind and brain are independent entities – the brain occupies World 1 and the mind World 2 – and that they acquire a mutual relationship through quantum physics" (Eccles, 1994, 27). However, Eccles is in no doubt that everything in this *mechanically* imagined interaction remains hypothetical: "Naturally we know nothing about the anatomy of this dialogue between the self and the brain" (ibid., 253).

This cannot be otherwise, since for observation of the psychophysiology, the chasm between the physical and physiological phenomena on the one hand and the psychological phenomena on the other still exists. This gulf naturally also applies to the ontological monism discussed above in its variant of physicalism (e.g. as per Roth or Searle) or to the psychophysical unity of emergentistic systems theory as postulated by Fuchs: "Conscious experience would then correspond to the highest emergent level of integration of the brain process, even if systems theory is really unable to say anything about the ontological nature of this 'leap' to the personal level" (Fuchs, 2005, 6). But what is meant here by "unity" or "correspondence"? Initially, simply that e.g. "every psychological disposition or disorder simultaneously represents a neurobiological disposition or dysfunction and vice versa" (ibid., 7). But this *"correspondence"* of physical and psychological phenomena which is actually present phenomenologically and has long been accepted as *parallelism* in its occurrence (Linke & Kurthen, 1988) does not result in its ontological *unity*. Because, in accordance with the phenomena and laws, this is actually not the case. Epistemologically there is nothing to prevent the recognition of the physical and psychological aspects as equally real. But how is their relationship to be understood in terms of causality, if both are real but do not "interact" in the sense of Descartes' and Eccles' dualistic interactionism?

---

109 According to Eccles' theory "every occurrence of a human intention brings about the firing of a group of neurons in the supplementary motor field in such a way that the 'correct' motor cortical neurons then discharge impulses via the different known pathways of the pyramidal tract and prompt the desired movement" (Eccles, 1994, 112). The mind itself is seen as a "non-material field" which, at a quantum physics level, "changes the likelihood of the emission of vesicles from presynaptic vesicle lattices". In this hypothesis Eccles sees a way of not having to disturb the law of conservation of physics because, according to the quantum physicist Margenau (1984) whom he cites, quantum physics provides aspects of immateriality: "[...] Some fields, such as the probability field in quantum physics, contain neither energy nor matter" (Eccles, 1994, 113–114).

And how is it then possible to maintain the justified demand for a unified (monistic) scientific view of the world?

5.2 The problem of ontological monism and psychophysical causation

For the unreflected experience of daily life it seems obvious that not only do bodily conditions affect our mental life but, conversely, mental occurrences can produce physical processes. In terms of ontological dualism, body and soul are perceived as two different entities which affect each other. It is the same in medical practice: "In medicine, which is a practical science, a strictly monistic interpretation of the brain/consciousness problem leads to a particular problem. We observe e.g. that a person sees something which angers them. This causes the person to turn red, in other words vasodilation of the facial skin, to an increase in blood pressure etc. If you wish to explain these physiological processes then it is difficult to avoid the interpretation that not only is subjective experience caused by brain function but it in its turn affects physical processes" (Handwerker, 1995, 195).

Consequently, i.e. for primarily practical empirical reasons and generally based on Engel (Engel, 1977), psychiatrists in particular always defend "an integral but nevertheless multi-dimensional and many-layered 'biopsychosocial' picture of the human being" against the reductionist molecular biological concepts of many brain researchers (Tretter, 2007, 504). From a system theory perspective this is also clearly viewed as a multi-dimensional "causative model" (Tretter, 2007, 498) but not in a linear causal way. According to Thomas Fuchs, for the "interactions" between the emergent layers of consciousness and neurobiology there is "never external causality or dualistic interaction between two worlds" but only a *"transformation"* within the overall psychophysical system in a bottom-up direction (e.g. induction of consciousness processes by psychotropic drugs) or in a top-down direction (e.g. induction of biochemical changes by psychotherapy) (Fuchs, 2005, 7). But what does this actually mean? The question is really how this "transformation" within the psychophysical unity, whose parts actually appear to observation as *two separate and different entities*, is *caused*. The transformation concept is therefore initially of no further help and the question of causality remains.

The body/soul problem thus poses the paradoxical question of how the causal relationship of the epistemologically clearly *different* and *dualistic* but *equally real* entities of physical and psychological phenomena can be understood in a *unified (monistic)* common scientific conception, without claiming these phenomena to be existentially the same (monistic) – which would contradict empiricism – and without resorting to an interaction which contradicts the laws of nature and therefore the unified overall view.

It appears that this question can be solved if we admit that there is not just *one* form of ontological monism and not just *one* type of causality. The current dominant ontological monism is – from an epistemological viewpoint – only directed at the percept or manifestation. Only here does it seek reality. Monism is then represented in such a way that what manifests as dualism is *declared to be* one. But for this to be achieved, one of the two (this usually happens to the psychological, non-spatial, inner element) must either be denied or added to the other in some way (usually therefore the physical, outer, spatial) or – due to the fact that they occur in parallel – included as a unity with it. In all such cases the prior dualistic character is of course not overcome and the declared monism not really achieved.

If we seek reality not only in the perceptual, manifest but, in the sense of ontological universal realism, also accept that its lawfulness is part of its reality, then monism appears to be achievable while retaining the existing diversity of the world of phenomena. For the *common* (monistic) element of the *different* (dualistic) phenomena no longer has to be enforced on the *side of the phenomena* – which is not possible – but actually occurs on the *side of the law* of what is cognised, even though the laws of the psychological and physical phenomena have their own distinct and specific *content*. But in terms of their *form*, all are *laws* in the same way, being made up of the same "substance" as it were. This, in Hegel's words, is pure spirit.

So the question of the relationship of the physical and spiritual does not appear for the first time with the brain/mind problem but already with the physical itself (cf. Section 3.2). And here all that can initially be said based on what can be observed epistemologically can be summed up as: the spatial phenomena of the physical *prove to be* constituted from their laws, the temporal processes of the living realm *prove to be* determined by theirs, the psychological phenomena *follow* psychological laws and the purely spiritual phenomena which are reached by thinking, the actual content of the laws, *follow* purely logical laws i.e. *are* of a logical nature. The fundamental element (the laws) of every level, the physical, organic, psychological and spiritual, can be reached in the same way by thinking, but the relevant phenomena can only be reached by very different observation methods and on different hierarchical levels of being. In terms of its *phenomena* the total being is therefore not dualistic but *multi-faceted*, in terms of its *lawful essence*, it is *monistic*. It is also through this lawful aspect that the constitutive or functional *relationships between* the disparate phenomena of the different levels of being can be found. This also applies to the physical and psychological phenomena. The entire world of phenomena therefore appears as being organised by the actual unifying element, the spirit. Spirit and phenomenon are only *two* (epistemologically dualistic) for our cognition which is separated into perception and thinking: in themselves they are *one* (ontologically monistic), as the spirit is the constitutive being of the phenomenon (cf. Section 2.5). And this is recognised in the actual act of cognition which relates the lawful essence and the phenomenon to each other (epistemological monism) (cf. Section 2.4). This is what characterises the other type of monism dealt with here, which was originally argued by Aristotle in his universal theory of form and matter and his view of cognition (Zeller, 1908, 172–189), by Goethe in the modern age and later by Steiner[110] and epistemologically justified by the last-named (Steiner, 1985b).

This monism includes other forms of causality than the outer ones beyond which Descartes and Eccles were unable to see with reference to the body/soul problem and which are rightly rejected for this area of being. The acknowledgement of physical and psychological phenomena as being equally real is

---

110 In the sense meant by Steiner, *monism* is initially the basic scientific approach of explaining *all phenomena* of the world by using an empirical principle of cognition. Steiner, however, distinguished different forms of monism. In his *Introductions to Goethe's scientific writings* he firmly opposes the *one-sided materialistic form "of that monism* which wants to found a unified view of nature – comprising both the organic and the inorganic – *by* endeavouring to trace what is organic back to *the same laws (mechanical-physical categories and laws of nature)* by which the inorganic is determined" and points out that Goethe likewise wants to wholly reject the mechanistic way of explaining the organic realm (Steiner, 1988b). Steiner and Goethe likewise reject the monistic explanation of consciousness and the life of the soul through their neurophysiological correlates. – Steiner himself argued a scientific point of view of a "one-world theory, or monism", as he described it in detail in his main philosophical work *"The Philosophy of Freedom"* (Steiner, 2011, 23, cf. also Chapters 2 and 7). However, this does not take place in a one-sided materialistic or spiritualistic form, but in a higher synthetic one which unites the material and the spiritual in an integrated view. Steiner's monism arises from his concept of knowledge and reality, i.e. from the ontological idealism or realism of universals which follows from the theory of knowledge. The full reality of a thing contains not only that which we can *perceive* of it, but also the *law* of what is perceived which we only grasp as a concept with our thinking. "It is due, as we have seen, to our organization that the full, complete reality, including our own selves as subjects, appears at first as a duality. The act of knowing overcomes this duality by fusing the two elements of reality, the percept and the concept gained by thinking, into the complete thing. [...] The world is given to us as a duality, and knowledge transforms it into a unity. A philosophy that starts from this basic principle may be called a monistic philosophy, or *monism*. Opposed to this is the two-world theory, or *dualism*. The latter does not assume just that there are two sides of a single reality which are kept apart merely by our organization, but that there are two worlds absolutely distinct from one another. It then tries to find in one of these two worlds the principles for the explanation of the other" (Steiner, 2011, 94).

not the same as claiming that these contrasting phenomena act outwardly on each other. An interaction of *this kind* – which of course cannot actually be observed (cf. Section 5.1) – would be like an interaction in mechanics: a causa efficiens would be applied *externally* to a causa materialis, etc.

A hierarchically ordered system with emergent layers in the sense of the monism referred to here does not arise due to an *external* effect of the higher-ranking *phenomenon* on the lower-ranking one or vice versa, but through an *inner* effect of the *laws* as has already been described for the hierarchically ordered world of substances and the structures of physical bodies and for the emergence of living phenomena and their laws (cf. Sections 4.8–4.9). Accordingly, the effect of each of the higher-order laws appears on the one hand in the creation of the emergent phenomena formed by them and on the other in the consequences which this has for the corresponding lower-order level: the higher-level law *supresses* it; the manifestation of its – now lower-order – laws in the phenomenon is *sublated* in the manifestation of the higher-order law. This counteraction can naturally only be brought about by something which exerts a force or action, which is why the result of this counteraction in what is acted upon is an index for the effectiveness or reality of the higher-order entity. In the world of substance and physical/spatial structure, this is demonstrated by the disappearance of the submergent properties in contrast to the appearance of the higher-level emergent ones; in the transition from the non-living to the emergence of life in the creation of a highly-ordered organic structure, which works actively throughout life against the decay of the body into disorder and the inorganic realm (cf. Section 4.11). And in the transition from the merely living to the emergence of consciousness, it appears again in the suppression of what is specifically living and vital in contrast to the appearance of consciousness.

The causative reality of the psychological realm is therefore demonstrated in the phenomena in two ways: first directly, in the emergent phenomena of the psychological realm itself and, second, indirectly in the suppression of the properties of life in the organism, in order to enable the appearance of consciousness.

5.3 Consciousness versus life: organic and psychological activity

In the context of the relationship between the body and soul there is in fact a rarely acknowledged but very revealing antagonism between the organic or vital functions on the one hand and the workings of consciousness on the other. This antagonism exists phylogenetically, ontogenetically, physiologically and in relation to functional anatomy: the higher the development of the nervous system or the workings of consciousness enabled by its processes, the *lower* is its regenerative power in a certain respect. This apparently contradicts the well-known fact that the human brain which enables the highest functions of consciousness, also has the highest metabolic rate. Although it only constitutes 2% of the body by weight, the human brain consumes 15–20% of the cardiac output and over 50% of blood glucose. But this very active metabolism is a regenerative *reaction* to the achievements of consciousness.

This reciprocal relationship between consciousness and vitality is shown phylogenetically in the ability to regenerate, for example. Lower invertebrates such as the almost 20,000 (!) species of flatworms have no central nervous system but only cerebral ganglia which, in evolutionary terms, are similar to the vegetative ganglia in human beings and accordingly only enable a low level of consciousness. But as a result they possess a huge regenerative capacity which enables them to regenerate a whole organism from each part of their body if cut in two at almost any point[111]. The evolutionary significantly higher amphibians already possess a basic central nervous system with the corresponding higher development of consciousness in comparison to Planarians, but for this they have forfeited vitality: they can, however, still regenerate whole extremities (Gross, 1974). Mammals, which have the highest capacity for consciousness in the animal kingdom, have lost the capacity for this kind of epimorphosis (complete replacement of lost body parts) and retain only a remnant in the form of a localised ability to heal wounds.

This contrast between the organism's capacity to create consciousness and that of vitality is also demonstrated in the ontogenesis of the individual and in the structural and functional tasks of its organs. As is well-known, growth and development take place in a cranio-caudal direction, with the head and neurocranium in advance and the extremities lagging behind, so that the skull in a two-month-old embryo forms half of the entire length of the body, a quarter of it at birth and an eighth at around 24 years (Husemann, 1994, 26). The growth rate is highest in the early embryonic stage and declines exponentially after birth, so that the final organic size is reached first in the neurocranium and last in the extremities. However, the decline in growth vitality is linked with increasing brain maturity, and the functions of consciousness awaken one by one in association with this. In full-grown mammals and human beings the neurons of the central nervous system are generally not able to regenerate after injury, mainly due to the growth-inhibiting molecules of the surrounding neuroglia and myelin sheaths, such as e.g. the protein Nogo associated with the membrane which is only expressed in the CNS but not

---

111 Cf. here and also for what follows the explanations on the Planaria, one of the four classes of flatworms, and on limb regeneration in the Amphibia (Section 4.7).

in the peripheral nervous system (Grandpré & Strittmatter, 2001). For this reason the latter definitely does remain capable of regeneration.

Just compare this to the regenerative power of the liver, which is actually similar to the almost inexhaustible vitality possessed by planarian worms, for example after a partial hepatectomy: "Liver manages to restore any lost mass and adjust its size to that of the organism, while at the same time providing full support for body homeostasis during the entire regenerative process" (Michalopoulos, 2007, 286). The liver is therefore justifiably called an organ of "life" in English and German ("liver" or "Leber")[112]. However, this organ of life does not produce consciousness but with its vitality occurs more or less in a "sleeping condition".

But even the brain needs to have regular sleep in order to maintain itself and recover from the efforts of consciousness. Sleep deprivation during brain development leads to a sharp increase in nerve cell mortality from apoptosis and to reduced brain mass (Morrissey et al., 2004). During sleep, increased amounts of growth hormone are released even in adults (Van Cauter et al., 2000) and sleep deprivation generally impairs the healing of wounds (Gumustekin et al., 2004). Even in flies, changes to the synapses which are caused by stress during waking are repaired (Donlea et al., 2009; Roberts, 2009). Cognitive learning processes even cause DNA double-strand breaks in neurons, which are completely repaired during sleep (Superbielle et al., 2013).

It has also become apparent that *consciousness is paid for with a loss of vitality on functional grounds*. This was something already pointed out by the German philosopher Karl Fortlage (1806–1881) in his *"Acht psychologische Vorträge"* (eight psychological lectures) (Fortlage, 1869, 35):

> When we refer to ourselves as *living beings*, thus according ourselves a property which we share with animals and plants, this living condition is necessarily

---

[112] What was stated above about the principle of the life processes in general is demonstrated particularly clearly by this regeneration process in the liver: these life processes are only made possible because the physical process cascades are *orchestrated* in space and time by a complex *active* process of organisation at a higher-level whole, meaning e.g. that both growth-promoting and growth-inhibiting factors have to be used at the right time and the right place and that individual types of cells need to change their function in certain circumstances in order to serve the regenerating organ as a functioning whole. "Liver regeneration after partial hepatectomy is a very complex and well-orchestrated phenomenon. It is carried out by the participation of all mature liver cell types. The process is associated with signalling cascades involving growth factors, cytokines, matrix remodelling, and several feedbacks of stimulation and inhibition of growth related signals. [...] In situations when hepatocytes or biliary cells are blocked from regeneration, these cell types can function as facultative stem cells for each other" (Michalopoulos, 2007, 286).

understood as something that never leaves us, and continues in us constantly both in sleeping and waking. This is the vegetative life of nourishing our body, an unconscious life, a life of sleep. The brain forms an exception to this in that, in the waking intervals, this life of nourishment, this life of sleep, is outweighed by a life of *consumption*. In these intervals the brain is exposed to a prevailing consumption and therefore ends up in a state which, if it were to extend to the remaining organs, would bring about a complete exhaustion of life or to death.

In fact all physiological processes which functionally serve consciousness directly are not constructive but *destructive* in nature, a process which Steiner – with reference to this description by Fortlage – described as an organic *"laming down"*[113] (Steiner, 1996, 149). For example, in the case of visual perception, *breakdown* of the previously produced rhodopsin structure occurs; the action potential from the visual image transmitted all the way to the CNS constitutes a controlled *breakdown*[114] of the membrane potential previously created by the expenditure of energy; and the neurotransmitters which have to conduct the nerve impulses over the synapses are *neurotoxic*, so that e.g. the damage to the organ in a brain embolism is not caused purely by a lack of oxygen but also by the accumulation of neurotransmitters which are no longer metabolised (Olney, 1987).

It could be objected that all the achievements of consciousness are associated not with a reduction of metabolic performance of the corresponding areas of the brain but an *increase*, and in fact this is the basis used by modern imaging techniques such as positron emission tomography (PET) or functional magnetic resonance imaging (fMRI) for assigning consciousness

---

113 In his book *"Riddles of the Soul"*, by *"laming down"* Steiner generally means "laming of our life in mental pictures" (Steiner, 1996, 19); meaning the transition of the mental pictures from a living to a non-living, "lamed down" state, when they become the conscious mental pictures of our normal consciousness through the activity of the nervous system. However, in extending the use of this term, Steiner also explicitly uses this expression to denote the "consuming life" of the *brain* mentioned by Fortlage, to which Steiner says: "termed by me 'laming down' in this book" (Steiner, 1996, 149). In the overall context of the text there it is clear that this – what I have here called "organic" – laming is the reason for the laming down of the mental pictures through which they become conscious. The organic laming is identical with the breakdown discussed in this chapter as a requirement for consciousness. Re the relationship of living force-bearing mental pictures or ideas to non-living ones, cf. also Chapter 7.
114 Cf. the description by Roth: "The occurrence of an action potential is in fact 'free' as the associated depolarisation of the cell membrane occurs along the concentration gradient and only uses energy which was produced previously. In contrast, restoring the membrane resting potential of -70 millivolt after a burst of action potentials – through the work of the sodium/potassium ion pump – is energetically expensive" (Roth, 2001, 196).

functions to areas of the brain. We must certainly agree with this, but this metabolism is *reactive*[115], to *compensate for* the consumption caused by the achievements of consciousness (cf. Vanzetta & Grinvald, 1999). Leopold states (Leopold, 2009, 387, ePH):

> Housekeeping in the brain is a challenge, because neurons undergo sudden bursts of activity that *consume* energy and pollute their surroundings. Consider, for example, what happens in the cortex when we first direct our gaze to a bright stimulus. In the visual cortex [...] thousands of previously quiescent neurons suddenly erupt in a cacophony of activity, each generating hundreds of electrical impulses per second. In *response* to the metabolic consequences of such activity, fresh blood is directed towards neurons and glia in active regions, flushing out waste products, delivering nutrients and *restoring* the local milieu.

Particularly in the case of higher mammals and human beings, their – highly developed – consciousness therefore needs to be regularly interrupted by sleep in order to regenerate the brain due to the demands made on it by the devitalisation and breakdown inherent in the workings of consciousness.

In other words: although consciousness develops *thanks to* the living organism, it works *in opposition* to it. The brain, with the system of senses connected to it *enables* consciousness, but not as its *causal agent*[116] but its *opponent* (Heusser, 2000a). Human beings and animals do not experience *through* but *in opposition to* the body. This then leads to *"experience"* (*"Erleben"*) instead of *"life"* (*"Leben"*). And this experience is emergent in comparison to simple life and has its own laws and forces. Logically it is *these* which must be the cause of the psychological phenomena, and the physical-etheric organisation and its "laming down" is the *condition* for their occurrence. To this extent, as far as the body/soul question is concerned, there is justification not only for

---

115 It does not contradict the fact that there is also an *anticipatory* increase in the local blood flow to areas of the brain in which, due to prior conditioning (!) by sensory stimulating elements, an activation brought about by consciousness, i.e. an increased metabolic demand *caused by this*, are to be *expected* (Sirontin & Das, 2009).

116 This also applies to e.g. the physiology and psychology of the senses. Sense organs, nerve fibres and brain create neither the objective content of perception nor the content of the images inherently linked to these, and neither do they create the subjective psychological form in which this complex of perceptions and images appears as a conscious experience in the individual (cf. Section 2.1). The whole physical process cascade between the object and the central nervous system which occurs during the sensory picturing experience serves the *mediation* of this experience. It is the *material condition* for the conscious *external* experience of an objective perception linked to the subject's inner picture (Heusser, 2000a). The lawful *psychological* activity can be seen as the *causative* factor of the psychological phenomena of sensation and imagination.

epistemological but also *ontological* dualism[117] (or rather: a *multi*dimensional theory of causes). This is the sense in which Steiner's description of the difference between the plant and animal in his introductions to Goethe's scientific writings is to be understood (Steiner, 1988b, 62–67):

> The plant differs from the animal in its lack of any *real* inner life. This last manifests in the animal as sensation, arbitrary movement, etc. The plant has no such soul principle. It still consists entirely of its externality, in its form. [...] Something different occurs now, to be sure, in animal life. Life does not lose itself here in its external features, but rather separates itself, detaches itself from its corporeality and uses its corporeal manifestation only as a tool. It no longer expresses itself as the mere ability to shape an organism from within outward, but rather expresses itself within an organism as something that is still there besides the organism, as its ruling power.

But this "ruling power" of the soul lives so to speak *at the cost* of the living, which can also be seen as an indirect indication of the active power of the psychological realm.

5.4 The soul as a formative principle. The "astral" organisation

In the case of animals and man, the soul life not only has a physically *destructive* function which enables the emergence of *consciousness* but also a *constructive, formative and developmental* function in the body. The animal's body is not used by the soul merely as a tool for the fulfilment of the inner needs contained in this soul, but is also *formed for this purpose*[118], i.e. *according to these needs*. This means that the animal typus, in contrast to that of the plant, contains not only the laws of the *outer* spatial and temporal form, but also those of the inner, psychological organisation, as Goethe so beautifully expresses in his *Metamorphosis of Animals* (Goethe, 1915, 256):

> Truly's each creature itself its own purpose, for nature creates it
> Perfect; and it in its turn begets progeny that will be perfect.
> Organs and members are shaped according to laws everlasting,

---

117 This "dualism" or multidimensional theory of causes is, however, not to be taken in an external mechanical or extensive spatial sense as in the interactionism of Descartes or Eccles, but in an inner dynamic and intensive form such as comes about e.g. in the ontological idealism argued by Nicolai Hartmann, i.e. as an antagonism of active laws in the hierarchically arranged system of nature. This applies especially to the expression "besides the organism" in the following quotation from Steiner.

118 "For this purpose" is meant here in the sense of the "inner" teleology or teleonomy (cf. Section 4.3).

Even the oddest formation its prototype latent preserveth.
Thus each mouth is adapted to seize the right food and to swallow
That which is fit for its stomach, – the one may be tender and toothless
While there are others with powerful jaws; but one organ
Always for proper nutrition will cooperate with the others.
Also the feet to the needs of the body are wisely adjusted,
Some of them long, while others are short, in perfect proportion.

In contrast to the plant, the animal therefore has an inner psychological centre which is served by every organ, *but in accordance with which this organ also has to be created* so that it can in fact serve the needs of this centre. Steiner therefore commented on these thoughts as follows: "In the case of the animal, every organ appears as coming from that center; the center shapes all organs in accordance with its own nature. The form of the animal is therefore the basis for its external existence. This form, however, is determined from within" (Steiner, 1988b, 67).

"From within" naturally does not mean "from consciousness" but from the lawful "being" of the psychological realm which, on the one hand, occurs in a formative capacity – *without consciousness*, similarly to plants – and on the other is able to produce consciousness in opposition to created form, as described above. Adolf Portmann (1897–1982) called this inner authority which gives the animal its form rather imprecisely but appropriately "inwardness" (Portmann, 2006, 69):

> We must immediately separate this inwardness from consciousness. Consciousness is a special capacity which is one of many types of inwardness. In itself this extends far beyond the sphere of consciousness. Recognition of unconscious inwardness by psychoanalysis has required many battles. Biologists, even when they did not employ this word, have long devoted themselves to the explanation of the unconscious workings of inwardness. The whole subject of instinct is only one part of this large topic area.

And Portmann draws attention to the fact that this inwardness is *not a spatial* principle, although it is involved in the creation of the spatial organisation of the body, and that it must be active not only *in the brain* but in the *whole organism*: "No one is able to pin it to one place: even if we attach the greatest importance to one organ such as the brain, nevertheless all the parts of the whole are involved in the entire inwardness" (ibid., 69) such as, when a planarian which has been cut in two regenerates its body and produces e.g. a new ganglia system, but also when the human being is formed in the womb (ibid., 70, ePH).

> The embryo, which measures a tenth of a millimetre at the beginning of this development, performs once only what we saw the worm perform when regenerating its body: it also produces its own brain. This development makes it

possible for a conscious ego to arise in this basic structure. However, we are all agreed that *this* ego is not the creative source – the ego is a possibility which depends on the reality of the self.

This "reality of the self", i.e. the inwardness which is involved in forming the body, can be seen as the real reason for the enormous "physiognomical" morphology of the animal form. This is – to use Portmann's words – *in itself* already an outward "expression" of inwardness: it appears "in the service of portrayal", "self-portrayal and self-enhancement", and therefore of social behaviour. As this outward expression, it is therefore of great significance in the interrelationship of sentient beings (Portmann, 1982). The inward being of the soul is therefore given expression as a phenomenon which has become "physical" so to speak. This process is understandable from the viewpoint of ontological idealism in that the lawfulness of the inward realm brings its "form" in the living organism into effect as its "material". Thomas Aquinas expressed it as follows: "The soul is therefore the form of the living body" (Aquin, 1977, 57, d. an. 2, I. 412 a 27).

Man and animal are therefore not just physical and living beings, but also *permeated by inwardness, by a soul* at their different levels of evolution. Inwardness here should not be understood simply as the *conscious* psychological element which is known in common parlance or in psychosomatics as *psyche* or "psychic authority" (Adler, 2005, 10). Conscious mental activity is dependent for its awareness on the brain (cf. Section 5.3) and is therefore, if it is regarded as a real self-active entity at all, often "positioned" in the brain (Eccles, 1994). However, as psychoanalysis has clearly demonstrated, the psyche is largely unconscious or subconscious, and in terms of its "location"[119] must be thought of as much more extensive, specifically as the inward psychological organisation acting throughout the *whole* organism. It is *this* comprehensive organisation of inwardness – and not the "psyche" – which imprints the human or animal organism physiognomically and organically with a "soul" and gives the above-mentioned purposeful orientation of the organism towards the conscious inner life of the "psyche".

In order to make a clear distinction between this complete organisation of inwardness which causally acts in the body and the usual concept of the "psyche", Steiner chose a different terminology also borrowed from ancient nomenclature, i.e. the *"astral"*, the *"astral organisation"* or the *"astral body"* (Steiner & Wegman, 1996). According to this, animals and man have not only

---

119 As already discussed above, the soul and spirit are not located at any specific point in the body, as they are not of a spatial nature. However, the spatial body provides them with their consciousness and this spatial *reference* is perceived as a location, a site from whence the world is experienced.

a physical and an etheric body like the plants, but also an astral body[120]. This is the bearer of everything in the soul, whether unconscious or conscious, be it constructive or destructive to the organs. And the same applies by analogy to the human spirit which forms the core or the actual "ego" of the human soul and gives this its self-consciousness (cf. Section 5.5). The soul of the animal or human being therefore unconsciously creates its instruments in the body in order to be able to use these for its conscious life (ibid., 9):

> The astral body must build up its organization within the physical and the etheric; the ego must do the same with regard to the ego organization. But in this *building* there is no conscious development of the soul life. For this to occur a process of *destruction* must oppose the process of building. The astral body builds up its organs; it destroys them by allowing the soul to develop an activity of feeling within consciousness; the ego builds up its "ego-organization"; it destroys this, in that will-activity becomes active in self-consciousness.

It goes without saying that the formative function of the soul and spirit has to precede the one which creates consciousness, because only *that part* of the life of consciousness can be developed for which the necessary living physical organisation has been produced. As a result, the soul and spirit are already active in an unconscious form in the body *before* they can appear as a life of soul in consciousness. This was stated back in 1859 by Immanuel Hermann Fichte in *"Mental Philosophy"* (Fichte, 1860, 12–17, 38):

> In order to explain the *conscious* phenomena of the soul (which is the proper aim of psychology), I hold that we must start from its *unconscious* state [...]. The real or actual consciousness is based upon a potential one, i.e. upon a *middle condition* of the soul, in which it already possesses the specific character of *objective* intelligence, but without being conscious of it. [...] The whole of the preconscious state of the soul is essentially and specially a process of *thinking*, without, however, its thought as yet touching the threshold of consciousness. [This is an] intelligence which develops [...] forms in space, and which stands in manifest analogy with the art-instincts of animals, and the creative aesthetic faculty in man, which can well be called a *'formative imaginative activity of the soul'*. [...] It is in a peculiar sense the *'fancy'*, – intelligence in its instinctive operation, which shows itself already in these preliminary forms; but which also constitutes the bond between the conscious and the unconscious region of the soul, inasmuch as it can be traced downwards in regular succession from

---

120 The expression "body" is naturally not to be taken in a physical sense when used in "etheric body" and "astral body", but in the sense of "organisation" or "formation" of life forces or psychological forces in the sense intended here. Cf. Steiner: "By the word 'body' is meant whatever gives a being shape or form. The term body must not be confused with a bodily form perceptible to the physical senses. Used in the sense implied in this book, the term body can also be applied to such forms as soul and spirit may assume" (Steiner, 1970a, 28).

the highest power of artistic creation, through the intermediate steps of dream-life in its many forms of presentiment and of instinctive action, to the space-constituting intelligence of the soul, as seen in the organic processes.

Life becomes experience, formative power becomes "fancy" (ibid., 38). It must be borne in mind that the living formative activity cannot be brought about by the soul (by the astral body) alone, because in all living creatures the processes of life and development of form are carried out by the *etheric body*. In the animal this must join to the laws of the soul and in the human being to those of the soul and spirit, if a physical development *appropriate* to the animal or human being is to come about. And the metamorphosis from life to experience involves not only a shift of the *soul and spirit's activity* from the organic realm to consciousness, but also a shift of the *etheric living* activity which serves the soul and spirit, from the organic realm to the formation of consciousness. This is what Steiner gives as the real reason for the above-mentioned decline in vitality when consciousness arises in the course of phylogenetic and ontogenetic development and with reference to psycho-physical functional relationships (cf. Section 5.3): it is as though etheric forces are *removed* from the body, i.e. the organs which mediate consciousness, for the benefit of the functions of consciousness. This corresponds to Steiner's and Wegman's description in *"Extending Practical Medicine. Fundamental Principles based on the Science of the Spirit"* (Steiner & Wegman, 1996, 6):

> At the beginning of a human life on earth – most clearly so during the embryonic period – the forces of the etheric body act as powers of configuration and growth. As life progresses, a part of these forces becomes emancipated from activity in configuration and growth and is transformed into powers of thought, the very powers that create the shadowy thought world we have in ordinary consciousness.
> It is of the greatest importance to know that ordinary human powers of thought are refined powers of configuration and growth. A spiritual principle reveals itself in the configuration and growth of the human organism. And as life progresses this principle emerges as the spiritual power of thought.
> And this power of thought is only one part of the power of human configuration and growth that is at work in the etheric.
> The other part remains faithful to the function it had at the beginning of human life. Human beings continue to develop when configuration and growth have reached an advanced stage, that is, to some degree a conclusion, and it is because of this that the non-physical, spiritual etheric which is alive and actively at work in the organism is able to become power of thought in later life.

The etheric organisation can therefore be viewed as a kind of missing link in the body/soul problem and in psychosomatics. Psychosomatics lacks the *life* between the psyche and the soma, so it has no real concept of how the body and soul should interact (Adler, 2005). However, the concept of the etheric enables an understanding of the *phenomenology* of psychological and especially of psychopathological phenomena, which cannot be interpreted on a psychological and neurophysiological basis alone. An example of this is the physical disorders associated with schizophrenia such as a bodily sensation of a current and similar body-associated phenomena, which can be meaningfully interpreted using the concept of the etheric body (Andres, 1992).

The same applies to the well-known phenomenon of *neuroplasticity*, the transformation of neural structures as a result of activity. "Activity" here means *mental* activity, the exercise of *functions of consciousness* in sensory experiences, memory, voluntary motor functions, etc. While all these functions are associated with the destruction in the nervous system described above and assume that a corresponding physical structure has previously been built up (cf. Section 5.3), they in turn result in a regeneration or renewal of the structure that corresponds to the specific *functional requirements of that mental activity* and thus allow for new acts of consciousness of this kind on the basis of an improved physical structure and function.

In recent years, research on neuroplasticity particularly in the area of the hippocampus has shown that the different growth factors such as brain-derived neurotrophic factor (BDNF), insulin-like growth factor-1 (IGF-1) and vascular endothelial-derived growth factor (VEGF) are responsible for this new organisation which usually begins after a few days of repeated mental activity (practice) and can persist for weeks. These growth factors induce gene activation and protein production in the neurons, leading to the sprouting of neurons but also to angiogenesis and increased metabolism to the extent that these are required for maintaining the new structures (see overview in Cotman et al., 2007). The biological long-term effect of short-term events in consciousness can be induced by an activity-dependent epigenetic regulation of these kinds of growth factors. For instance, an epigenetic regulator protein (Gadd45b) has been identified in mice whose gene belongs to a gene family whose products are known to be responsible for reactive processes such as DNA repair and adaptive immune reactions. The neuronal activation triggers the Gadd45b gene whose protein leads to a DNA demethylation of growth factors such as BDNF and fibroblast growth factor-1 (FGF-1), which are in turn responsible for the neurogenesis (Ma et al., 2009).

This is the physiological and molecular biological level which, however, cannot be understood mechanically but only systemically: all these processes in the nerves, metabolism and vascular supply have to be precisely coordinated as a complete spatial and temporal event and therefore require an *active* systemic *organisation* at a *higher level* than the genetic and other events (cf. Sections 4.7–4.8). "Growth factors *orchestrate* most, if not all, of the brain responses to exercise through either direct or indirect effects" (Cotman et al., 2007, 468, ePH). It is clear that these growth factors themselves also have to be orchestrated for this purpose: "These growth factors work *in concert* to produce complementary functional effects, modulating both overlapping and unique aspects of exercise-related benefits in brain plasticity, function and health" (ibid., 466, ePH). A systemic living level (that of the etheric organisation) is at a higher level than the molecular one (level of physical organisation) and this living level in turn produces the structure for the psychological functional needs (astral organisation), in other words follows the laws of the living and the psychological which, in the case of human beings, is further permeated by their spiritual ego. The molecular biological events described can therefore be seen as an expression of the physical development activity of the soul and spirit whose aim is to build the physical instrument for the corresponding functions of consciousness.

The viewpoint of objective ontological idealism developed in the previous chapters and applied to chemistry, biochemistry, biology, morphology and psychophysiology leads to the conclusion that neither the human brain nor the remaining physical organisation can be viewed as a type of purely molecular biological and electro-physiological machinery in its functioning, as is almost always stated in the text books. Instead, molecular laws and forces are sublated and directed by biological laws and forces and these in turn are sublated and directed by the laws and forces of the soul and spirit.

5.5 Spirit versus soul: the emergence of self-consciousness and self-determination

The human being is not only a living physical being with a soul like the animals, but is also endowed with an individual thinking self-conscious and self-determining *spirit*. Everyday experience shows that the human being consists of body, soul and spirit, as Steiner illustrates with a simple example in his book *"Theosophy"* (Steiner, 1970a, 18–19). Particular attention should be paid here to the difference between soul and spirit:

> I cross a meadow covered with flowers. The flowers make their colours known to me through my eyes. That is the fact which I accept as given. I rejoice in the splendour of the colours. Through this I turn the fact into an affair of my own. Through my feelings I connect the flowers with my own existence. A year later I go again over the same meadow. Other flowers are there. New joy arises in me through them. My joy of the former year will appear as a memory. It is in me; the object which aroused it in me is gone. But the flowers which I now see are of the same kind as those I saw the year before; they have grown in accordance with the same laws as did the others. If I have informed myself regarding this species and these laws, then I find them in the flowers of this year again just as I found them in those of last year. And I shall perhaps muse as follows: "The flowers of last year are gone; my joy in them remains only in my remembrance. It is bound up with *my* existence alone. That, however, which I recognised in the flowers

of last year and recognise again this year, will remain as long as such flowers grow. That is something that has revealed itself to me, but is not dependent on my existence in the same way as my joy is. My feelings of joy remain *in me*; the laws, the *being* of the flowers remain outside me in the world."

Thus man continually links himself in this threefold way with the things of the world. One should not for the time being read anything into this fact, but merely take it as it stands. There follows from it that man has *three sides to his nature*. This and nothing else will for the present be indicated here by the three words *body, soul, and spirit*. Whoever connects any preconceived opinions, or even hypotheses with these three words will necessarily misunderstand the following explanations.

Steiner does not provide a definition of body, soul and spirit here but points out the three areas in which these manifest and which can be clearly distinguished from each other empirically, in which the properties of body, soul and spirit can be *experienced* (ibid., 19):

By *body* is here meant that through which the things in man's environment reveal themselves to him; as in the above example, the flowers of the meadow. By the word *soul* is signified that by which he links the things to his own being, through which he experiences pleasure and displeasure, desire and aversion, joy and sorrow in connection with them. By *spirit* is meant that which becomes manifest in him when, as Goethe expressed it, he looks at things as a 'so to speak, divine being'[121].

In terms of the above example of the flowers, this refers to the intelligibility of the so-called "eternal" laws as being independent of any specific spatial and temporal presence. What then becomes manifest is the spiritual content of the laws on the one hand and the spiritual activity which reveals this content on the other (ibid., 19–20):

In this sense the human being consists of *body, soul* and *spirit*.
Through his body man is able to place himself for the time being in connection with things; through his soul he retains in himself the impressions which they make on him; through his spirit there reveals itself to him what the things retain for themselves. Only when one observes man in these three aspects can one hope to be enlightened about his nature. For these three aspects show him to be related in a threefold way to the rest of the world.

Through his body he is related to the objects which present themselves to his senses from without. The materials from the outer world compose this body of his; and the forces of the outer world also work in it. And just as he observes the things of the outer world with his senses, so he can also observe his own bodily existence. But it is impossible to observe the soul-existence in the same way. Everything in me which is bodily process can be perceived with my bodily

---

121 Cf. the extensive quotation of this point in Goethe in Section 3.1, Note 47.

senses. My likes and dislikes, my joy and pain, neither I nor anyone else can perceive with bodily senses. The region of the soul is one which is inaccessible to bodily perception. The bodily existence of a man is manifest to all eyes; the soul-existence he carries within himself as his *own* world. Through the *spirit*, however, the outer world is revealed to him in a higher way. The mysteries of the outer world, indeed, unveil themselves in his inner being; but he steps in spirit out of himself and lets the things speak about themselves, about that which has significance not for him but for *them*. Man looks up at the starry heavens; the delight his soul experiences belongs to him; the eternal laws of the stars which he comprehends in thought, in spirit, belong not to him but to the stars themselves. Thus man is citizen of *three worlds*. Through his *body* he belongs to the world which he also perceives through his body; through his soul he constructs for himself his own world; through his spirit a world reveals itself to him which is exalted above both the others.

This is the world of spirit which determines the world of phenomena as its lawful essence and which in turn can be made consciously manifest by the human spirit in the form of ideas.

Evolution has progressed to the human being and, in him, to developing a spirit beyond the purely material, living and psychological levels. Dieter Wandschneider therefore states (incidentally just as did Hegel): "In it [the human mind] nature transcends itself as nature and gains at the same time the potential to reveal the essence underlying nature" (Wandschneider, 2005, 208). There is therefore no contradiction between the acceptance of naturalistic evolution as described by Darwin on the one hand and an understanding of the same evolution in the sense of objective idealism on the other (cf. Section 2.6). As Wandschneider says (ibid., 211):

> The laws of nature are rather like the logic determining the process of nature: they do not exist in time and space like a stone or an earthworm but possess *ideal* character. This ideal, then, is what was described above as the essence of nature underlying the natural being. In short, the concept of natural evolution implies that nature contains potentials that, in turn, stem from the laws of nature and thus form an ideal basis of nature – an inevitable metaphysical premise[122] for all natural sciences.
>
> [...] This much can be said: we are forced to assume something like a logic underlying nature in the form of natural laws, and this logic leads to the evolution of the faculty of cognition, which in turn is capable of directing itself toward nature and of penetrating and understanding it logically. All in all, this points to an (objective-) idealistic[123] ontology of nature.

---

122 However, this comment on the ideal and therefore spiritual basis of nature cannot be a *premise* but is the *result* of the theory of knowledge. Cf. Chapter 2.
123 Wandschneider states his standpoint of objective ontological idealism more precisely in the next sentence: "Mind you: *objective* here designates an idealism not

This allows us to acknowledge the empirically obvious *fact* of naturalistic evolution but still reject the claim that man is nothing but a more highly evolved animal. For even if the human being has arisen from animal-like earlier stages of himself, what makes him human is his spirit, an *emergent* principle which cannot be explained *from* the physical, living and mental nature of his preliminary forms but which gradually becomes a reality *in* these. This spiritual principle in its state *before* being realised can then be called "universale ante rem", or similarly a "spiritual ancestor" of the human being, just as its early animal-like forms are its "natural ancestor". Steiner formulates this as follows (Steiner, 1988a, 4):

> The evolution of the world is then to be understood in such a way that the preceding unspiritual, out of which the spirituality of man later unfolds itself, contains something spiritual above and beyond itself. The later, spiritualized sense-perceptibility in which man appears thus arises through the fact that the spirit ancestor of man unites himself with the imperfect, unspiritual forms, and, transforming these, then appears in sense-perceptible form.

Looked at in this way, the human spirit is *added* by degrees as a self-realising "universal" from the condition of "ante rem" into that of "in re" to the already realised "in re" laws of the physical, living and mental spheres and marks them with its typically *human* form through which the human being differs both morphologically and psychologically from the animal.

This explains why the human organism is based on the same morphological basic typus as the higher mammals on the one hand but on the other assumes a *unique position* amongst these which is specifically human. Because the typus concept – as Goethe wrote to Knebel when he sent him his treatise on the discovery of the intermaxillary bone in a letter dated November 1784 – leads to the consequence "that one can in fact *not find the difference between the human being and the animal in anything singular*. On the contrary, the human being is very closely related to the animals. *The agreement of the whole* makes every creature into what it is, and the human being is human just as much due to the form and nature of his upper jaw as through the form and nature of the last joint of his little toe" (Goethe, 1958a, ePH). Steiner has shown that Goethe's discovery of the intermaxillary bone in human beings was *a result* of his thoughts on the morphological connections in the realm of nature which he conceived in discussion with Herder (Steiner, 1988b). In his *"Outlines of a Philosophy of the History of Man"*, Herder had already pointed out that

---

> of Berkeleyan but of Hegelian type, for which there are, I think, good reasons and to which we owe probably the most well-thought-out philosophical concept of nature that occidental philosophy has brought forth" (Wandschneider, 2005, 211).

the upright stance required in the human body is the fundamental requirement for the reasoning activity of the human being (Herder, 2014; Steiner, 1988b). In fact the entire anatomical and functional organisation of the human being is designed in such a way as to enable the specifically human i.e. *spiritual* and *culturally creative* functions to develop in thinking, speaking and action (Kipp, 1948; Poppelbaum, 1981; Schad, 1985b). The specifically human elements of the anatomical characteristics are already visible in the embryo (Schad, 1982b); and it can be shown that the higher mammals, apes and primates in particular develop an embryonic morphology which is more like that of *humans*, such as e.g. in the morphology of the skull (Schindewolf, 1972), which is also anatomically designed for an *upright stance* e.g. in terms of the ratio of the width to the depth of the thorax and the position of the foramen magnum at the base of the skull (Verhulst, 1999). This is lost as the animal grows (e.g. in the skull morphology[124] and through the relative reduction in the width to depth of the thorax, corresponding to quadrupedalism), so that the morphology of the adult becomes completely typical of the animal. From a paleontological angle, ontogenesis of the skull in humans and anthropoid apes indicates "a base form which must have been more anthropoid than the recent anthropoid apes" (Schindewolf, 1972, 272). The naive Darwinist theory that human beings are descended from apes can therefore be countered by the empirically justifiable hypothesis that apes can be viewed as *offshoots* or even as a "retrograde evolution" (ibid., 272) of phylogenetic primary forms of the *evolutionary line to the human being* (Verhulst, 1999). Although these must be thought of on the animal level, this does not equate to the present-day forms of animals.

In terms of ontological idealism, the evolution of man should be thought of as the gradual imprinting of the specifically human spiritual principle into those pre-human organic forms through which they were transformed into the higher emergent *human* organisation. This in turn is the precondition

---

124 Cf. the description by Schindewolf (Schindewolf, 1972, 266): "The great similarity to the human being in this juvenile stage gradually recedes during ontogenesis and gives way increasingly to forms typically characteristic of the mammals. In a reversal of the proportions, the mature skull then displays an absolute dominance of the facial skull with its extremely elongated snout in contrast to the brain capsule with its completely retarded growth. The forehead is receding, the frontal bone has lost its obvious curvature. The temporal line is raised above the calvarium and lies on a sagittal crest; there is also a prominent nuchal crest. The eye sockets slope backwards and are protected by a supraorbital boss. The foramen and condyles lie on the posterior edge of the base of the skull and are angled backwards and downwards. The rows of teeth have assumed an elongated rectangular form with almost parallel sides and the canines are enormously enlarged".

for the human spiritual activity through which they work their way up from purely *natural* beings to *cultural* ones. "In this manner the biological principle of survival is 'unhinged': cultural principles take its place (which may well be analogous to the principle of survival) in the course of *a new form of evolution, the cultural evolution of man*" (Wandschneider, 2005, 211, ePH). Looked at in this way, cultural evolution can complement natural evolution in a certain way (Wandschneider, 2005, 208):

> Now, man himself is a child of nature. Nature thus appears to have brought forth, in humanity, a being that is capable of understanding nature [...]. In the form of man, evolution has developed the organ of cognition that nature itself lacks. Only the activity of man's cognition can make visible what nature is in its essence. From this perspective, mind appears as the completion and perfection[125] of nature, which, as we have said, also manifests in the completely new possibilities of technology in contrast to those of "biologically developed" nature.

But this involves a responsibility, because simply releasing the spiritual man from the bonds of nature does not guarantee that he will form a harmonious relationship to nature. Through cognition human beings bring to light the laws of nature. Through their actions, in the creation of culture and technology, they bring to light new ideas and laws and make them reality e.g. as works of art or machines. Whether the relationship of the creations of culture are in harmony with those of nature is not always examined by any means. However, as a spiritual being with the faculty of intellectual judgement, the human being is responsible for his actions, as indicated by Wandschneider (ibid., 208):

> As mind, man has the ability to oppose nature and to pervert and destroy it, yet nature has completed itself only in the human mind. Both results come from the same root: the ability to understand the underlying essence of nature. Animals can perceive only the outer appearance of nature. But man's ability to understand the essence is simultaneously his liberation from natural limitations and the negation of naturalness.

---

125 Steiner describes the human task of cognition of nature specified here as "to bring the world process to a conclusion" (Steiner, 1988a, 101): "If man were a mere sense being, without spiritual comprehension, inorganic nature would certainly be no less dependent upon natural laws, but these, as such, would never come into existence. Beings would indeed then exist that perceived what is brought about (the sense world) but not what is bringing about (the inner lawfulness). It is really the genuine and indeed the truest form of nature that comes to manifestation within the human spirit, whereas for a mere sense being only nature's outer side is present. Science has a role of universal significance here. It is the conclusion of the work of creation. It is nature's coming to terms with itself that plays itself out in man's consciousness. Thinking is the final part in the sequence of processes that compose nature."

5.6 Soul and spirit: Intelligence in animals and humans

Creative feats of intelligence are also often attributed to animals and questions raised about the important difference between human beings and animals due to the former's mental ability. This applies in particular to the limited ability of chimpanzees to use tools creatively, to learning associated with specific situations, to the symbolic use of words and to the small logical feats associated with this[126]. However, these achievements generally require a natural or artificial *conditioning* and the *triggering* by sensory objects which are actually present; and it has remained a matter of debate as to whether this constitutes *pure i.e. abstract* feats of intelligence. After decades of research on animal intelligence, James and Carol Gould conclude as follows (Gould & Gould, 1994, 190–191, ePH):

> Certainly chimps in the wild understand similarity and difference, analogies, cause and effect, and a great deal more besides. They are able to formulate fairly complex and subtle plans. *The issue is whether they accomplish their thinking and planning solely by manipulating visual images* – actual or mental pictures of objects – *or if they also have more abstract conceptual tools available.* [...] Perhaps there is a middle ground, in which generic category icons are used in visual thinking. Concepts would be manipulated mentally in some way, and thus would probably have some kind of representation independent of a specific canonical example. A model based on *something like mental icons – generic category images* that might act as a mental alphabet for objects and actions and characteristics – would help explain the ease with which chimpanzees form and use conceptual categories, as well as their limited ability to learn language. Certainly there is no compelling evidence that language-trained chimpanzees are quicker at solving real-world problems than their untutored peers; abstract words and symbolic thinking, so essential to much of our culture-based behavior, do not seem to be powerful intellectual tools for chimpanzees except in situations specifically designed by researchers to make word-use necessary.

But "mental icons" or "general category images" would be *pictures*, possibly mental images. While these *contain* laws as a picture, the lawfulness does not appear in a pure, abstract non-pictorial form. From an epistemological viewpoint, pictures are mental *perceptions, psychological phenomena,* whose *meaning* first has to be found and understood by thinking[127]. This would point to the fact that even the most intelligent higher mammals such as chimpanzees only experience intelligently structured representations and can behave

---

126 For example, by indicating the relationship between two objects such as corks and bottle tops as "the same" or "not the same" in which, when questioned, the animal had to point to appropriate symbolical "words" in the form of plastic shapes (Gould & Gould, 1994, 189).
127 Compare what is said in Section 2.3, Note 24 on Kekulé's pictorial experiences.

intelligently as a result, but that they cannot, unlike the human being, make manifest *what is intelligible as such, i.e. as pure concepts*, the pure *spiritual* content, which underlies the inorganic, the organic and their own psychological world. They lack the faculty of abstraction, an intellectus sensu strictu, an independent spiritual faculty, which is necessary to experience what is intelligible as such, i.e. pure thoughts, abstract ideas, and the laws of nature.

This appears e.g. in the intelligent use of tools which is ascribed to higher mammals and some species of birds in particular and which has been studied e.g. in the anthropoid apes in a variety of experimental settings. Through associative learning, the animals are in fact able to create and use simple tools (e.g. made from branches) to get hold of food. People therefore often believe that the animals "understand" these actions and Tomasello also makes these kind of statements in places in his most recent book (Tomasello, 2014). However, Tomasello makes it clear that this "understanding" consists of a mental representation of intelligent *"antecedent-consequent relations"* (which as is well known can also be attained by conditioning) and not in a *logical* understanding of the *causal relation of forces* which apply to the use of the tool (ibid., 16–17). Tomasello in fact states elsewhere (Tomasello, 1999, 23):

> But, in addition, the understanding of intentionality and causality requires the individual to understand the mediating forces in these external events that explain "why" a particular antecedent-consequent sequence occurs as it does – and these mediating forces are typically not readily observable. This understanding seems to be unique to humans.

In 2008 a comprehensive review by Penn and his team re-analysed the evidence for a wide range of "human-like" cognitive abilities in various species of animals at a functional level and compared this with human intelligence. The authors draw the following conclusions (Penn et al., 2008, ePH):

> Across all these disparate cases, a consistent pattern emerges: Although there is a profound similarity between human and nonhuman animals' abilities to learn about and act on the perceptual relations between events, properties and objects in the world, only humans appear capable of reinterpreting the *higher-order relation* between these perceptual relations in a structurally systematic and inferentially productive fashion. In particular, only humans form *general categories* based on structural rather than perceptual criteria, find *analogies* between perceptually disparate relations, *draw inferences* based on the *hierarchical or logical relation* between relations, *cognize the abstract functional role* played by constituents in a relation as distinct from the constituents' perceptual characteristics, or postulate *relations involving unobservable causes such as mental states and hypothetical forces*. There is *not simply a consistent absence of evidence for any of these higher-order relational operations in nonhuman animals; there is compelling evidence of an absence.*

Penn et al. therefore contradict the famous statement by Charles Darwin: "Nevertheless, the difference in mind between man and the higher animals, great as it is, is certainly one of *degree and not of kind*" (Darwin, 1871, 105, ePH), stating instead: "We argue that Darwin was mistaken: the profound biological continuity between human and nonhuman animals masks an equally profound *discontinuity* between human and nonhuman minds" (Penn et al., 2008, ePH).

So, although a series of intelligent types of behaviour have been brought to light in recent decades by extensive research in neurobiology, animal behaviour and animal psychology in higher animals such as anthropoid apes, birds, dolphins and elephants, human beings seems to differ *significantly* – and not only in degree – from the animal, through their rational intelligence. Leading authors in this area agree that there remains a vast distance between the feats of intelligence of these animals and humans, and that only the human being can communicate rational thoughts (Gazzaniga, 2008; Suddendorf, 2013; Tomasello, 2014), which is of course due to the fact that only the human being can *understand* rational thoughts.

In addition, it must not be overlooked that even the most intelligent animals can at best produce feats of intelligence corresponding to those of human babies and small children below the *age of three years* (Suddendorf, 2013). But this is the age at which the child's *self-consciousness or ego-consciousness* first *awakens* in its mental life (Largo, 2000), a consciousness which is necessary for the formation of independent thoughts; and the faculty of pure intellectual (rational) abstraction only starts to develop gradually during the school years (Stamm & Edelmann, 2013). This development is biologically dependent on the corresponding increase in the myelinisation of the neurones and is only fully available after puberty, but can then be developed still further (Stern, 2006). The specifically human form of self-conscious intelligence is therefore completely absent in the animal.

But what exactly is the difference between this self-conscious human intelligence and that of the animal? Intelligence is generally taken to mean a mental ability to act in accordance with logical intelligible laws. From this point of view it is initially of no importance whether this behaviour is conscious or not, i.e. whether these laws are understood or not. For instance, the building achievements of beavers, swallows and wasps or the flight acrobatics of seagulls are intelligent, in that they operate in accordance with structural or aerodynamic laws, however without this implying that the animals have a logical understanding of these laws. A similar consideration applies to the pre-scientific construction of dwellings by human beings e.g. the igloos of Eskimos or the art of throwing the boomerang by Australian Aborigines. Largely unconscious instinct can therefore be "intelligent" as can the clearly more conscious emotions (see e.g. Daniel Goleman's *"Emotional Intelligence"*;

Goleman, 1996). More or less conscious spontaneous or learned actions can also be intelligent without having to rest on intellectual insight into the logical content of the intelligence used, such as e.g. the above-mentioned use of tools by animals based on mental representation of antecedent-consequent relations or with throwing the boomerang. Intelligence is inherent in all these types of actions, but is not conscious. This basically applies even to purely organic or inorganic events such as e.g. the morphogenesis of plants or the formation of crystals in accordance with mathematical and geometrical laws[128]. It is only mature *human thinking* which can be *conscious* of all these laws underlying the inorganic, organic and mental realms by bringing them forth in itself, a mental feat of which the animal is obviously not capable.

But what distinguishes this intellectual achievement of the human being from that of the animal? Its *noetic – i.e. spiritual – quality*, its *activity*, its *full consciousness and self-consciousness* and its *objectivity* and *universality*. This activity is *noetic* in that it consists of nothing other than the fully conscious penetration of the purely ideal – i.e. spiritual – content of the relevant laws. It is *active, fully conscious and self-conscious* because this thinking penetration of laws is only possible through the fully self-conscious intellectual activity of the individual. And it is *objective* and *universal* because it penetrates the *universal, general objective* laws which disregard all individual particular cases to which these laws apply and only contains their generality which, moreover, applies to all thinkers in the same manner (Steiner, 2011, 75–76):

> Our thinking is not individual like our sensing and feeling; it is universal. [...] There is only one single concept of "triangle". It is quite immaterial for the content of this concept whether it is grasped in A's consciousness or in B's.

By producing the universal spiritual activity to penetrate the objective spiritual content, the individual is *spirit*. The intellectual activity of the spirit is direct manifest intelligence. Human spirit *is* intelligence.

The spirit *transcends* the soul because, as we have seen, the *soul* is characterised by *subjective inner experience* or *inwardness* which is present in a similar manner in the human being and the animal (cf. Sections 5.1–5.4). In contrast to the spiritual realm, the processes and content of the realm of the soul always have the character of this subjective inwardness, i.e. the *subjectiveness* and *singularity* of the affected person. Drives and emotions, for instance, always belong to the individual subject. Likewise, the mental representations of the outer world perceived by the senses contain the specific perspective of the subject in contrast to the general universal character of pure

---

[128] This fact has given rise to the expression "intelligent design" which is justified within certain limits. For a discussion of the difficulties of the creationist concept linked to this expression see Section 4.6.

ideas as José Luis Bermúdez describes using the example of visual perception: "It is plain that each of us sees the world from an egocentric perspective. In particular, we see objects as located on an egocentric frame of reference centered on the eyes" (Bermúdez, 1997). This egocentric perspective is preserved in the sensory representation. Although this is also a thought like an idea, it is a thought in pictorial form, not a thought in the form of an idea. Tomasello calls it "imaginastic or iconic schematizations of the organism's (or, in some cases, its species') previous experience" (Tomasello, 2014, 12). Even if, like James and Carol Gould, you grant chimpanzees *generic* category images (Gould & Gould, 1994, 190–191), as noted above, these are still only images and not abstract, logically comprehensible ideas. In addition, mental representations *arise of their own accord* in consciousness e.g. when your glance focuses on an object. Acquired antecedent-consequent relations likewise arise in consciousness basically associatively – as a result of associative learning – as postulated by Tomasello (Tomasello, 1999, 22–23; Tomasello, 2014, 16–17) i.e. under suitable conditions of perception the corresponding sequence of representations appear in consciousness *by themselves*, whereas the logical understanding of the reason for a sequence of this kind only occurs through *active* intellectual effort. Naturally this applies not only to the associative mental images of animals but also to those of human beings.

In summary, it is possible to distinguish between the *intelligence of the soul* of human beings and animals and the *intelligence of the human spirit*. The intelligence of the *life of the soul* is *passive* in a certain sense, i.e. it arises spontaneously in consciousness as a result of the context, and *not self-conscious, concrete, i.e. pictorial, subjective and individual,* and its inner lawfulness remains *hidden* beneath the cloak of subjective inner experience as it were. The intelligence of the human spirit is *active, self-conscious, purely intellectual, objective* and *universal*, and is mentally directly *manifest*. The soul *contains* inherent, but not directly apparent intelligence: the spirit *is* intelligence. It is only the spirit that can make the soul's intelligence (and that of the organism) – but also its own intelligence – consciously manifest. For, through its universal nature, it transcends itself in its physical, psychological and spiritual forms of manifestation, whereas the soul remains "imprisoned" in its subjective inwardness[129].

---

129 Compare Thomas Nagel's description of these facts in *"Mind and Cosmos"* (Nagel, 2012, ePH): *"Consciousness* presents a problem for evolutionary reductionism because of its *irreducibly subjective* character. This is true even of the most primitive forms of sensory consciousness, such as those presumably found in all *animals*. The problem that I want to take up now concerns *mental functions such as thoughts, reasoning, and evaluation that are limited to humans*, though their beginnings may be found in a few other species. These are *the functions*

This allows an explanation of a series of specific properties which make the human being something unique compared to the whole of the animal kingdom and which Michael Gazzaniga describes in his book *"Human. The Science behind what makes us unique"* (Gazzaniga, 2008). This includes the control of thoughts, feelings and actions, episodic memory and planning the future, self-awareness and self-reflection, speech, aspects of imitation and social learning, and theory of mind. The reviewer of this book in *Science*, Ralph Adolphs, wrote: "Trying to find a single theme that ties all these together or subsumes them is daunting" (Adolphs, 2009). However, it is not daunting when one thinks of the human spirit with its universality and its ability of logical insight and reasoning. The *spirit* is the central thread in this list of abilities.

Gazzaniga's inability to notice this can well be attributed to his reductionist assumption which holds this spirit to be an illusion i.e. simply a production of the brain (Gazzaniga 2011). However, this does not prevent him from giving a very good empirically based description of specific human characteristics.

---

that have enabled us to transcend the perspective of the immediate life-world given to us by our senses and instincts, and to explore the larger *objective reality* of nature and value" (ibid., 71). "Just as *consciousness* cannot be explained as a mere extension or complication of physical evolution, so *reason* cannot be explained as a mere extension or complication of consciousness. To explain our *rationality* will require *something in addition* to what is needed to explain our consciousness and its evidently adaptive forms, something at *a different level. Reason can take us beyond the appearances* because it has *completely general validity*, rather than merely local utility" (ibid., 81). "This, then, is what a theory of everything has to explain: not only the *emergence* from a *lifeless universe* of reproducing *organisms* and their development by evolution to greater and greater functional complexity; not only the *consciousness* of some of these organisms and its central role in their lives; but also the development of consciousness into *an instrument of transcendence* that can grasp *objective reality and objective value*" (ibid., 85). Nagel tries to find an explanation for the fact of emergent evolution, from the inorganic to living organisms to animal consciousness all the way to human reason with its transcendental, objective and universal spirituality. He rejects materialistic reductionism and theistic creationism on account of their internal contradictions and inadequacies. Instead he seeks an emergentistic explanation which he calls *"natural teleology"* (ibid., 91). "The teleology I want to consider would be an *explanation* not only of the appearance of physical *organisms* but of the development of *consciousness* and ultimately of reason in those organisms" (ibid., 92). Nagel only *demands* an explanation of this kind but does not succeed in finding it. In the previous Sections 4.5, 5.1, 5.5 and others, the attempt has been made to show that this kind of explanation arises from the epistemologically justifiable fact that the emergent properties of life, soul and spirit can be explained by the hierarchical organisation of causally active laws. Cf. also Section 5.9 on the emergent perspectives of body, life, soul and spirit.

He merely has no causal explanation for these characteristics because he denies the reality of what is specifically human, the spirit. However, if we can acknowledge the spirituality and universal nature of the human ego – or at least assume it hypothetically – then this gives rise automatically to the characteristics mentioned, as presented in what follows.

Control of thoughts, feelings and actions is only possible when an inner activity centre is able to *rise above* the mental activities of thinking, feeling and volition and to actively control these from a higher sphere. The animal with its purely subjective inwardness is unable to do this. Likewise, *episodic memory and planning the future* are only possible if the centre of consciousness is capable of *transcending* the current psychological experience and of entering into a sphere of experience which is detached from space and time. In this detached state it then looks back on a past episode or situation or imagines a future episode or situation. Significantly, both take place detached from current experience of the present situation, because when imagining conditions which are no longer there or not yet there, the conditions which *are* there are not imagined. This is apparently not possible for animals. What is characteristic of animal behaviour corresponds to a dependency on the immediately present sensory or physical experience, as though the animal could not rise above this.

It is similar with the capacities for *self-awareness* and *self-reflection*, because these also require a transcendence of the self which a creature which lives entirely in the individual self, like the animal, cannot do. This is why the animal is not able to put itself into the place of another creature in the same way as a human being can, resulting in the differences between the human and the animal mentioned by Gazzaniga in terms of certain *aspects of imitation and social learning*, as well as *theory of mind*. For, to become aware not only of the physical or psychological behavioural expressions of the other person during the social interaction and to respond with your own subjective inwardness, but to place yourself in the *inner intentions* of the other person which underlie this behaviour, once again assumes the ability for self-transcendence and – tied to this – the ability to enter the *mind of another*, something only to be found in the human spirit. This ability lies in the universal nature of human thinking. In his *"Philosophy of Freedom"* Steiner says of this:

> It must, however, not be overlooked that only with the help of thinking am I able to determine myself as subject and contrast myself with objects. Therefore *thinking must never be regarded as a merely subjective activity*. Thinking lies *beyond* subject and object. It produces these two concepts just as it produces all others. [...] *Thinking is thus an element which leads me out beyond myself and connects me with the objects.* (Steiner, 2011, 49–50, ePH)

My self-perception confines me within certain limits, but my thinking is not concerned with these limits. In this sense *I am a two-sided being*. I am enclosed within the sphere which I perceive as that of my personality, but I am also the bearer of an activity which, from a *higher sphere*, defines my limited existence. (ibid., 75, ePH).

In contrast to the animal, the human being is in fact a *two-sided being of soul and spirit*. This also explains the phenomenon described by Michael Tomasello and verified by a wide range of empirical research, that anthropoid apes, even when cooperating, can only apply their own *"individual intentionality"*, but the human being can, in addition, develop *"shared intentionality"* with his own kind and therefore *"common intentionality"* (Tomasello, 2014). "Shared" or "common intentionality" means that two or more individuals have *the same intention, know from each other that they have this same intentionality*, communicate this *cooperatively* with each other (which for humans is possible through verbal and non-verbal *speech*) and are able to implement this through joint action. In contrast, the cooperative behaviour of animals develops primarily through *"competitive interactions and imperative communication"* (ibid., 150). The social cognition of animals forms in accordance with this (ibid., 31, ePH):

> In virtually all theoretical accounts, great apes' skills of social cognition evolved mainly for competing with others in the social group: being better or quicker than group mates at anticipating what potential competitors might do, based on a Machiavellian intelligence [...]. And indeed a number of recent studies have found that great apes utilize their most sophisticated skills of social cognition in contexts involving competition or exploitation of others as opposed to contexts involving cooperation or communication with others [...]. Great apes are all about cognition for competition. Human beings, in contrast, are all about (or mostly about) cooperation.

This makes sense if we bear in mind that the animal's fundamental experience is that of *individual subjectivity* i.e. *the soul* and that in human beings this is permeated by the *trans-subjective spirit*. In contrast to the animal, the human being – apart from his specific objective thinking ability through which he can *understand* the intentions of others – therefore develops a *specific social perceptive ability* through which he can *perceive* the intention of the other. In 1917 Steiner first described three senses beyond the initially highest ones which the human being shares with the animals (the sense of sight and hearing): 1. the sense for words, 2. the sense for thoughts, 3. the ego sense (Steiner, 1996, 127). This enables an actual *perception* of the manifestations of speech, thoughts and the self of the other in social cognition, not merely their hypothetical assumption by *analogy* as is usually assumed in the "Theory of Mind"[130].

---

130 The *word or speech sense* is, however, not known by this name in current social cognitive science but its existence is basically proven by neurobiological and neuropsychological research (Peveling, 2015). The tenability of Steiner's concept of

Man is of course a being with a soul as are the animals, something described very clearly by Charles Darwin based on the emotional and physiognomical modes of expression e.g. for fear, anger or joy/pleasure and since confirmed by research on emotional expression (Darwin, 1998). The same is true for the intelligence inherent in subjective animal consciousness. The human being has no special position in this respect. But, in addition to this, the human being is a *spiritual* being with the ability to transcend this consciousness, to understand himself and the world objectively through his self-conscious and self-produced intelligence, to give his life *sense and meaning* and, in so doing, form a relationship to the spiritual foundation of the world through *science, art and religion*. It is the spirit which is responsible for the transition from nature to culture in human affairs.

Just as the form of the human body follows the basic typus of the higher mammals but raises this to a "special morphological position of the human being" (Goerttler, 1972), so is the "psychological typus" of the human being which he shares with the animals raised to a completely personal individuality by the influence of his spiritual self. It is this spiritual self which gives the human soul its innermost identity[131].

5.7 The importance of the spiritual factor in health and medical care

The spiritual aspect of the human personality is apparent in all facets of life, as can be seen in matters related to health, and therefore in medical care. For example, the ability to *cope* with a disease is something which only human beings possess, as it assumes *cognitive* abilities like self-reflection, information

---

the sense of thinking is currently being tested on the basis of empirical research. An *ego sense* was also adopted in 1922 by the German philosopher Max Scheler (Scheler, 1973) and, following on from this, taken up into the current scientific debate (Gallagher & Zahavi, 2012). In the line of reasoning adopted by Steiner, Scheler and Gallagher & Zahavi, the important point is that the thoughts and the ego of the other person are *not constructed in our own subjective selves through analogy* as is usually hypothetically assumed in the Theory of Mind hypothesis, but *objectively perceived* on this other person, and that this can become empirically clear through an analysis of the phenomena of the perception field.

131 This is probably the reason why all expressions of the human soul have such a unique character, even in small things of life such as everyday emotions which are basically similar to those of the animals. The "laugh" of an animal which is more like a "grimace" is certainly not comparable with the human smile as an expression of his most personal inner being. Steiner actually describes it as follows: "A person who really observes the soul knows that the animal cannot weep but at the most howl, nor can it laugh but only grin" (Steiner, 1981b, 96).

gathering, the control of emotions connected to illness and conscious changes in behaviour. It is unthinkable that an animal could "cope" with its illness in this way. Of course animals can also react to stressors with defensive behaviour which is called "coping". This consists of a more passive or more active i.e. more aggressive reaction of the animal to an internal or external stressor, with a varying influence on its ability to cause illness. For example, mice which are inoculated with tumour cells develop more lung metastases if they are subjected to social stress. Those which show a more passive coping mechanism have more lung metastases, increased corticosteroid levels and reduced immune functions in comparison with the more active animals (Azpiroz et al., 2008; Koolhaas, 2008). However, this form of spontaneous, more or less effective form of coping which also exists in humans, has to be distinguished from the aforementioned cognitive form of coping which can only be observed in humans. On the other hand, as soul and spirit are not separated but form a unity in the human being, it is unlikely that only emotional or only rational factors play a role in human coping.

For example, it is known that the psychological (mostly emotional) state can have an effect on the course of disease in human beings. In an analysis of 477 publications on breast cancer from 1974 to 2007, Montazeri and colleagues have shown that psychological factors such as psychosocial stress and depression can reduce later quality of life and sometimes the survival rate in breast cancer (Montazeri, 2008). Conversely, psychosocial intervention (which always involves elements of typical human social cognition and some forms of rational discourse) can reduce emotional stress in women with breast cancer. Evidence that psychotherapy can prolong survival in these patients is attributed in part to immune functions such as the natural activity of killer cells (van der Pompe et al., 1996). Cognitive behavioural therapy can improve mood, tiredness and physical fitness in chronic fatigue syndrome (O'Dowd et al., 2006) and be predictive for later improved quality of life and reduced health costs for status after stroke (Darlington et al., 2009). This illustrates that the essential spiritual core of the human soul is a relevant factor in the genesis and therapy of disease, something which does not apply to animals in the same way.

Some forms of psychotherapy are directly aimed at this spiritual core of the human being, such as e.g. the *logotherapy* initiated by Viktor Frankl (1905–1997) (Frankl, 2005). In his "dimensional" anthropology, Frankl distinguished the following aspects of the human being: 1. the physical dimension (body), 2. the psychological dimension (psyche), and 3. the *noetic dimension (spirit)*. According to Frankl, the spirit can raise itself above the psycho-physical: only this latter becomes ill, not the spirit. However, this is aimed existentially at the fulfilment of *meaning*: a lack of meaning can lead to illness and the purpose of logotherapy (lógos: Greek for meaning) is to discover this – naturally ego-related, individual – meaning.

The concept of *salutogenesis* founded by Aaron Antonovsky (1923–1994) is also linked to the spiritual ego function of the human being (Antonovsky, 1979). According to this concept, the human being can better maintain his psychophysical health in the face of external stressors if he can develop an inner *sense of coherence* based on the *comprehensibility* of the world and on the *manageability* which results from this, but especially on the *meaningfulness* or *significance* of the person's own life and biography. This can only be achieved by the intelligent spiritual core of the human being, never by the soul of an animal.

Steiner also holds that the spiritual core of man is not something that can become ill. The expression *"mental illness"* is inadequate according to Steiner, because "the spirit is always healthy, and in fact cannot fall sick in the true sense of the term. To talk of mental illness is sheer nonsense. What happens is that the spirit's power of expression is disturbed by the bodily organism, as distinct from a disease of the spirit or the soul itself. The manifestations in question are symptoms, and symptoms only." (Steiner, 1948, 176, ePH). This is why anthroposophical psychiatry and psychotherapy also take account as far as possible of the relationships between the physical and soul/spiritual functions in diagnosis and therapy. Therapy is therefore designed to be multimodal and integrative with elements related to the body such as conventional psychopharmaceuticals, anthroposophical medicines, external applications, movement therapy (e.g. curative eurythmy) and other therapies (Treichler, 1981a; Walter, 1955; Bissegger et al., 1998). Anthroposophically oriented psychoanalysis and psychotherapy deliberately work on strengthening the spiritual functions of the self (Vandercruysse, 1999) and include questions on meaning and destiny i.e. spiritual and transcendental aspects (Priever, 1999).

Anthroposophical medicine and psychotherapy also include a consideration of *biography* in which the development of the spiritual individuality of the person is expressed. There is no sense in talking about "biography" in the case of animals. The question, "What will become of you?" can be answered very easily about a new-born cat: the same as its parents, in other words a cat, though with some variation in its physical appearance and its emotional behaviour. It will completely fulfil the purpose of its *species*. This question is impossible to answer when applied to a new-born human child. Certainly the child will also conform to its species as a *physical and emotional* being and will express the characteristics of its sex, family and ethnic background more or less clearly. But above and beyond this the *spiritual individuality* of the human being – something which is decisive for human development and biography – will develop which, *of itself,* has nothing to do with the parents, nation, race, etc. Steiner expressed this clearly in his *"Philosophy of Freedom"* in the chapter *Individuality and Genus* (Steiner, 2011, 200–201):

> A human being develops qualities and activities of his own, and the basis for these we can seek only in the person himself. What is generic in him serves only as a medium in which to express his own individual being. He uses as a foundation the characteristics that nature has given him, and to these he gives a form appropriate to his own being. If we seek in the generic laws the reasons for an expression of his being, we seek in vain. We are concerned with something purely individual that can be explained only in terms of itself. [...] Anyone who judges people according to generic characters gets only as far as the frontier where people begin to be beings whose activity is based on free self-determination[132].

For medical diagnosis and therapy aimed at the whole person it is thus never enough to consider simply the combination of the "bio-psycho-social" dimensions because these still lack *the key element of being human, the spiritual dimension*. "If we would understand the single individual we must find our way into his own particular being and not stop short at those characteristics that are typical" (Steiner, 2011, 202–203). Medicine based on anthroposophy therefore works towards the development of a diagnosis and therapy which corresponds to the physical, living, emotional *and spiritual* dimensions (Sieweke, 2008; Steiner & Wegman, 1996), and which tries to take account of the lawfulness of the soul and spiritual development of the human being in the course of their life in relation to the physical and bodily development in a differentiated way (Treichler, 1981b). As a result, *biographical work* has developed as a special branch of anthroposophical psychotherapy which always includes spiritual and transcendental points of view (Burkhard, 2005).

Spirituality is also an essential element of a fully "human" medicine. As a spiritual being with the capacity for transcendence (cf. Section 5.6), it is in man's nature not to be satisfied with the world of physical phenomena and emotional experience but in addition[133] to seek a relationship to the

---

[132] This does not mean that the spiritual ego is always in a state of free self-determination. It is *not* in such a condition e.g. as an infant, while asleep or in states of restricted consciousness, but also not in all those situations in life and the normal daily routine when the ability to judge has (still) not been reached or is limited in some way. But the human ego is in the process of developing this self-determination: "No human being is all genus, none is all individuality. But every person gradually emancipates a greater or lesser sphere of his being, both from the generic characteristics of animal life and from domination by the decrees of human authorities" (Steiner, 2011, 203).

[133] This is what Faust is referring to in his statement to his famulus, Wagner: *"One impulse art thou conscious of, at best;/ O, never seek to know the other!/ Two souls, alas! reside within my breast,/ And each withdraws from, and repels, its brother./ One with tenacious organs holds in love/ And clinging lust the world in*

spiritual in nature and the universe. This he does through science, art and religion. The thinker "seeks the laws of phenomena, and strives to penetrate by thinking what he experiences by observing"; the artist "seeks to embody in his material the ideas that are in his I"; and the religious believer "seeks in the revelation which God grants him the solution to the universal riddle" (Steiner, 2011, 22). The importance of spirituality and religion for medicine has become increasingly clear over the last decade (Heusser, 2006b; Büssing et al., 2006), especially for ethical medical issues (Heusser & Riggenbach, 2003); for coping with illness in the case of patients with chronic and serious illnesses (Büssing et al., 2005; Büssing et al., 2009; Büssing et al., 2013a; Büssing et al., 2014); for the relationship of health care professionals to their patients (Büssing et al., 2013b; Büssing et al., 2015), and in palliative care where it has been recognised that the spiritual needs of some patients in the face of incurable diseases, dying and death can often not be adequately met by the provisions of a technical materially-oriented medicine and a secularised culture (Efficace & Marrone, 2002). This is why palliative care nowadays speaks of four forms of suffering: physical, psychological, social and *spiritual suffering*; and meeting the needs arising from these is unquestionably counted amongst the aims of palliative and hospice care (Bausewein et al., 2007, 6).

This development clearly shows that spirituality and the search for meaning are increasingly recognised as an integral element of modern medicine (Breitbart, 2002). However, what is still missing is the recognition of the human being as a true being of soul and *spirit* and the conceptual integration of the spirit into medical anthropology and the theory of medicine. Even leading representatives of biopsychosocial medicine currently speak only of a "psychological entity" in the human being and its functions (Adler, 2005) but without defining the nature of this entity in more detail or in fact acknowledging it as a reality. Biopsychosocial medicine also lacks any differentiation between the entities of soul and spirit, although this has a tradition at least in the history of German thought[134]. However, the task of human medicine is to develop a medical anthropology in theory and practice appropriate to the human being. This, however, requires a scientific method which is not focused

---

its embraces;/ The other strongly sweeps, this dust above,/ Into the high ancestral spaces." (Goethe, 2005, lines 1110–1117).

134 Beckermann in his *"Analytische Einführung in die Philosophie des Geistes"* (analytical introduction to the philosophy of mind) summarises this as follows: "In German a distinction is usually made between spirit and soul, with the realm of rational thinking and action assigned to the spirit and the realm of feelings and intuition assigned to the soul" (Beckermann, 2001, 4). (For the sake of completeness it should be added that the fully conscious thinking spirit can naturally also possess intuition, cf. Section 2.3 and Chapter 7).

on reductionist conjecture but on a theory which grasps the lawfulness *of the phenomena* and does not limit scientific method to physical phenomena but also extends to include the phenomena of life, soul and spirit, as this work attempts to demonstrate by means of the epistemology as defined by Goethe and Steiner. The same applies to the question of freedom.

5.8 The question of freedom

The question of freedom in human thinking, will and action has occupied western philosophy and science since Plato as well as religious thinkers in Buddhism, Christianity and Islam over and over again (see overview in: an der Heiden & Schneider, 2007). While human freedom – inasmuch as it is based on the cognitive faculty of human reason and is therefore an achievement of the human *spirit*, something which distinguishes man from the animals – was considered possible by a series of thinkers from ancient times until the 20th century, in the last few decades this human freedom has been increasingly rejected in the wake of the results of neurophysiological research. In Germany, public debate on this is dominated by the theories of Wolf Singer and Gerhard Roth (Geyer, 2004), whose thesis Singer summarises as follows: "We are determined by circuits: we should stop talking about freedom" (Singer, 2004). Singer in particular points out that the neurosciences provide "increasingly convincing proof that human and animal brains scarcely differ, that their development, structure and functions obey the same principles" and maintains that the behaviour of animals and human beings is ultimately "subject to the deterministic laws of physiochemical processes" and that therefore "the contention that behaviour is materially conditioned is equally applicable to human beings". Further, because the neural processes underlying even the higher cognitive powers and behaviour can now largely be elucidated thanks to the increasing refinement of neurobiological measurement methods, while not holding the emergent cognitive powers themselves to be the same as the physicochemical interactions in the nerve networks, he nevertheless sees these as "causally explicable" by them (ibid., 35–37). This means that for him and those who think like him, the freedom experienced by the human being becomes an illusion and his spiritual ego simply a "cultural construct" (Singer, 2001) or a "virtual actor" produced by the brain (Roth, 2001).

It appears that nothing much has changed in the basic concepts of psychophysiological research since the materialistic theories of the 19th century "despite the mountain of data" which the neuroscientists have accumulated in the meantime (Breidbach, 2001, 17). Brain, consciousness, spirit and freedom and therefore the key elements of the human being become products and illusionary epiphenomena which are explained by the physicochemical structures and processes which themselves are self-organised and selected by evolution.

However, this answer to the question of freedom is not conclusive: it is merely a consequence of the epistemological and ontological *premises* of reductionism. There is a different answer to the question of freedom which emerges from the physicochemical, biological, evolutionary and psychophysiological facts looked at from the viewpoint of empirical ontological idealism. The emergence of higher structures and functions does not equal their causal explanation from the structures and functions of lower levels, but their conditionality on and compatibility with these: emergent phenomena and laws are epistemologically of equal value to submergent ones and cannot be reduced to these ontologically. They thus have their *own state of being* and therefore their *own causation*. As described above, this applies to the understanding of substance, the concept of the organism, the relationship of the body and soul and the issue of consciousness.

With reference to the question of the spirit and freedom, this means that the spiritual characteristics which are emergent compared to the animal and the resulting human capability of freedom cannot be *causally* attributed to the physicochemical interactions of the nerve cells. Sentences such as: "From a neurobiological viewpoint consciousness is a state of special neuronal activity" (Roth, 2001, 204); or, "In developing an ego the brain is generating a virtual actor" (ibid., 204); or: that the cognitive functions are "causally explainable" as arising from the physicochemical neuronal interactions (Singer, 2004, 36) are simply wrong. These are not findings but hasty opinions including, according to Singer's own epistemological premises: "We can only know what we can *observe*, classify through our *thoughts* and are able to imagine" (ibid., 30, ePH). However, it is impossible to *observe* the *generation* or *causation* of consciousness, ego and cognition from neurobiological processes. The only scientifically correct statement would be to describe the relevant neurobiological processes as being verifiably essential *conditions* of consciousness, the manifestation of the mental self and cognition. In this realm science can initially observe only the physiological processes on the one hand and the psychological and cognitive ones on the other *in themselves* and grasp them in accordance with their laws. Between the two is the gulf already mentioned. In fact, both Singer and Roth are aware of this. Roth concedes the objection, that a purely neurobiological explanation of human action is a category error (Roth, 2006, 19, ePH):

> As a brain scientist it is important to take these arguments seriously. The fact is that, in his research, the brain scientist can only take account of anatomical features and physiological processes which in themselves do not divulge any information whatsoever on any kind of meaning. This applies in equal measure to events at the level of individual cells and to measurements of complete brain processes using functional magnetic resonance imaging. A functional and therefore meaningful interpretation only results from a *comparison* between

the structures and functions of the brain on the one hand and the reactions or verbal statements by the subject or patient on specific experiences on the other.

If we admit this, then it is essential to be aware of the inadmissibility of asserting the existence of causality where none can be observed. Roth is also ultimately aware of this, something he concedes (Roth, 2001, 205, ePH):

> Of course, this does not answer the question of what the philosophers see as being explained by this. *It certainly does not provide an explanation of the uniqueness of the personal experience of consciousness,* – this trait which is so dear to the philosophers of mind. It may be that this is an *unbridgeable gap in our explanations* which cognitive neurobiology will have to accept (the philosophy of mind likewise): after all, not everything has to be explained by one theory.

But this is exactly the crux of the matter: how can the *characteristic feature* of consciousness, ego and cognition, the experience of self and of freedom, be explained in relation to neurobiology? Neurobiology can explain neither its content nor its form. This admission by Gerhard Roth naturally undermines the credibility of his reductionist theory. But a reductionist theory is not needed, either epistemologically or ontologically. *From an epistemological* angle, psychophysiological and neurocognitive research is actually not possible without the multiperspectivity of the empirical point of view. In line with Michael Pauen "we not only want to know something about the 'neuronal basis' of mental phenomena but wish to investigate *these phenomena themselves*" (Pauen, 2001, 86). Helmut Schwegler calls this *aspect pluralism:* "Various empirical approaches are taken equally seriously and initially recorded in domain theories which can then be further linked together in a large number of ways" (Schwegler, 2001, 79). As regards the qualitatively clearly different mental and neuronal characteristics, Pauen speaks of an epistemological *property dualism* (Pauen, 2001). *Ontologically* it would be only consistent to acknowledge property dualism or aspect pluralism also as *ontological dualism* or *pluralism* without making a *speculative* interactionistic dualism or pluralism from this in the sense meant by Eccles.

For in *empirical* research it is initially only a matter of acknowledging what is *actually experienced* in the different observational realms as physical, psychological and spiritual phenomena, as the relevant *facts*, irrespective of the nature (physical, psychological, spiritual) of these facts and irrespective of how these different facts might then *relate* to each other. Because physical, psychological and spiritual facts are epistemologically of equal value for cognition, i.e. in the *form of experience* in which they occur (even if each does so in a different realm of experience, cf. Sections 2.2 and 5.1), it is therefore quite justified to acknowledge these facts as being *of equal value ontologically*, even if they are not *of the same type*. For *reality*[135] is a matter of empiricism, not theory. The actual soul and spiritual aspects of the human being can be grasped empirically and ontologically as reality just as much as the neurophysiological and other biological processes necessary for their occurrence, in fact initially without the need to explain the causal relationship between them.

In this respect the fact of the human experience of freedom is also a "manifest, indubitable experience", as Peter Bieri expressed it (Bieri, 2001, 19). Even for Singer the experience of the "attributes of our humanity" in the inner "first person perspective" is an *indisputable fact*. He includes in this all the inner emotional and spiritual experiences which we can have, such as perceptions, experiences of ourselves and feelings and also the "spiritual mental dimension [...], which enables us to judge freely about ourselves, to evaluate and to decide" (Singer, 2004, 33). In addition, he admits that the emotional and spiritual experiences are felt to be *no less real* than the physical sensory ones amongst which the neural processes studied by the researcher must be included: "We experience these immaterial phenomena as being just as real as the manifestations of the world of things which surrounds us. [...] We experience the phenomena which we call spiritual or psychological or emotional as realities of an immaterial world of whose existence our personal experience, however, allows just as little doubt to arise as our sense perceptions allow of the existence of the material world" (ibid., 33). This is an empirically justified ontological statement which is consistent – at least in terms of the epistemology put forward here. We have "the conviction supported by our personal experience [...], that we participate in a spiritual dimension which is independent of the material world and ontologically different" (ibid., 36).

But this ontological difference – conceded on empirical grounds – is not generally accepted by neuroscience and especially not by Singer: a "conflict for our self-conception [arises] between two, what appear to be equally convincing, equally correct but incompatible images of man" (Singer, 2004, 37). But why is this incompatible "for our *conception* of ourselves" when it is obviously *not* incompatible for our *experience* of ourselves? Because the *theory* applied to the experience *cannot* accept what this experience clearly shows: the ontological equivalence of physical, emotional and spiritual realities. In the scientific "longing for a satisfactory relationship of the self-conscious ego to the general world picture" (Steiner, 1973a, 349), this is what underlies the temporary impossibility of coming to a coherent answer which combines

---

135 In agreement with Steiner and following the overview of various concepts of reality in Hügli and Lübcke (Hügli & Lübcke, 2003), *reality* here is seen (both generally and specifically) as the lawfully determined facts which are directly or indirectly perceptible in the empirically given object or subject (cf. Section 2.5).

these different classes of facts in a consistent unified picture of the world and man. Because this is what neurophysiology is justifiably looking for (Singer, 2004, 39).

The reason for the presumed incompatibility initially lies mainly in the fact that unity in the phenomena present is looked for on the side of *perception* instead of on that of the *concept* or *law*. This is associated with the fact that, in the nominalistic sense, the laws are held to be something subjective, which only have meaning for the human need for order and not for objective reality. But unity will never be found on the side of perception. Because it is not only the physical and mental perceptions which are different or "incompatible" but the physical perceptible sense qualities themselves. That which relates all the viewpoints to each other and can therefore examine the unity which runs through everything is the *conceptual side*, the cognition which has to find the ideal lawfulness of the different phenomena and their interrelationships. The unity of the disparate entities is then an ideal one but can nevertheless be acknowledged from the viewpoint of ontological lawful realism as being real[136].

A further element of accepted theory is the opinion that outer objective physical phenomena are more real than inner subjective psychological or spiritual phenomena, and a view incorporating both must "be believed" (ibid., 37). But both kinds of phenomena appear as empirically and ontologically *equivalent*, as Singer himself stated, and so the experienced fact of freedom, for example, is no more a question of belief than the neuronal processes. There is therefore no proper justification for Roth and Singer when explaining psychophysiological connections in giving clear preference to the external third person perspective instead of the inner first person perspective, in fact in wanting to replace the latter with the former (Roth, 2004; Singer, 2004).

The consequence of an unjustified "replacement" of the first by the third person perspective is statements such as "cognitive achievements of human brains" (Singer, 2004, 39) which contradict actual empiricism. It is not the brain but the human *spirit* which produces observable cognitive achievements. The brain only provides the material conditions for this. To be epistemologically correct requires a clear distinction between the independent pursuit of both perspectives on the one hand and their relationship to one another on the other, something mentioned by Roth himself. If this is not done, then it is easy to assign a determinism to the *spirit* which actually belongs to the *brain*. But the aim of neurocognitive research cannot be the denial of the obvious experience of freedom but the development of concepts which *clarify* the *experienced* compatibility of spirit and brain in a unified and consistent scientific anthropology. This is also the task of neurocognitive

---

136 Cf. also the discussion of the problem of monism in Section 5.2.

research, as defined by the epistemological attitude of Goethe and Steiner (Steiner, 1973b, 21):

> For the free and active working, straight from the inner resources of the human being, is a perfectly elementary experience of self-observation. It cannot be argued away; rather must we harmonise it with our insight into the universal causation of things within the order of Nature.

This is by no means trivial. The neuroscientists rightly draw attention to two main problems which have so far made it impossible to find a harmony of this kind and which have therefore led to the dispute over human freedom. First is the unsolved issue of the energetic interaction of spirit and brain and second the neurobiological knowledge that "actions in which we *feel* ourselves to be free are prepared and executed in the brain" (Roth, 2006, 11). Singer formulates the first problem as follows (Singer, 2004):

> Furthermore, there is the particularly awkward problem of causation for which we likewise know of no possible solutions. If this immaterial spiritual entity exists which takes possession of us and gives us freedom and dignity, how is this to interact with the material processes in our brain? For it must influence the neuronal processes so that what the spirit thinks, plans and decides can also be carried out. Interactions with the material realm require the exchange of energy. If the immaterial is to summon up energy in order to influence neuronal processes, then it must have access to energy. But if it possesses energy, then it cannot be immaterial and must be subject to the laws of nature. [...] If the premise that interpretations of the world have to be consistent in order to be correct is valid, then there are three options: our personal experience is deceptive and we are not what we think ourselves to be; or our scientific descriptions of the world are incomplete; or our cognitive abilities are too limited to experience the unifying element behind the apparent contradiction.

As personal experience is described by Singer himself as being ontologically of equal value to that of experience of the outer world (e.g. the processes in the brain), the latter must be just as deceptive as the former. The downgrading of personal experience in favour of the experience of the brain which is undertaken by Singer, Roth and others to solve this problem cannot be justified on this basis. The other option, that "our scientific descriptions of the world [could be] incomplete" is not even explored by them. But a major problem appears to occur here. Singer's formulation *assumes* that the cause and effect relationship between spirit and brain must be thought of as an *inter*action or an *exchange* of energy, as is the case in mechanics elsewhere and, further, that energy can only be *material*, forgetting that natural laws are of a spiritual nature. In short, a mechanistic and nominalistic mode of thought is adopted.

The same problem appears different from the perspective of ontological realism adopted here. If *material* energy and substance are to be "*subjected* to natural laws", they *obey* something *spiritual*, the law. This means that the spiritual is fundamental even for the lowest elements of matter (cf. Sections 3.1 and 3.2). The emergent laws of the hierarchically organised world of substance, the organismic structures of time and space in microorganisms, plants, animals and human beings, the emotional and spiritual characteristics of the living element in animals and the additional spiritual characteristic in the human being are all of a *spiritual* nature. Seen in this way, there is not merely empty matter at work in the justifiably postulated self-organisation of the brain (Stephan, 2001), i.e. even during the *formation* of the organism and brain in phylogenetic and ontogenetic development as is often believed (Cruse, 2004), but constitutive *spirit*, and what is more, *starting on the material plain*. From the perspective assumed here, the material level is then augmented by the levels or the lawful action of life (the etheric body, cf. Section 4.11), the soul (the astral body, cf. Section 5.4) and the human spirit (the innermost "I", cf. Section 5.5). These all contribute to the creation of the human brain. Their laws do not exist *externally* to the brain in the sense of an "interaction" or an "exchange of energy" but act in it *inherently* as *its* laws.

And because the human soul and spirit *use* the brain in the elaboration of their *purely mental activity*, what was previously built up is destroyed again in a certain sense (cf. Section 5.3). So although the human experience of freedom develops *thanks to* the brain, in a certain sense it also acts *in opposition* to it. This is not something that involves an *external mechanical* interaction with the exchange of energy but occurs through the development of its inner mental formative activity which is then no longer available for organic formative processes. Looked at like this, the thought of a mental agent which, despite the brain being determined by natural laws, is capable of freedom, is compatible both with the facts of evolutionary history and neurobiology and with the incontestable fact of the human experience of freedom.

The second problem concerns the preparation by the brain of acts of will which are felt to be "free", which is often put forward by Gerhard Roth in particular as proof against the freedom of the will. What exactly is meant by this?

In 1983 Benjamin Libet showed that, in experimentally controlled movements, consciousness of desiring the relevant movement only appeared *after* the movement had been initiated at the neuronal level by the creation of what is known as a "readiness potential" in the contralateral motor cortex (Libet et al., 1983; Libet, 1985). This result was essentially confirmed in 1999 by Haggard and Eimer (Haggard & Eimer, 1999), although it is not the readiness potential in general but the *lateralised* readiness potential in particular which these authors link to the onset of awareness of the urge for movement. Roth also pointed out that the cortical readiness potential is in turn preceded by other stimuli in the basal ganglia which have no conscious correlate, and therefore came to the conclusion that (Roth, 2001, 195):

> These – and many other – results are incompatible with the assumption that an autonomous spirit uses the brain as an instrument in order – according to Eccles – to realise itself in the material world. On the contrary, they show that consciousness is the *end result* of very complex interactions of many brain centres, of which the majority basically function unconsciously and the spirit does not set the brain activity in motion as is assumed by interactive dualism which views the brain as an instrument of the spirit.

This interpretation can be countered by the argument that an autonomous spirit or a self-acting soul is possible not only in the sense of Eccles' interactive dualism, and that it must first be established whether the acts of will examined in these experiments could in fact be considered as *free*. Because to feel yourself to be an agent or to know yourself to be a free agent – or in other words: will and free will – are not identical. Libet's subjects were told to move a finger within a specified time window as soon as they felt the urge to do so and then to indicate at what time they became aware of this "urge" using a rotating clock hand. An animal could equally well be aware of an urge to make a movement, followed by the corresponding movement; but nobody would describe a conscious urge *of this kind* as free. Singer nevertheless makes a distinction between *free and unfree will*, the distinction being "the differing *degree of consciousness of the motive* which leads to decisions and actions. We obviously assume that motives which we can bring into consciousness and subject to a conscious deliberation are subject to free will, while motives of which we are not able to be consciously aware are obviously not subject to free will" (Singer, 2004, 51, ePH). "Decisions and acts which we base on the conscious consideration of variables, in other words on the *rational* treatment of content which is capable of consciousness, we judge to be free" (Singer, 2004, 60, ePH). In this sense Libet's urges which become conscious are in no way free. Haggard and Eimer explicitly point out that the urges examined by them and Libet are *not* based on a conscious rational consideration of this kind: they are "events pertaining to the implementation of a specific movement, rather than more abstract representations of action occurring at processing stages prior to selection of a specific movement" or, in other words (Haggard & Eimer, 1999, 132):

> While the LRP [lateral readiness potential] may bear a causal relation to W [will] judgement, the LRP is a relatively late event in the physiological chain leading to action. In our terminology, LRP onset represents *the stage at which representation of abstract action is translated into representation of specific movement*. Thus, the LRP onset is *not the starting point of the psychological processes that culminate in voluntary movement*, but *it may be the starting point of conscious awareness of our motor performance*.

So what is the *psychological* process which, in the words of Peter Bieri, presents a "manifest, indubitable experience of freedom"? This is a question which can and must be answered from *psychological* observation, from an inner perspective, initially without the intervention of neurophysiology. Because free activity of the soul and spirit must first be empirically clearly *defined* as such, in contrast to actions which are not free. Only then is it possible to look for the neuropsychological correlate of the experience of freedom. On this topic, i.e. the investigation of the question of freedom based on psychological observation, in his main philosophical work *"The Philosophy of Freedom"*, Steiner undertook the most detailed empirical analysis to this day of the experience of freedom (Steiner, 2011). One key element of this is summarised below.

In a single act of will Steiner distinguishes between *motive* and *driving force* (Steiner, 2011, 125ff.). The motive is either a concept or a mental picture, i.e. a thought in a general or particular form i.e. one related to a percept[137]. The motive provides the *aim* of my will, the driving force is what "determines me to direct my activity towards this aim" (ibid., 126). The motive is "the momentary determining factor of the will", the driving force is "the permanent determining factor of the individual", i.e. "the will-factor belonging to the human organization and directly conditioned by it", which results from the "individual make-up of the person" which Steiner also therefore called "the characterological disposition" (ibid., 125–126). This therefore determines whether I wish to make a mental image or an idea into the motive i.e. into the aim of my action. The same mental image or the same concept can motivate different people very differently, depending on their subjective disposition or the same person depending on their inner state. The characterological disposition contains the individual's instinctive and emotional inclinations, but also their previous experiences, habits, acquired mental pictures and concepts, in fact everything that, as a "more or less permanent content of our subjective life" (ibid., 126) can provide the driving force for making a current mental image or idea into the motive for an actual desire for action.

---

137 With reference to what was explained in Section 2.3, the content of the *concept* (the general thought) becomes known through mental insight, i.e. through conceptual intuition. "A *mental picture* is nothing but an intuition related to a particular percept; it is a concept that was once connected with a certain percept, and which retains the reference to this percept. My concept of a lion is not formed *out of* my percepts of lions; but my mental picture of a lion is very definitely formed *according to* a percept. I can convey the concept of a lion to someone who has never seen a lion. I cannot convey to him a vivid mental picture without the help of his own perception. *Thus the mental picture is an individualized concept*" (Steiner, 2011. 90).

In relation to the type of driving forces Steiner therefore distinguishes four possible stages of willing:

1. The *drive*, through which perception "translates itself directly into willing, without the intervention of either a feeling or a concept" (ibid., 127), as may occur in the satisfying of physical needs. This type of willing which is characterised by "the immediacy with which the single percept releases the act of will" (ibid., 127) is also present in conventional habitual social behaviour, when we let an action follow the perception of an event or a person as a *reflex* or through acquired social *tact* "without reflecting on what we do, without any special feeling connecting itself with the percept" (ibid., 127).
2. The *feeling* which can be linked to perceptions. "When I see a starving man, my pity for him may become the driving force of my action" (ibid., 128).
3. *Thinking and forming mental pictures*. "A mental picture or a concept may become the motive of an action through mere reflection" (ibid., 128). This can happen if you remember e.g. what you have learned or done previously in such a situation. Steiner therefore calls this driving force of the will *practical experience*. Over time, as a result of habit, this can turn into purely tactful behaviour, "when definite typical pictures of actions have become so firmly connected in our minds with mental pictures of certain situations in life that, in any given instance, we skip over all deliberation based on experience and go straight from the percept to the act of will" (ibid., 128).
4. *Pure thinking*, or *practical reason*, i.e. "conceptual thinking without regard to any definite perceptual content" (ibid., 128), so that we do not will something and take action due to *those* considerations which are linked to memories, examples, authorities or what we have learned, in other words to practical *experience* but due to *a purely conceptual insight into the ideal content* of a concept or idea to be gained through intuition[138]. The inventor of a completely new device, for example, is in this kind of situation with the intellectual conception (intuition) of its idea. Only once the conception has thoroughly *illuminated* his reason will he realise it, i.e. put it into practice in a physical form which is accessible to perception.

---

138 "We determine the content of a concept through pure intuition from out of the ideal sphere. Such a concept contains, at first, no reference to any definite percepts. If we enter upon an act of will under the influence of a concept which refers to a percept, that is, under the influence of a mental picture, then it is this percept which determines our action indirectly by way of the conceptual thinking. But if we act under the influence of intuitions, the driving force of our action is *pure thinking*. As it is the custom in philosophy to call the faculty of pure thinking 'reason', we may well be justified in giving the name of *practical reason* to the moral driving force characteristic of this level of life" (ibid., 129).

Mental images or concepts become the aims (motives) for action when the subject consciously or unconsciously links these types of driving forces with them. Drives are aroused naturally by mental images (of enjoyment to be obtained, etc.), feelings likewise by mental images but also by general concepts (e.g. as enthusiasm for environmental or political ideas) and practical experience is likewise linked to mental images or general concepts which were acquired in the past and which are therefore available internally. This includes moral commandments or normative ethical principles, the categorical imperative as defined by Kant.

In contrast to this, for practical *reason* it is solely a matter of the *insight gained from a person's own thinking* (conceptual intuition) into the ideal content of the goals which they aim for. This is independent of whether this content was first gained through conceptual intuition or was previously thought or learned or was in fact acquired by the subject through a normative ethical principle or commandment. "As soon as I see the justification for taking this content as the basis and starting point of an action, I enter upon the act of will irrespective of whether I have had the concept beforehand or whether it only enters my consciousness immediately before the action" (ibid., 129). The point is whether the thinking individual is *currently* able to understand the motive.

The highest form of willing is therefore that in which the motive is a pure, objectively understandable idea and the driving force is the subject's own pure thinking. From an ethical standpoint i.e. in terms of maxims of action, the individual subject is therefore left fully to himself on the one hand and – *simultaneously (!)* – placed fully in the objectivity of the goals of his action on the other. For at this level of willing it is *my* thinking which can clearly understand the meaning of the objectives of action and which can creatively intuit *my own ideas* when faced with a specific situation for action, without being bound by a normative ethical road map or a categorical imperative. At the same time, this thinking is directed in a completely supra-individual manner to the objective content of the idea which, like all thoughts, is understandable for others. "It is clear that such an impulse [pure thinking] can no longer be counted in the strictest sense as belonging to the characterological disposition. For what is here effective as the driving force is no longer something merely individual in me, but the ideal and hence universal content of my intuition" (ibid., 129). In his *"Philosophy of Freedom"*, Steiner called the viewpoint which takes as its objective the development of this highest intellectual form of willing *ethical individualism* (ibid., 135).

As defined by Steiner, *it is only this willing based on pure thinking* which can be called *free*: "The mere concept of duty excludes freedom because it does not acknowledge the individual element but demands that this be subject to a general standard. Freedom of action is conceivable only from the standpoint of ethical individualism" (ibid., 138–139)[139].

This allows the experience of freedom which Peter Bieri calls a "manifest unquestionable experience" (Bieri, 2001, 19) to be defined more exactly. Why is it impossible to doubt this experience? There is no possibility of doubt because we ourselves consciously produce the *thinking activity* which brings forth the thoughts of the objective, judges it as meaningful and therefore sees it as something to be realised, and because the *content of thought* which we then produce and understand lies before our inner eye in total clarity. We are completely certain of the existence of our thinking because it is something we produce ourselves. We cannot doubt it *because* we create it. René Descartes based his entire philosophy on the certainty of this experience of thinking. Steiner puts it in the following words (Steiner, 2011, 37):

> The feeling that he had found such a firm point led the father of modern philosophy, Descartes, to base the whole of human knowledge on the principle *I think, therefore I am*. All other things, all other events, are there independently of me. Whether they be truth, or illusion, or dream, I know not. There is only one thing of which I am absolutely certain, for I myself give it its certain existence, and that is my thinking. Whatever other origin it may ultimately have, may it come from God or from elsewhere, of one thing I am certain: that it exists in the sense that I myself bring it forth.

The neurophysiological correlate of the experience of freedom can therefore only be that of pure motive-forming thinking. The correlate of the cognitive creation of intention and control of action is located in the

---

139 Ethical individualism should not be confused with an asocial way of life of the individual subject which is common for instance in impulsive forms of volition The thinking necessary for ethical individualism enables not only insight into the subject's own motives but also knowledge of the motives of others and the connection between the two: "I differ from my fellow man, not at all because we are living in two entirely different spiritual worlds, but because from the world of ideas common to us both we receive different intuitions. He wants to live out *his* intuitions, I *mine*. If we both really conceive out of the idea, and do not obey any external impulses (physical or spiritual), then we cannot but meet one another in like striving, in common intent. [...] Were the ability to get on with one another not a basic part of human nature, no external laws would be able to implant it in us. It is only because human individuals *are* one in spirit that they can live out their lives side by side. The free person lives in confidence that he and any other free individuals belong to one spiritual world, and that their intentions will harmonize. The free individual does not demand agreement from his fellow human beings, but expects to find it because it is inherent in human nature" (Steiner, 2011, 139–140).

frontal lobe of the brain, specifically in the prefrontal cortex (overview in Goschke, 2006). The prefrontal cortex is in turn part of a complex neural system which includes other cortical and subcortical areas which are responsible for less conscious psychological functions such as emotional and unconscious motivational processes. Pure thinking itself is psychologically prepared by non-thought processes, such as the *need* (in other words a feeling) of wanting to understand some fact through thinking[140]; and feelings then follow the purely mentally – i.e. freely – created motives[141]. These are in fact important for turning the created motive into actions (Damásio, 1994).

It is similar for actions which are not free, those which are instrumental as links in a chain of actions for realising a higher-order, freely chosen motive. When I drive to the library with the freely chosen intention of fetching a book from there, then the individual movements of handling the steering wheel which serve the realisation of this aim, the movements of climbing out of the car, going up the steps, etc. have to be classified not as freely willed but as half-conscious or reflex-like in terms of the lowest of the above-mentioned levels of willing. In the real sense meant here, in "free" action, free only applies to the component of the free creation of the motive and control of the action carried out in pure thought which, for its execution, requires the prefrontal cortex, but not to the semi-automatic individual actions which serve as instruments for the execution of this motive and which only require the involvement of subordinate centres of the brain. In fact, the mental experience disappears from consciousness during the action itself, i.e. in the process of moving the limbs we are no longer aware of what is happening there as a psychological act of will[142].

---

140 Cf. *"The Philosophy of Freedom"*, Chapter II: *The Fundamental Desire for Knowledge* (Steiner, 2011, 21ff.).
141 "While I am performing the action I am influenced by a moral maxim in so far as it can live in me intuitively [i.e. purely in thought]; it is bound up with my *love* for the objective that I want to realize through my action. I ask no one and no rule, 'Shall I perform this action?' – but carry it out as soon as I have grasped the idea of it. This alone makes it *my* action" (Steiner, 2011, 135–136).
142 This is also why e.g. in his classification of the emotional life, Franz Brentano completely omits the will and only discusses the formation of mental images, judgements and feelings (Brentano, 1911). Cf. Steiner's detailed discussion on the topic in: (Steiner, 1996). Steiner distinguished the three main forms of activities of the soul and how conscious they become as follows: 1. thinking and forming mental pictures: *fully conscious*, 2. feeling: corresponds to the degree of consciousness of *dreaming*, and 3. willing in moving the limbs: that of *sleeping* (Steiner, 1996).

The impulses for action studied by Libet and by Haggard and Eimer with their readiness potential belong to the last two, the subordinate ones, and *presuppose*, both psychologically and physiologically, the superordinate free ones of the prefrontal cortex, if they are to be interpreted in the context of the question of freedom at all. For an action which we can describe as free, the insight into the motive and into its justification must *already be carried out* by reflection *before* the action which belongs to it can be started. *Looked at purely for itself*, the "urge felt" to move the finger which Libet investigated does not belong to the experiences of freedom under any circumstances and can therefore not be used to refute freedom. According to Thomas Goschke's explanation, Libet's original interpretation of his findings rests "on a mistaken idea of the causal role of conscious[143] intentions [...], according to which conscious intentions are the *immediate triggers* of individual deliberate movements" (Goschke, 2006, 140). In contrast to this, an alternative conception arises from a more complete interpretation of the psychophysiological findings, "according to which conscious intentions would be better viewed as *modulated boundary conditions*, which [...] *configure* cognitive, sensory and motor systems in a particular way" (ibid., 140), i.e. *define* their characteristics. Goschke therefore claims (ibid., 140):

> Viewed in this way, the causative action of conscious intentions does not consist in them "initiating" individual movements, but in the fact that they modulate the readiness with which particular actions are activated in reaction to particular stimuli. When interpreting Libet's findings this means that the subjects actually formed the causally relevant intentions long before the moment when they carried out the individual movements – i.e. at the beginning of the experiment when they received the experimental instructions and agreed to carry them out.

In order to discuss the question of freedom, we must therefore return to these initial "causally relevant intentions" arrived at through insightful reflection. Because an action can only be called free if it is executed in accordance with a freely created motive. This motive must be clearly formulated and understood *before* the action is initiated. Psychologically and physiologically the question of freedom must be directed towards the execution of *pure thinking* upon which the creation of a motive is based.

Pure freely activated thinking therefore arises, dynamically speaking, like an island of certainty from the continuous flow of diverse mental activities

---

143 It would be more precise and correct to replace "conscious intentions" here and "conscious" in the following quotation by "free intentions" and "free". Because even the "urge" to move the finger studied by Libet is of course conscious. It is apparent from the final sentence in the passage quoted below that, with "conscious intentions" Goschke means those formulated on the basis of mental insight, in *contrast* to what is meant by Libet.

which are experienced in varying degrees of consciousness and in a more or less active or passive form but which by no means lay claim to the psychological status of freedom. "Even decisions which are felt to be conscious are always prepared and influenced by a wide range of unconsciously negotiated processes" (Singer, 2004, 52).

However, on the *psychological* level, it should not be concluded from the fact that pure thinking is embedded in other activities of the soul that the experience of freedom referred to here is *caused* by the activities of the soul which precede it or work in the psychological subconscious in parallel to it. Observation of the soul shows clearly that e.g. the *urge for knowledge* which precedes thinking may well be a need which arises spontaneously (i.e. not freely) in response to a particular perception, but that this need only supplies the *instigation* for the subsequent thought and judgement activity. As thinking individuals we experience this as *freely caused* in the above-mentioned sense (cf. Steiner, 2011, 21ff.). "Free *from*" other activities of the soul such as feelings etc. because thinking prevails over these and supresses them as long as it can maintain itself; and at the same time "free *to*" in that it intends only the content of thinking and nothing else. Pure free thinking is emergent compared to these other activities of the soul[144] and it is only this which is experienced as being caused solely by the human ego. Inner observation shows "that *only* in the thinking activity does the I know itself to be *one and the same being with that which is active*, right into all the ramifications of this activity. With no other soul activity is this so completely the case. For example, in a feeling of pleasure it is perfectly possible for a more delicate observation to discriminate between the extent to which the I knows itself to be *one and the same being with what is active*, and the extent to which there is something passive in the I to which the pleasure merely presents itself. The same applies to the other soul activities" (ibid., 44–45).

In a similar way the *neurophysiological correlate* of the unfree activities of our mind which precede or accompany pure thinking cannot be seen as the *cause* of the neurophysiological correlate of pure motive-creating thinking. Because here, in the bio-physiological realm, it is not linear causality in the sense of a mechanical chain reaction in which the prior event in time is the *cause* of what follows which applies, but *self-organised action and the occurrence of emergent conditions* which actualise their laws *out of themselves*. What *precedes this actualisation in time* and what *underlies* it as a

---

144 These other activities of the soul include *having* thoughts or the spontaneous appearance of thoughts *of their own accord* in the field of our awareness. This kind of thinking or imagining is not empirically experienced as a production of our ego, as the thinking referred to here which is actively caused by it (Steiner, 2011, 45).

material and functional substrate is not the cause, but the *necessary condition* and *instigation* (cf. Sections 4.7–4.8 and 5.2–5.4).

In terms of the question of freedom, only the causal relationship of pure thinking to *its* neurophysiological substrate will therefore be taken into consideration in what follows. What was presented above in more general terms for consciousness also applies here to pure self-aware thinking: first, there is no observation which indicates that the *brain* gives rise to the free decision. The claim "that the so-called 'free decisions' are also made by the brain itself" (Singer, 2004, 57) is epistemologically wrong. These kind of decisions can only be observed as being made by the spiritual ego of the individual. Secondly, it really cannot be disputed that the cognitive acts which are felt to be free basically rest on the same type of neurophysiological processes as do those that are unfree (ibid., 61). However, this would mean that the inner execution of such acts rests physiologically on destruction in the same way as those other, unfree processes (cf. Section 5.3). So, while I think *with* my brain, I simultaneously think *in opposition to* my brain. The brain is like the ground on which and, in a certain sense, *against* which our ego with its *independently created* activity moves. This means that the action of our spiritual processes – and therefore the freedom of the human will – is not incompatible with the fact of physiological processes. It would then follow that it is not the experience of certainty of freedom but its causation by the brain which is the "illusion" and "social convention", i.e. that of certain neuroscientists.

Furthermore, actual observation of pure thinking on the one hand and the processes of the brain on the other clearly show two different types of "objects" which are each subject to different laws. The brain functions in accordance with *physiological* laws, pure thinking in accordance with *logical* laws (Steiner, 2011, 36):

> This transparent clearness concerning our thinking process is quite independent of our knowledge of the physiological basis of thinking. Here I am speaking of thinking in so far as we know it from the observation of our own spiritual activity. How one material process in my brain causes or influences another while I am carrying out a thinking operation is quite irrelevant. What I observe about thinking is not what process in my brain connects the concept lightning with the concept thunder but what causes me to bring the two concepts into a particular relationship. My observation shows me that in linking one thought with another there is nothing to guide me but the *content* of my thoughts; I am not guided by any material processes in my brain. In a less materialistic age than our own, this remark would of course be entirely superfluous. Today, however, when there are people who believe that once we know what matter is we shall also know how it thinks, we do have to insist that one may talk about thinking without trespassing on the domain of brain physiology.

The epistemological *equality* of thinking and the brain and the resulting ontological equality which even Singer admits likewise means the ontological recognition of the human spirit with its capability of freedom. This does *not* deny that this capability of freedom is dependent on the brain. On the contrary, it could even be conceded that the brain shows an expression of the thinking activity *before* this thinking activity becomes *conscious* as something free which I myself have activated. This is similar to the sense experiences and the initiation of movement studied by Libet and others which becomes conscious *after* the brain activity which is linked to it with a *time delay* of a few hundred milliseconds (Roth, 2001, 194–197). From the viewpoint adopted here, a time delay of this kind is actually to be expected, without calling into question the reality of the experienced freedom. Admittedly this appears paradoxical and will therefore be explained in more detail below.

Something which is first in *time* is not necessarily first *principally* and *causally*. This is only so in mechanics where the cause of a manifestation is always present *before* the effect and this leads to a continuous chain of observable processes. Neither the experiences of consciousness in general nor those of freedom in particular can be proved as being *created* by those neural processes which precede them in time in *this* way. First there is no observable transition from the neuronal to the mental events and therefore no empirical proof of the claimed creation.

Secondly, neuronal processes are only *conditions* for the emergence of consciousness: the latter is emergent compared to them and realises itself in them – as does everything emergent in comparison to its basis – by its own laws. This self-realisation is clearly observable in the case of pure thinking whose creation by the active ego is an existentially certain experience for the thinker.

Thirdly, it is not claimed here that the spiritually freely active ego is an entity "which *precedes* the neuronal processes" (Singer, 2004, 57, ePH) or that free will is *"unconditional"*, i.e. independent of "prior conditions" (Goschke, 2006, 107). The point is merely that of distinguishing the *conditionality* from the *causality* and pointing out that neural conditionality is compatible with mental efficient causation, a view which is discussed as "compatibilism" by other authors in the current debate (cf. Goschke, 2006). Furthermore, it is said that when acting out of freedom, the full *insight* into the motive and its purpose precedes the action. But it is obviously out of the question that the thinking ego experiences itself as "prior" to the neural processes. From an internal perspective the ego does not experience any neural processes, but only its thinking activity and the content of thought which this brings about. It is also aware of how this thinking activity must forcefully overcome other experiences of consciousness which arise, such as when it is disturbed by external noises while concentrating on solving an arithmetic problem or by looking forward to the imminent evening meal. The *very fact* that every fibre of the thinking referred to here must be produced consciously by the thinker in order to appear in the first place, gives the thinker a guarantee that nothing else can play into the content of this experience of thinking and freedom than what he himself has produced, i.e. thinking content and thinking activity, in other words nothing from e.g. the physical or emotional organisation.

Psychophysiological experiments can also be expected to produce results showing that, at the neural level, the neural correlate of this focused execution of thinking activity will appear in place of the corresponding correlate of other experiences of consciousness. However, the inner perspective does not reveal the nature of the connection between thinking and its own physiological correlate and it cannot therefore be claimed that thinking must precede the occurrence of its neural correlate. This would only apply in the interactionist dualism of Descartes or Eccles where it is claimed "that the 'immortal autonomous spirit' acts upon the brain and triggers the observed correlative brain processes, just as the pianist elicits notes from a grand piano by pressing the keys" (Roth, 2001, 194). Only if interactionist dualism were correct would the activation of thinking precede its physiological correlate just as the activation of the finger movement precedes the depression of the piano key.

But this is not the case if the brain/spirit relationship is seen in the systemic emergentistic view of ontological idealism presented for discussion here. In contrast, the occurrence of physiological events would then be compatible with the corresponding experience of freedom *afterwards*, *although* in the above-mentioned sense the experienced free spiritual activity must be causally explained by itself and not by its neurophysiological correlate.

After all, the *law* of an emergent phenomenon is to be found *in itself*, independently of its manifestation and, as the *principle* or *"being"* of this manifestation, *precedes* it (as universale ante rem) just as the idea of a sculpture precedes its manifestation. It is then unimportant whether the idea precedes its appearance *in time* but that it is *first in principle*. And *it remains this first in principle* during the entire course of its process of actualisation, in other words, never ceases to be first in principle during its entire activity. But just as in the sculpting of the work of art in plasticine or clay from the realisation of the idea movements or forms emerge even *before* the work of art *appears* – movements which cannot be causally explained by the material – this occurs in an analogous sense in the *self-actualisation* of emergent laws in nature. For example in the organism, when the embryo has not yet revealed the developmental law of the mature form but nevertheless presents an expression of this at a subordinate level. The appearance of thinking which manifests in the act of thought can be thought of in a similar way.

This approach can be reconciled with the empirical psychophysiological facts just as well as the reductionist view of present day neurophysiology, but in comparison actually better, inasmuch as it is not forced to deduce consciousness, spirit and freedom from the brain by disregarding the observational gap between physiological and psychological empiricism. When Singer claims: "Decisions arise in the brain as a result of self-organisation processes" (Singer, 2006, 86), this is once again epistemologically wrong. The formation of decisions is not observed *in the brain* but *in the spirit*. This, inasmuch as it is free, "organises" or actualises itself. What is then observed – in a different way – in the brain as a *physiological* process of self-organisation would be viewed *as a consequence* of the spirit's self-actualisation in the sense meant here. In this interpretation, as a neural process of *becoming aware* of thinking, what may possibly come first *in time* is not the neural *cause* of thinking and the experience of freedom based in it, but the *consequence* of the essence (lawfulness) of thinking itself which is manifest in the neural substrate. Both – the neural preparation and the awareness of thinking which follows this – are thus a form of actualisation of the first in principle. However, the first in time, the physiological preparation, is a *condition* for what follows. Consequently, the thinking which *emerges* in the realm of consciousness cannot be the causal reason for the brain processes as is assumed by the type of interactionistic dualism represented by Eccles. Neither can the first in time, the preparatory brain process, be the cause of thinking as believed by Roth, Singer and many neurological scientists; but both have their origin in the essential nature of thinking which is the cause of its own manifestation. This essence therefore has a double role which may explain why the emergence, i.e. the manifestation of thinking as a conscious activity is correlated not with creation but with destruction from an organic angle. Steiner was the first to point this out in his *"Philosophy of Freedom"* (Steiner, 2011, 124):

> The essence that is active in thinking has a twofold function: first, it represses the activity of the human organization; secondly, it steps into its place. For even the former, the repression of the physical organization, is a consequence of the activity of thinking, and more particularly of that part of this activity which prepares the *manifestation* of thinking.

From this point of view the brain does not serve the creation of an illusionary "spirit" but the *appearance of consciousness* of the real thinking human spirit active "in the brain" (ibid., 124–125):

> An important question, however, emerges here. If the human organization[145] has no part in the *essential nature* of thinking, what is the significance of this

---

145 What is meant here by Steiner is the "psycho-physical organization of man" which contains everything of a physiological and conscious or unconscious psychological nature, which could subject thinking to something other than its own laws and therefore make the experience of freedom an illusion (ibid., 123).

organization within the whole nature of man? Now, what happens in this organization through the thinking has indeed nothing to do with the essence of thinking, but it has a great deal to do with the arising of the ego-consciousness out of this thinking. Thinking, in its own essential nature, certainly contains the real I or ego, but it does not contain the ego-consciousness. To see this we have but to observe thinking with an open mind. The I is to be found within the thinking[146]; the "ego-consciousness" arises through the traces which the activity of thinking engraves upon our general consciousness, in the sense explained above. The ego-consciousness thus arises through the bodily organization.

And the ego-consciousness is the consciousness of oneself as a thinking being who experiences his thinking activity as it is produced and the content of thinking contained in this. Only when this is thought in full consciousness can it become the motive of free will. It is the *consciousness* of the insight into the content and purpose of a motive in the form of a pure thought and produced entirely independently by the human spirit through pure thinking which constitutes the freedom of human will. Freedom is neither an "unconditional", uninitiated and undetermined will of an elevated hypothetical spirit playing on the piano of the body as defined by Eccles, nor a social convention and illusory creation of determinism of a piano which plays itself as defined by Roth and Singer, but that form of the human will which displays this as a fully conscious activity of the self-aware human spirit in its experience of its own reality. The thinking ego does not owe the brain its being and activity, but its *consciousness* and, *indirectly* through this, its freedom.

5.9 A comprehensive basis for medical anthropology: body, life, soul and spirit

Human medicine is not simply a matter of physics, chemistry and biology and, despite many similarities between animals and humans, not a matter of zoology either, but – to intentionally express it with a tautology – a matter of *human anthropology*. While this does contain aspects of physics, chemistry biology and to a certain extent "zoology", it also goes beyond these due to the specifically human element. By focussing the scientific attention of the last 200 years on the molecular and biological basis of the organism and consciousness, the organism's material structure and function and its inner relationship beyond the species boundary have been explained to an extent never seen before, something which has also lead to enormous progress in medicine. On the other hand, the *human element as such* has largely disappeared from the field of view or has been challenged as being something special and unique.

---

146 Compare what was said with reference to Descartes in this section about the experience of existence of the ego as a thinking entity.

The negative results of this development mentioned in the introduction must likewise not be overlooked.

A solution to this problem in terms of medical anthropology requires us to *pose the question of the being of man anew*: and the empirical ontological idealism put forward here and the view of knowledge and reality upon which it is based could make a relevant contribution to this. What is important here is that the various phenomena of physics, chemistry, biology, morphology, psychology, etc. which can be found by observation and which are emergent in their natural hierarchy should not be reduced to the underlying levels but be treated as of equal value by science. This means initially considering each of these realms of phenomena on their own through observation and thinking and only then – in an additional effort of cognition – explaining their mutual dependencies.

The outcome of this approach is that every realm of phenomena can be ascribed its own characteristics and laws which cannot be derived from those of other phenomenal realms; that reality comprises not only the phenomena, but also the laws which determine them; that the mutual dependence of the realms of phenomena does not need to be a cause but can also be a condition; that causality need not only mean external causation but can also mean self-causation. In short, that different forms of activity can be acknowledged and all phenomenal realms of reality must be seen as epistemologically and ontologically of equal importance. By integrating the findings from these areas, a comprehensive scientific view of the human being in their complex physical, living, emotional and spiritual reality arises, a view that can justifiably be called an "integrative" or "holistic" approach. The medical picture of the human being as anthropology i.e. as a science of the human being, therefore gains a new meaning in both theory and practice which can contribute to the change in direction in medicine which many people today are looking for.

As determined by the phenomena, the integrative or holistic approach mentioned is a systematic hierarchically ordered one. Systematic concepts have been repeatedly presented in the medicine of recent decades. They generally contain layered theoretical models which primarily postulate a hierarchy of the *physical* structure of the human system and which, with variations, comprise roughly the following divisions: *elementary particles* → *atoms* → *molecules* → *macromolecules* → *organelles* → *cells* → *organs* → *organism* → *family* → *ethnic groups* → *humanity* (Feigl & Bonet, 1989; Medicus, 2006). In addition to this there are sub-divisions which cover the *physical, psychological and social* aspects of the human being such as e.g. the bio-psycho-social model (Adler, 2005). Of course it can always be argued that for every division it is possible to create sub-divisions (von Uexküll & Wesiak, 1998, 98). Other classifications take account not only of the psyche but the thinking spirit of the human being (body → psyche → spirit) (Frankl, 2005) or differentiate the body into a level of matter (Greek: hýle) and one of life (Greek: bíos), i.e.

(hyle → bios → psyche → logos) (Danzer, 2013); and, recently, emphasis has been placed on the religious side of the soul and spirit for medicine (model with physical, psychological, social and religious constitution of the human being) (Bausewein et al., 2007; Büssing et al., 2009). Models like these are not derived from theory but from *empirically* derived perspectives of the patients and of applied medicine and also lead to practical therapeutic applications (cf. Section 5.7). So far, however, there has been almost no discussion of the implications of this multiperspective view for the fundamental understanding of the human being, or it is discussed as part of the body/soul debate, usually with the result that soul and spirit are not recognised as having any real status due to a reductionist *theory*.

However, if you consider cognition in the sense meant here as the penetration by thought of the *actual* phenomena and effects of inherent laws (cf. Chapter 2), this leads to a differentiated synthetic view of the human being and his relationships to the environment in which not just *one* (the material) form of causality and reality can be accepted, but *different ones* which are nevertheless related to each other. *Systemically* in human beings there are initially four main emergent levels which can be distinguished, through which the human being is related to the other realms of nature and is himself differentiated as a hierarchically ordered complete being. These are the main realms of physical substance, life, the soul and the spirit which are shown in a simplified form in Table 1 and which have been described throughout history and since the 20th century by several well-known scholars in addition to Steiner (Steiner & Wegman, 1996): Max Scheler (Scheler, 1947), Nicolai Hartmann (Hartmann, 1964), Helmut Plessner (Plessner, 1975) and Gerhard Danzer (Danzer, 2013) (see also Heusser, 1999b; Heusser, 2013).

Tab. 1: Emergent realms of being and systemic levels in nature and man

Mineral	Plant	Animal	Human being	Systemic realms
			Spirit	"I"
		Soul	Soul	Astral body
	Life	Life	Life	Etheric body
Matter	Matter	Matter	Matter	Physical body

All four realms of nature consist of *matter* including the plants and the physical body of animals and man. In terms of what was explained in Section 3.3 about the holistic character of substance, even this physical level turns out to be hierarchically arranged.

But plants, microorganisms, animals and human beings are not just physical systems but also *living* ones with the emergent characteristics of growth, metamorphosis, nutrition, breathing, defence, excretion, healing, reproduction and others. Due to the empirical reasons discussed, the systemic cause of this can be identified as a special class of specific organisational laws and their forces, for which Goethe coined the terms organic "formative drive" or "formative force" (Goethe, 1950j, 237, 240) and Steiner usually referred to as "etheric" forces, and which can be made causally responsible for the functional and structural forms of the living realm as I have attempted to show above. When the activity of this organisation of forces ceases, death and decay of the body into the mineral realm follow (cf. Section 4.11).

Animals and human beings possess the further emergent characteristics of the *realm of soul* with the phenomena of consciousness: sensation, pain, desire, joy, fear, instinct, drive, intention, voluntary movement, etc. These psychological phenomena – even from the way in which they arise as inner psychologically experienced qualia – show themselves to be of a completely different kind to the physical phenomena and accordingly follow other laws than these. Steiner refers to this as an "astral" organisation (cf. Section 5.4). The actualisation of the psychological laws and forces is *dependent* on the presence of the corresponding living and material basis, but is not *caused* by these, in a similar way to the relationship of the etheric forces to the physical ones.

The human being differs from the animals by the additional and emergent fact of his individual *spirit* which gives his soul the capacity for *thinking* and therefore the capacity for insight, freedom and culture, which are not accessible to the animals (cf. Section 5.6). Hegel aptly remarked that: "It is *thinking* which makes the soul, with which an animal is also endowed, into spirit" (Hegel, 1969, 13). The human spirit is the entity in man which can bring to light the spiritual basis of material, living and ensouled nature in the form of laws, and in this way can also know itself. This is what is experienced as the innermost determining essence of the soul, as the actual ego or the spiritual individuality of the human being.

Each of these emergent levels has its own phenomena and laws, where the laws of the higher system levels can only be actualised when the relevant subordinate system levels and other conditions are met. To this extent the actualisation of the superordinate system characteristics is *dependent* on the subordinate ones, but it is *not caused* by them, but has its own independent existence. The external expression of this independent existence and action is the fact of self-organisation at all these system levels. Self-organisation, however, is only possible through *self-activity*. The philosopher Nicolai Hartmann (1882–1950) aptly expressed this interrelationship of the four levels of existence of the physical, organic, psychological and spiritual existence of the human being as a complete hierarchical structure (Hartmann, 1964, 177–183):

> It [the nature] clearly shows the four major levels of physical, psychological, organic and spiritual existence.
> Thus organic nature rises above the inorganic. It does not float freely on its own, but requires the conditions and laws of the physical material level: it rests on these, even though these are in no way sufficient to constitute the living. Psychological existence and consciousness likewise depend on the supporting organism, which is essential for their existence in the world. And the great historical phenomena of spiritual life are dependent on the psychological life of the individuals who are its bearers. From one level to the next, across every break, we find the same relationship of support or conditionality "from below" and at the same time the independence of that which is supported with its own specific form and laws.
> This relationship is the true unity of the real world. In all its manifoldness and heterogeneity, the world cannot do without unity. It has the unity of a system, but the system is composed of levels. The structure of the world is one of layers. The point here is not the unbridgeable nature of the gaps – because it may be that these only exist "for us" – but that new laws and categorical forms arise which, while subject to the lower ones, are nevertheless demonstrably unique and independent of them.

Each of these layers is of equal importance epistemologically and ontologically and each must be cognised in principally the same way: their manifestation must be observed (possibly with the help of instruments) with "outer" senses and their laws must be brought to light with the "inner sense" of the human spirit, thinking. In this respect cognition must of necessity assume a multi-perspective approach if it wants to grasp the complex reality of the human being and of nature in their differentiated constitution as physical, organic, psychological and spiritual entities. This is the way in which Goethe viewed his comprehensively conceived morphology (Goethe, 1950I, 415):

> *Morphology*
> Is based on the conviction that everything that exists must also give intimation and evidence of itself. We shall apply this basic law from the first physical and chemical elements to the most spiritual expression of the human being.
> We shall turn immediately to what has form. The inorganic, the vegetative, the animal, the human all intimate themselves, it appears, as what they are to our outer and inner senses.

An attempt has been made here to show that a differentiation of medical anthropology according to body, life, soul and spirit is not only compatible with the *available empirical facts*, i.e. the phenomena and laws of matter, the biological and psychological processes and the free spiritual activity of the human being but, following a logical line of enquiry, actually comes about through them. In addition to the 20th century scholars such as Max Scheler and Helmut Plessner already mentioned, Gerhard Danzer, a leading German professor in psychosomatics at the Charité in Berlin, once again provides a comprehensive description of the human being on these four levels: body matter ("hyle"), life ("bios"), soul ("psyche") and spirit ("logos") in his recent book on medical anthropology, psychosomatics and person-centered medicine, *"Personale Medizin"* (Danzer, 2013).

This view is also implicitly implied by someone who is currently the main reductionistic interpreter of humanity, Wolf Singer, when he distinguishes dead substance, living substance, soul and spirit based on actual experiences (Singer, 2006, 86) and, at the same time, assigns these experiences an epistemologically and ontologically equal status (cf. Section 5.8). But his reductionistic thinking habit prevents him from drawing the obvious conclusion from this equality and hence prevents him from coming to a recognition of the human being as a material *and* living, psychological and spiritual being in the real sense. He writes (ibid., 86):

> The subdivision of the world into the levels of non-living matter, living organisms and psychological and spiritual processes only reflects the coexistence of descriptive systems for distinct experiences. However, the existence of different descriptive systems does not mean that the phenomena referred to in them cannot be related to each other. For example, it could behave like the different views of an object. Reduction or explanation would then simply mean the production of links between phenomena which have received a different description from differing positions. However, this is a process which is carried out unchallenged and with great success in the natural sciences.

This must be agreed with by and large. But differing "descriptive systems" only have meaning, when different systems are also *perceived*. And in this case it is epistemologically not simply a matter of the description but of grasping the different laws of what has been perceived. Furthermore, the *linking* of the levels of body, life, soul and spirit is not identical with "reduction or explanation" because a higher emergent level, in accordance with its content, is not causally explainable by or reducible to the subordinate levels. In addition, not only the *causality* but also the *conditionality* of the higher system levels on the lower ones would be a possible relationship between them and, in fact, one which could allow the superordinate and subordinate levels an equal identity. Singer can obviously not see or admit this. He seems to be so blinkered by thinking habits that he can only conceive of a consistent image of the human being under the *presupposition* of *reductionism*: "However, if you accept that our views of the world are constructions of the brain, then the conflict between the reductionist approach of modern brain science and the spiritual positions appears to be soluble" (ibid., 86). However, this does not lead to a solution of the "conflict", because the spiritual as a product of the material is only *declared* but – as demonstrated – not verified. In addition, it is only an apparent conflict: it arises because of the wish to attribute the manifold *world of phenomena* to *a single* type of this in order to reach a unified view of man and the world, while overlooking the fact that the real and unifying element which reveals the relationships between the diverse elements is present in the *lawfulness* of this multiplicity. This oversight is the result of nominalism.

What is more, it is incomprehensible why, in view of the epistemological *equality* of the brain and spirit, the spirit should be reduced to the brain. Because the other option, of attributing the brain to the spirit, would then be equally justified. Even this is implied in Singer's comments, although he is obviously unaware of it. In other words, if Singer makes our world view, i.e. our consciousness of the world, a mere construct of the brain, then the brain itself must be a mere construct of the brain rather than reality. Because there is no other way of knowing about the brain than by becoming *conscious* of it through perception and thinking. Singer does not seem to notice that his theory cancels itself out. Singer's line of reasoning is an example of how nominalism and reductionism do not lead to a consistent medical anthropology which corresponds to actual empiricism and to the daily experience and self-conception of the patient. On the other hand, the multi-perspective rational empiricism practised by Goethe in his scientific work and established epistemologically by Steiner could contribute to solving this problem.

This applies not only to our understanding of the human being but, as is implicit in what has already been said, to the human being's *relationships to the world*. For the approach presented here also opens up the perspective that the human being not only requires to have a connection to his environment and to the universe in a *physical* respect, as commonly assumed, but that relationships could also be possible for his etheric organisation of formative forces, his soul and his spirit in a similar manner. In other words: it can be supposed that the universe consists not only of the physical bodies and forces with which we are familiar but also contains etheric, astral and purely spiritual forces and occurrences to which the human being has different relationships. However, some empirical scientific material has already been collected for the realm of the etheric forces and their importance for the biology of plants, animals and man (Wachsmuth, 1932; Wachsmuth, 1965; Adams & Whicher, 1980; Bockemühl, 1985; Edwards, 2006; Endres & Schad, 2002; Zürcher, 2010) which cannot be discussed further here. But beyond the living etheric realm these kind of perspectives might also be important for the human soul and spirit, for example for the broad-ranging questions which arise in connection with the spiritual and religious needs and hopes of human beings, questions whose relevance for medicine has become increasingly obvious, for example in relation to the question of the existence of the human soul and spirit independently of the body after death and also before birth or conception (Steiner, 1969b; 1970a; 1972; Heusser, 2003).

From this viewpoint medical anthropology is the branch of science dealing with human beings which attempts to grasp their physical-organic-psychological-spiritual wholeness and relationship to the outside world with reference to medical needs. It therefore unavoidably encounters questions on the reality of the immaterial in the human being, nature and the universe which can no longer be solved by natural scientific means alone. The question is, whether and how an *empirical scientific* approach to the solution of these kinds of questions is in fact possible. This is what anthroposophy, as a spiritual science, is concerned with.

5.10 Summary

A further application of the principles of cognition and reality as defined by Goethe and Steiner is formed by the realm of the soul and spirit as a phenomenon. The property of *consciousness i.e. the psychological* as a phenomenon is just as emergent in relation to the purely living organism as is the spatial-temporal organisation of the living in relation to pure substance or the physical body. Substance and organismic organisation therefore turn out to be necessary *conditions*, but not *causes* of the psychological realm. This is illustrated by the problem of psycho-physical causation. The common conviction that the psychological is *caused* by the brain cannot be held on epistemological grounds because the physical and organic phenomena of the brain and the – totally different – psychological phenomena of consciousness are separated by a chasm for observation. No causation can be observed. In contrast to substance and the living organisation of the brain, the psychological appears through its own phenomena, laws and activity, and thus as its own reality. This is not contradicted by the fact that consciousness can only be manifest under the *preconditions* of the living, functioning brain. On the contrary: the fact – still accorded too little attention – that *consciousness is accompanied organically by breakdown and devitalisation* which need to be regenerated between the conscious phases, is rather an argument in support of the possibility that the psychological processes might rest on an activity of their own type. For the purpose of the development of consciousness, this must exist in a certain contrast to the organic-living realm. On the other hand it can be assumed that the soul is also involved in an unconscious way in the development i.e. the regenerative reconstruction of its organic basis. For the brain's structures and life functions have to be organised *according to their psychological function, in other words in accordance with psychological laws* as is presumably the case with the example of activity-dependent neuroplasticity. A distinction must be made between the soul realm, when it is active in *organising* the development in the body, and the soul realm which becomes *conscious* to varying degrees in relation to bodily breakdown.

This soul realm is referred to generally as the *"psyche"*. The "psychological" therefore only represents a part of a much more extensive activity of the soul. This more extensive activity is often referred to by Steiner by the term *"soul"* or *"astral" organisation*.

Furthermore, due to the self-aware intelligence of the human being, a phenomenological distinction can be made and applied between the soul and spirit, because the spirit (the actual "I" of the human individuality) is another emergent non-reducible entity which can be contrasted with the purely psychological realm of the soul. The soul is subjective individual inwardness with inherent but not fully conscious qualities of intelligence, as can also be seen in the feats of intelligence of animals. The spirit transcends this subjective inwardness and its intelligence is able to consciously penetrate the objective universal laws in reality and put them to its service. Only through the spirit does the human change from a being of nature to one of culture. It is only the spirit which is accessible to questions of sense and meaning and it is only the spirit which can be assigned the capability of freedom. The only actions which can be viewed as being free are those whose motive has first been formed and understood through fully conscious thinking. In this regard I have tried to show why the readiness potentials found by Benjamin Libet and others which appear in the brain before the appearance in consciousness of a spontaneously intended action, are no argument against human freedom.

The chapter ends with a fourfold concept of the human being which recognises the physical, organic, psychological and spiritual aspects of the human being based on natural scientific and psychological empiricism as real interacting organisational elements and makes these the basis of medical anthropology.

6 From anthropology to anthroposophy

6.1 The question of the reality and cognition of the spiritual

In Chapter 2 it was shown that the purely spiritual realm is known not only to philosophy and the other humanities but is something also familiar to science, i.e. in the laws of nature brought to light by thinking. And empirical reasons were given for claiming that this spirit is objective, both for inner intellectual contemplation and also due to the fact that outer nature objectively complies with the relevant natural laws. The natural law was not viewed merely as something abstract in the subject's mind but as an actual component of reality itself. Further, it was argued that this reality must be an active one, otherwise the phenomena could not in fact be subject to its lawfulness. With reference to this and following classical and modern universal realism – and in particular, Goethe's and Steiner's view of cognition and reality – a case was made for empirical ontological idealism.

This view, however, is associated with a major problem. For *science initially does not recognise an "active" spirit*, but only the pure though abstract spirit of human thoughts. And because a nominalist interpretation leads to the general belief that this is merely subjective or even a product of the material brain which has nothing to do with the objective world, science usually makes no attempt to look for or recognise a "spiritual" element in nature, the universe or mankind. "Spirit" is therefore left to belief, irrational feelings or speculative assumption.

However, the "spirit" which we are talking about here is the brightest, clearest and most precise entity that a scientist can ever experience. It is only this spirit which makes science into science: the complete, comprehensible, ideal lawfulness. This applies to all laws accessible to science, at all emergent levels of the material realm, of the living phenomena of organisms, of psychological phenomena and of the spiritual activity of the human being. It also applies to statistics, i.e. to laws which are not manifested with predictable explicitness but only with statistical probability, for of course statistics and mathematics themselves are ideal laws and only comprehensible as such. Looked at like this it is the *spirit* which enables the scientist to really look *into* nature. Without this spirit he can only look *at* its outer appearance.

This is what is meant by Goethe's well-known objection to Albrecht von Haller when he criticises his poem:

> *Into the core of nature*
> *No earthly mind can enter.*
> *Happy the mortal creature*

To whom she shows no more
Than the outer rind.

This corresponds to the nominalist attitude which only looks at the *outside* of the natural phenomena and thinks that the thoughts about nature belong only to the thinking spirit. However, for the ideal realism subscribed to by Goethe, the matter presents itself in such a way that, while to observation the human spirit is confronted by the phenomenon like a "rind", he nevertheless penetrates the being or "core" when grasping the law, so that in cognition he always has the whole. So Goethe voices the criticism (Goethe, 1994, 237):

'Into the core of nature' –
O Philistine —
'No earthly mind can enter'
But have the grace
To spare the dissenter,
Me and my kind.
We think: in every place
We're at the center.
'Happy the mortal creature
To whom she shows no more
Than the outer rind.'
For sixty years I've heard your sort announce,
It makes me swear, though quietly;
To myself a thousand times I say:
All things she grants, gladly and lavishly;
Nature has neither core
Nor outer rind,
Being all things at once.
It's yourself you should scrutinize to see
Whether you're center or periphery.

But the lawful spirit which is inherent in nature in this form, appears to thinking simply in the abstract lifeless form of scientific *thoughts*. And, irrespective of whether you consider the contents of these thoughts to be subjective, as in nominalism, or objective as in realism: in view of this abstract spirit, it appears absurd to speak of an *active* spirit in human organisation and in nature. It is easy to see why, under such circumstances, the existence of an active spirit should be called into question.

But despite this, if the relationship of the law to its phenomenon is thought through carefully, it appears that the laws cannot be simply something abstract and powerless in terms of their objective validity outside in nature, but must be *active*. For, if the phenomenon is actually organised in accordance with the law, it must also have been *created* in accordance with this law. So, the law outside in nature – as universale in re – cannot be the powerless abstraction as appears to our mind – as universale post rem – but must contain an *active force*, otherwise the order could not be created[147]. This is pointed out by Steiner in his essay *"The Abstractness of Our Concepts"* in *"Riddles of the Soul"* when reflecting on the Viennese philosopher Vincenz Knauer (1828–1894) (Steiner, 1996, 118):

> The human being forms concepts about sense-perceptible reality. For epistemology [the science that investigates our knowing activity] the question arises: How does what man retains in his soul as a concept of a real being or process relate to this real being or process? Is what I carry around in me as concept of a wolf equivalent to any reality, or is it merely a schema, formed by my soul, which I have made for myself by noting (abstracting) the characteristics of one or another wolf, but which does not correspond to anything in the real world? This question received extensive consideration in the medieval dispute between the Nominalists and the Realists. For the Nominalists, the only thing real about a wolf is the visible substance, flesh, blood, bones etc., present in this one particular wolf. The concept "wolf" is "merely" a mental summation of characteristics common to the various wolves. The *Realist* replies to this: Any substance you find in a particular wolf is also present in other animals. There must be something else in addition that orders substance into the living coherency found in a wolf. This ordering real element is given through the concept.

As presented in Chapter 2, "concept" here means the law grasped by thinking of the order in nature, in this case in the wolf. Steiner elaborates this point as follows (ibid., 118–119):

> One must admit that Vincent Knauer, the outstanding expert on Aristotle and medieval philosophy, said something exceptional in his book, *The Main Problems of Philosophy* (Vienna, 1892) when discussing Aristotelian epistemology:
>
>> A wolf, for example, does not consist of any material components different from those of a lamb; its material corporeality is built up out of the lamb flesh it has assimilated; but the wolf does not become a lamb even if it eats nothing but lamb its whole life long. *What makes it into a wolf, therefore, must obviously be something other than hyle[148], sense-perceptible matter; and indeed it must not and cannot be any mere thing of thought, although*

---

147 The relationship of this problem to the currently much-discussed subject of *self-organisation* in nature is examined in Sections 3.3 and 4.3.
148 *Hyle* (Greek for matter) and *morphe* (Greek for form) constitute the whole of something, according to Aristotle. There is no matter without form. In the above example if you ignore the form of the whole which constitutes the wolf or the lamb, then the remaining matter continues to be subject to its own form e.g. to its structural law which is the same in the case of a corpse. This decays, because it is no longer subject to the higher law which in the living animal organises the matter into the form which constitutes the specific whole (or parts of the whole, if you descend to the level of the organs or cells).

233

it is accessible only to thought, although it is accessible only to thinking and not to the senses; it must be something working [productive] and therefore actual, – something very real.

In the state of universale in re, the organising law can therefore be thought of logically as an *active* idea, although as universale post rem it is only experienced as a powerless abstraction. This should not be taken merely *historically* as a requirement of Aristotelianism or medieval philosophy, but *principally*, i.e. as a consequence of the logical relationship of that which *provides* the law to that which *receives* the law. If the latter is *really* to be organised by the former, then this, i.e. the organising law, must have been actively realised in that which receives the law. In other words, the organising law itself has to possess energy[149]. This applies to all systematic formative laws in physics, biology and psychology.

This is the reason why the physicists David Bohm (1917–1992) and David Peat (born 1938) introduced the concept of *"active information"* into the scientific debate (Bohm & Peat, 1990). Information is therefore an *active* entity, in the first place as a causal explanation for the occurrence of emergent properties in quantum physics. The active information is thus assigned a real status, in addition to matter and (material) energy. David Peat states (Peat, 1999, 49–53):

> Toward the end of the 1980s David Bohm introduced the notion of Active Information into his Ontological Interpretation of Quantum Theory. His idea was to use the activity of information as a way of explaining the actual nature of quantum processes and, in particular, the way in which a single physical outcome emerges out of a multiplicity of possibilities. Initially this idea, of information as a physical activity, was tied to Bohm's particular theory but [...] it is possible to go further and elevate 'information' to the level of a new physical concept, one that can be placed alongside Matter and Energy. [...] Information, therefore, allows a distinction to be made between what could be called raw or 'unformed' energy and a more subtle energy, an activity which can be identified with information. This information acts on raw energy to give it form. [...] Form has associated with it the idea of a Gestalt, of global patterns, perception and non-locality. [...] It is really correct, for example, to speak of a 'field' of information, since information does not fall off with distance, neither is it associated with energy in the usual sense. Possibly the notion of field should be widened or, at the quantum level, we should be talking about pre-space structures, or about algebraic relationships that precede the structure of space and time.

These thoughts are fully compatible with the ontological idealism subscribed to here which sees the law itself as something active: in the state of universalia ante rem the formative laws precede their effect in the matter ("pre-space

---

149 Cf. the explanation on the relationship of organism and mechanism in Section 4.4.

structures", "precede the structure of space and time"). In this respect they are not fixed to any "place" ("non-locality"). Looked at like this, information as an active law may indeed be referred to as a "field" in analogy with the physical field concept, but one which is not spatially restricted, in contrast to the physical fields of physical energy ("a 'field' of information, since information does not fall off with distance, neither is it associated with energy in the usual sense").

This makes it clear from another angle why neither a vitalistic morphogenetic field nor mechanistic morphogenetic substances can hold true as an effective causal explanation of the formative processes in the organism, but Goethe's typus idea certainly can (cf. Sections 4.7 and 4.9). Because the "information" for this creation, the formative or organising law as such, actualises itself, and this can only happen when the law possesses the required effective force, in other words when it is *active*, to use David Peat's term. This is the idea of the wolf as "something active, therefore actual, therefore eminently real" as defined by Vincenz Knauer.

Because "information" or a law according to which an object is created (universale in re) is simultaneously the same one through which the creation is grasped by the cogniser (universale post rem), it is clear that the "active information" provides both the ontologically connecting element between spirit and matter and the epistemologically connecting element between object and cogniser. As the formative law it is the spirit which is inherent in the phenomenon and as a conceptual law it reveals the idea, the meaning, to the cogniser[150] of the relevant phenomenon. This idea can be communicated by verbal or other means. For this it needs to be understood by thinking activity in the consciousness which is enabled by the corresponding brain function. From the viewpoint of ontological idealism we can therefore agree completely with Peat when he supposes the active information to be the element which connects the form, meaning, biology, neurosciences, consciousness research,

---

150 In terms of objective ontological idealism it is obvious that "information" must be understood here in the sense of the *content* of its *meaning* and not in the sense of Shannon and Weaver's information theory (Weaver & Shannon, 1949). "Shannon and Weaver's 'Information Theory' is concerned with the way data – bytes of information – travel along a telephone line or within other transmission systems. The 'meaning' of a particular message is irrelevant, what is significant is only the mechanism of its encoding and decoding, plus the relative roles played by noise and redundancy" (Peat, 1999, 49). This point is highlighted here because Shannon and Weaver's information theory is usually taken as the basis of the *physiological information concept* and because the difference and relationship between information *content* which is what we are talking about here and information *carrier* which is what Shannon is referring to is often not dealt with adequately in the physiology books.

spirit and matter with each other: "Information connects to concepts such as form and meaning, which are currently debated in a variety of fields from biology and the neurosciences, to consciousness studies and the nature of dialogue. It may well provide the integrating factor between mind and matter" (Peat, 1999, 49).

But the problem remains that the *active effective* element of the information detaches itself from the *percept* and only produces a postulate *inferred by thinking* from what has been created. Of course this applies initially in a certain way to *all activity* whose existence – from a scientific or psychological viewpoint – must be *postulated* on the basis of what has been created (Stuart & Klages, 1984, 12), be this in physical[151], chemical, biological or so-called psychosomatic interactions, in fact in the whole realm of science whose findings provide the elements of medical anthropology.

In his book *"Von Seelenrätseln"* (Riddles of the Soul) Steiner now classifies all these types of science – quite independently of medicine – as *"anthropology"*[152]. This expression refers to all sciences based on perception and conceptual thinking arising from normal consciousness, in other words, the natural sciences, psychology and philosophy. The spirit can only be experienced in these in its abstract powerless form[153].

The reality of this action is a fundamental problem for "anthropology", as Steiner describes following the above-cited text by Vincenz Knauer on the action of the wolf law (Steiner, 1996, 119):

> But how, in the sense of a merely anthropological investigation, could one wish to attain the reality indicated here? What is communicated to the soul by the

---

151 For instance, electromagnetic, radioactive and gravitational forces cannot be perceived with the senses. Nevertheless their existence and the way in which they act can be logically postulated from what has been caused by them. The postulation of specific organic formative forces (etheric forces) in organisms is analogous (cf. Section 4.11), something frequently overlooked.

152 "To avoid continuous, long-winded paraphrases, I would like to use the word 'anthropology' from now on to designate that approach in science which bases itself on sensory observation and the intellectual processing of such observation, asking the reader to permit me this uncommon usage" (Steiner, 1996, 7).

153 Steiner attributes the abstract form of experiencing normal consciousness to its dependence on the physical-bodily organisation. "Anthropology" as a scientific "subject" in this sense is "only accessible as what is experienced in ordinary consciousness through the bodily organisation" (ibid., 106). This includes not only the physical phenomena but also psychological ones in the sense meant by Franz Brentano, in the broadest sense everything which can be experienced and scientifically studied as elements of the soul and spirit in line with what was discussed on the physical conditionality of consciousness in Chapter 4, all the way to logic, epistemology and philosophy (ibid., 30–32).

senses does not produce the concept "wolf". But what is present in ordinary consciousness as this concept is definitely not something "working" [productive]. Through the power of this concept the assembling of the sense-perceptible materials united in a wolf could certainly not occur. The truth is that this question takes anthropology beyond the limits of its ability to know.

The reality of a thing in the way meant here is only *present* when it is made accessible through suitable means of *perception* and is grasped in accordance with its law (cf. Sections 2.4–2.5). However, as the acting energy as such eludes sense perception and thinking, anthropology arrives at a *limit to knowledge* in a certain sense. This is why the existence of forces as such has been the subject of argument from time to time in the history of science, for example recently in 1963 by J. J. C. Smart (cf. Hacking, 1996, 64ff.). However, even if the existence of forces in general is admitted, the possibility of detecting them directly empirically can still be doubted.

But the limit to knowledge which is defined by this does not need to be an absolute one. If a way could be found of perceiving not only what has been created but also what *produces this*, then this boundary could be crossed at least at this point. The search for the discovery of this point and for a crossing of the boundary which exists for observation between that which is created and that which effects this marks one of the movements in the history of philosophy and science in Europe in the 19th century, a movement which can also be called the search for a transition from "anthropology" to *anthroposophy*. As this movement is barely known nowadays in the shadow of the huge scientific and anthropological development which has taken place, but clarifies the position of anthroposophy in the context of science, it will be sketched out in a few lines in this chapter. From the viewpoint of the history of science it can provide an understanding of what Steiner described in his book *"Von Seelenrätseln"* (Riddles of the Soul) more from the angle of the theory of science and the theory of psychology (Steiner, 1996, 3–7):

> In the first essay on anthropology and anthroposophy [...], I seek to show briefly that the true natural-scientific approach not only does not stand in any contradiction to what I understand by "anthroposophy", but that anthroposophy's spiritual-scientific path must even be demanded as something essential by anthropology's means of knowledge.
> Out of experiences that are not just personal to him, the advocate of anthroposophy believes himself justified in stating that human activity in knowledge can be developed further from the point at which those researchers stop who want to base themselves only upon sensory observation and intellectual judgement of such observation. [...] In this sense anthroposophy believes itself able to begin its research where anthropology leaves off.

The above-mentioned endeavours of the movement within European philosophical and scientific history are expressed by Goethe in his *"Faust"* in a unique poetic form (Heusser, 2000a).

Faust is portrayed in this work as the representative par excellence of the science which has developed in the west. He has completed all four of the classical faculties and can therefore survey the universe as a whole: he is in possession of the knowledge which a scholar of his era was able to achieve and is "cleverer than all the dandies, doctors, masters, writers and clerics".

But he has therefore become all too undeniably aware of the limitations of all this scholarly knowledge (Goethe, 2005):

I've studied now Philosophy
And Jurisprudence, Medicine,
And even, alas! Theology,
From end to end, with labor keen;
And here, poor fool! with all my lore
I stand, no wiser than before! (354–359)
And see, that nothing can be known!
That knowledge cuts me to the bone. (364–365)

So he feels that he has reached a *limit to knowledge* and this has become an almost unbearable situation, an existential crisis for him. What exactly is it that is causing his suffering?

Alas! in living Nature's stead,
Where God His human creature set,
In smoke and mould the fleshless dead
And bones of beasts surround me yet! (414–417)

His entire science based on sense perception and thinking does not grant him access to the *living essence* of active creative nature, but only to what has already been created, what has come to an end and is dead. The *activity* of the laws which, in terms of Goethean morphology, have created the finished solid form is not accessible to this science. But it is this *activity* which is the basis of everything that exists, which Faust seeks with a fervour which verges on despair.

That I may detect the inmost force
Which binds the world, and guides its course;
Its germs, productive powers explore,
And rummage in empty words no more! (382–385)

It is the *view of the active force* which underlies everything created which Faust, as the most developed university scholar of his time, seeks but not in a *theoretical*, merely intellectual way but in an empirical, experiential, fully *participatory* one.

Faust's colleague and famulus, Wagner "the soulless sneak" (521), still has no notion of *this* kind of striving. The spirit for him is not something real and active which he might look for in nature as an *experience* but something purely ideal which he finds separated from nature in books and which he knows only in the form of *finished thoughts*:

I've had, myself, at times, some odd caprices,
But never yet such impulse felt, as this is.
One soon fatigues, on woods and fields to look,
Nor would I beg the bird his wing to spare us:
How otherwise the mental raptures bear us
From page to page, from book to book! (1100–1105)
Pardon! a great delight is granted
When, in the spirit of the ages planted,
We mark how, ere our times, a sage has thought,
And then, how far his work, and grandly, we have brought. (571–573)

It is Wagner who later on in the phial "from many hundred matters, We by alloy – for alloy is everything – Compound the human matter throughly" would like:

What man mysterious in Nature once did hold,
To test it rationally we make bold,
And what she erst constrained to organize,
That do we bid to crystallize. (6858–6861)

When Wagner can view only the *matter* as something active, but not the *idea*, the organising law through which the matter is arranged in the organism, then the organism does not arise through "organisation" for him, but through "alloy", i.e. through an *outer* meeting of substances. These substances should solidify into a whole in the manner of inorganic matter, so that even the highest is present, i.e. "a brain, that thinks quite splendidly" (ibid., verse 6867).

In his scientific quest Wagner attends only to the sensory material reality: he is only familiar with spirit in the form of human thought to which he cannot really attach any active reality. Faust, however, who not only *also* knows this reality and this spirit, but has explored them to their limits so to speak, searches in despair, even violently, at this boundary for a way of crossing it in order to come to an experience of this *creative* and *spiritual* force which is at the root of all creation, but which remains hidden from ordinary cognition based on sense perception and thinking. This is why he addresses the famous lines to Wagner:

One impulse art thou conscious of, at best;
O, never seek to know the other!
Two souls, alas! reside within my breast,
And each withdraws from, and repels, its brother.

One with tenacious organs holds in love
And clinging lust the world in its embraces;
The other strongly sweeps, this dust above,
Into the high ancestral spaces. (1113–1120)

Goethe does not mean this in a trivial or poetic manner: it is the expression of a deep longing which is behind the cultural creativity of his era in the most diverse forms and which can also be discovered in science's striving for knowledge in the universities. On the one hand this is the striving to develop knowledge of the human being and nature on a purely empirical scientific foundation with a thoroughness not previously known, as has been the case particularly since the time of Du Bois-Reymond. On the other hand it is the attempt to found and develop an *empirical spiritual science* with the same sound basis, so that access can be gained to the initially hidden *activity* in the human being, nature and the universe. This is the striving to partner a scientific *anthropology* with a scientific *anthroposophy*, and not only in the sense of Steiner's understanding of these expressions, but in terms of the nomenclature of the time, from representatives of that second scarcely-recognised movement. This will be discussed below using two examples.

6.2 The anthropology and anthroposophy of I. P. V. Troxler

The first example is the doctor, philosopher and politician Ignaz Paul Vital Troxler (1780–1866) from Beromünster near Lucerne (Heusser, 1984)[154]. He made a name for himself in politics early in life as an opponent of the restoration in the post-Napoleonic era. For example, the establishment of the two-chamber system for the national parliament in the Swiss Federal Constitution of 1848 goes back to Troxler. Troxler studied philosophy and medicine in Jena and then practiced as a doctor in Vienna and Switzerland for a time. His patients included Ludwig van Beethoven and Heinrich Pestalozzi. In Jena he attended Hegel's first course of lectures, was one of Schelling's favourite pupils and soon wrote his own theory of medicine essays in the style of Schelling's natural philosophy of the day.

However, his basic philosophical position soon differed from or rather extended beyond Schelling's view. As Schelling's natural philosophy of the time dealt with bringing to light the "absolute" in human reason as an *idea* underlying the human spirit and nature, the ideal and the real, the subjective and the objective, in order to create the "identity" between spirit and nature,

---

154 A more detailed description of the life, work and importance of Troxler and references and information on further literature on the topics dealt with in this chapter are given in (Heusser, 1984), particularly concerning his philosophy, anthropology, anthroposophy and theory of medicine.

this no longer satisfied Troxler. In contrast to Schelling he wanted not only to *think* the spirit in nature, but to learn *to observe it empirically*.

This need for an empirical contemplation arose for him particularly in relation to the problem of life. In his 1806 paper with the telling title *"Über das Leben und sein Problem"* (On life and its problem) (Troxler, 1925) he described "life" as an active "something" which, while its existence was a requisite of the empirically perceptible facts created by it, never displayed itself to empirical observation. Life was therefore the "unknown" (ibid., 36), indeed for the intellectual contemplation of reason a "nothing" (ibid., 28) and that was its problem. The task was therefore to make it *accessible to perception* as he suggested in his next publication in 1808. A *sense* needed to be developed which Troxler called a "vital sense" (Troxler, 1808, 2) in keeping with the object in question. And the – for the time being only postulated – science based on this kind of perception he referred to not as "biology" but *"biosophy"*, as already indicated by the title of the publication.

However, as a doctor, Troxler was clear that medical anthropology required a solution not only to the riddle of life, but also to that of the soul and spirit. So for him the human being was not simply a physical-material body with living, feeling and spiritual characteristics, but a fourfold being consisting equally of the physical *body* ("Körper"), the *living body* ("Leib") and in addition the *soul* ("Seele") and *spirit* ("Geist"), as he described in his 1812 publication *"Blicke in das Wesen des Menschen"* (views into the being of mankind) (Troxler, 1921)[155]. Troxler also considered these four entities to be the elements through which the human being is related to nature: "All realms of nature appear again in the human being: *matter* has the nature of the earth, *living body* that of the plants, *soul* that of the animal nature, *spirit* that of the human being" (Troxler, 1936, 192)[156]. According to these basic

---

155 Goethe first read this book in which Troxler describes the fourfold division of the human being into matter, living body, soul and spirit in November 1812 and counted it amongst the most important publications of the year, although – despite brilliant parts – it tended to confuse people's heads rather than setting them right (Honegger, 1925).

156 In this "tetraktys" as Troxler called the fourfold division of the human into matter, body (life), soul and spirit using a Pythagorean expression, Troxler saw the real and original image of the human being which had previously been known to the ancients as "sarx", "soma", "psyche" and "pneuma" and had also been represented in this form by the apostle Paul. Due to advancing knowledge, this image of man needed to be discovered in a new way and, in particular, developed as a basis for medical anthropology (Troxler, 1921). Modern-day psychosomatics is only aware of the "psyche" and the "soma". But *this* "soma" renounces life, something which Paul still attributed to it, and actually means the physical body "sarx" (Greek for flesh), which finally ends up in the "sarcophagus" (that

premisses he attempted to sketch out a medical anthropology in which the development, physiological functions and pathological processes were not only to be understood simply as the result of physical interactions but as an expression of a harmonious or disharmonious interaction of material-physical, bodily-living, soul and spiritual forces in an organ or organ system (Heusser, 1984).

What Troxler had in mind was the development of a cognition which would be able to examine the processes of this supersensible system of forces in their realm using the same principle as applies to cognition in the realm of scientific anthropology: using the appropriate *perception* and through *understanding* what has been perceived (Troxler, 1942). For Troxler this gives rise to the prospect of a higher form of science which, in his *"Naturlehre des menschlichen Erkennens oder Metaphysik"* (natural theory of human knowledge or metaphysics) in 1828 and in his three-volume *"Logik"* in the following year he called *Anthroposophie* (anthroposophy) (Troxler, 1944; Troxler, 1829). This was to be a spiritual scientific anthroposophy exactly following the natural scientific model (Troxler, 1944, 114):

> But a time will come, and it is not far away, when anthroposophy will explain the natural phenomena of the realm of spirit in the human being to the spirit, just as physics explains the rainbow to our sight and the Aeolian harp to our ear.

The consequences for medicine of a development like this would be that not only sensory-physical empiricism but also the *spiritual experience* would need to be taken into account when finding the causes of illness and for explaining pathological processes, something Troxler mentioned in one of his early writings (Troxler, 1806, 27 and 21):

> He who is not able to reach and observe such illnesses at their origin with a spiritual view and to follow them from there, will certainly not recognise them in the corpse – even though he rummage around in the entrails and look, smell, taste, feel and touch, he will at most find the residues of processes which have ceased. A type of much higher research is therefore required in order to find the cause itself. Doctors will actually have to rise far above their previous views on the aetiology of such diseases.

It must be realised that at the time, shortly before the dawn of the modern scientific age with Du Bois-Reymond and others, Troxler's ideas met with a high level of interest amongst many of his contemporaries and that he was

---

which accepts the "sarx"). A phenomenological distinction between substance and life, as has already been attempted above could, however, lead to a differentiation of the "somatic" into a physical and a living organisation and likewise the "psychological" in the human being can be differentiated into the psychological and the spiritual (cf. Section 5.9 and Heusser, 1999b).

someone who was well-known and highly respected beyond his immediate circle. His *"Logik"* published in 1829, for example, lead to his appointment as professor of philosophy at the University of Basel in 1830. When Hegel died in 1831 in Berlin, Troxler was suggested as a possible successor. And when the University of Bern was founded in 1834, Troxler was appointed as its first professor of philosophy. Troxler also had to deliver the opening address at the University of Bern because, as the authorities who assigned him the task wrote: "you are the person who will perform this not unimportant act in a spirit which is most suitable to the celebration and the importance of the day" (Widmer & Lauer, 1980, 151). In the same winter semester of 1834/35 Troxler held his lecture series on philosophy which he published in the spring (Troxler, 1942). He describes here amongst other things, how philosophy as a science existing solely in *thought* had reached its peak in the systems developed by Schelling and Hegel. In particular, philosophy had reached its "highest and final stage of development" in Hegel's system of "pure thinking which is not attached to the phenomenon" (ibid., 290). The prospective "reform of philosophy" would no longer be found in "this or that system" but required "a specific higher organ of consciousness and cognition" (ibid., 81). The task was to develop an *anthroposophy* which was based on spiritual *perception* and the *understanding* of what was perceived, just as in the case of the sciences based on sense perception[157]. In this sense anthroposophy was considered as a requirement of anthropology right at the founding of the University of Bern.

6.3 The anthropology and anthroposophy of I. H. Fichte

The second example illustrating the emergence of anthroposophy in the history of European science as a necessity of scientifically based anthropology is Immanuel Hermann Fichte (1796–1879), son of the famous philosopher Johann Gottlieb Fichte. I. H. Fichte studied philosophy in Berlin, was

---

157 Troxler called the perceptive faculty for spiritual reality a "superspiritual sense" (übergeistigen Sinn) and the ability of the human spirit to understand what has been perceived supersensibly the "supersensible spirit" (übersinnlichen Geist) (Troxler, 1942, 6th lecture). While these designations are rather awkward they make sense inasmuch as the spiritual faculty of perception involved in ordinary consciousness for scientific anthropology knows "spirit" only in the form of abstract human thought and because, in *this* respect, the perception of the active element of the natural laws cognised as thought can be designated as being "above" the spiritual. Likewise, the judging spirit which otherwise concerns itself with concepts of the sense world, in anthroposophical cognition where empiricism is concerned, has to deal with the "super" sensible, and therefore "supersensible spirit" (cf. also Steiner, 1988a, 123, Note 3).

examined by Hegel and others for his doctorate in 1818, was later temporarily professor of philosophy and pedagogy in Bonn and held the chair of philosophy at the University of Tübingen from 1842 until his retirement in 1863.

One of the main subjects of I. H. Fichte's philosophy was the question about the nature and reality of the human spirit. On this point he was firmly opposed to Hegel. His criticism of Hegel was that the latter only allowed the spirit in its *general* form accessible to pure thinking, not as a *particular, individual* being which is how the human spirit experiences itself (Fichte, 1876, 127–128):

> To him the spirit, just like nature, is in itself a general being without individuality. The question therefore arises as to how at least the appearance of something individual can arise in it. In truth the matter is completely the reverse: in reality spirit and soul are only present as something individual and only thus an object of research.

Fichte's intention was to make scientific statements in psychology like those in the natural sciences, i.e. "in the way of *direct* observation" (Fichte, 1860, 29). And because the individual spiritual element of the human being can be actually *experienced* – and not merely *thought* – Fichte felt himself justified in claiming that: "The human soul is an individual substance" (ibid., 15), in which the expression "substance" is of course not meant physically but in a figurative sense as spiritual.

According to Fichte the human spirit can therefore be defined empirically as belonging to its own kingdom in the classification of the realms of nature (ibid., 13):

> The soul is, therefore, in regard to this original, inexplicable, and otherwise unattainable property of consciousness, a being sui generis; it forms, in the whole series of things, a step of existence for itself.

Fichte also criticised the hypotheses of the materialism of his time that consciousness and the human spirit could ultimately be explained by matter and its movement in the human brain. He therefore also rejected the material interpretation of Darwinian teaching (Fichte, 1876, 81–82):

> This, with its irreproachable intention of proving the development of the higher from the lower in perfect continuity, formulates the sentence: that life and the organisms on the earth could have arisen from the purely mechanical action of the physical and chemical forces while excluding any system or purpose. And this initially unsubstantiated claim appears to have turned into a kind of creed with us, providing the yardstick for the free-thinking scientific researcher.

In contrast to this view, Fichte draws attention to the *new* properties on each of the higher stages of development and organisation in nature which cannot be derived from the lower ones, in other words, what is nowadays referred to as "emergence" (ibid., 82):

> However, we are at first met by the previously refuted fallacy that something different and new can arise from a composition of elements which is not itself present in any of those elements. This conclusion would contradict the law of logic "that nothing can come of nothing" which also applies to nature. However, this whole line of argument that "the hypothesis is conceivable" these kind of things can be "assumed" etc. bears the stamp of such subjective uncertainty and objective improbability, that it is difficult to reach the conclusion that it has really solved the problem.

Fichte, however, also applies the sentence "that nothing can come of nothing" to itself. He explains the actual spirit which becomes manifest not from the material conditions of its appearance but *from real but latent i.e. potential states of itself which precede its appearance* (Fichte, 1860, 12–13):

> In order to explain the *conscious* phenomena of the soul (which is the proper aim of psychology), I hold that we must start from its *unconscious* state, or, what is the same thing, must *go back* from the soul as a developed subject, to its undeveloped and primary essence. [...] The real or actual consciousness is based upon a potential one, i.e. upon a *middle condition* of the soul, in which it already possesses the specific character of objective intelligence, but without being conscious of it. It is from the conditions of this preconscious existence that our *actual* consciousness must be explained, and out of which it must gradually be developed. This then is the first point in our investigation: the genesis of consciousness out of the conditions of a *prior or preconscious* existence of the soul.

The pre-conscious states of the soul include those in which it expresses itself in the body through drives and instincts which in themselves are intelligent, but in which the intelligence has not yet manifested in a pure self-conscious form. Instinctive actions are "only an unconscious kind of thought" (ibid., 32). And the emergence of consciousness is then explained *causally* from these early forms and not from the material-bodily conditions which are necessary for making consciousness possible. This applies to both the phylogenetic and ontogenetic development of the human being (ibid., 19):

> The mind is, in its fundamental constitution, an *a priori* being furnished with a system of impulses and instincts *out of* which it works itself gradually into self-consciousness, in order to create what we may term the *empirical* form of [consciousness]. According to this, mind must be regarded, in the most special sense, as existing *prior* to experience – existing, that is, in all its individuality *previous to* its own conscious states, and as being the producing cause of them.

What is very interesting is Fichte's idea that the soul and spirit in their preconscious condition are not only active in the body in the form of drives and instincts, but also *as the force which itself creates the body* and are thus involved in the creation of the physical conditions which enable consciousness. This would explain e.g. the fact that the body is formed according to the soul and spirit and carries their impression or their physiognomy, so to speak: "So from where else but directly from the soul can the individual character of the body come?" (Fichte, 1859, 55). And Fichte tries to prove this point on empirical grounds, particularly in his *Anthropology* from 1856 (Fichte, 1856). For the organic process through which this physical development comes about is effectively an intelligent one in which all individual organic processes have been coordinated into a particular whole and are also maintained in the face of disturbances. This takes place with an "individual self-activating rationality but at the same time with a power which does not reflect consciously but acts in a real artistic formative manner" (Fichte, 1876, 483). So, for Fichte, "the whole of the preconscious (organic) state of the soul is essentially and specially a process of *thinking* without, however, its thought as yet touching 'the threshold of consciousness'" (Fichte, 1860, 16). This is a thinking "which, due to its manifest analogy with the art-instincts of animals and, in its highest form, with the creative aesthetic faculty in man, may be called an 'imaginative activity of the soul which develops forms in space'" (ibid., 17).

This "formative power" (ibid., 34) i.e. the "geometrising activity" (ibid., 37) of the soul is the essential requirement for creating the bodily condition for the later development of consciousness. And in the development of *conscious* imagination and intelligence there occurs a "suspension" (ibid., 39) of the soul's powers which have previously been at work in the body, i.e. a liberation from this work.

Fichte's postulate of this preconscious activity of the soul in the body rests on a way of reaching a conclusion as is customary in science and psychology: "From the actual facts of mind in its conscious state, we draw conclusions respecting its preconscious capacities, and its whole substantial nature. [...] This circular procedure is however no other than that which meets us in all inductive processes based upon experience" (ibid., 14).

The conclusions which Fichte draws in this way are far-reaching and have consequences for his basic ideas on the relationship of body and soul. It follows that, "the soul, which shapes and maintains its (outer) body must necessarily precede it in some sense, which naturally still needs to be considered in more detail" (Fichte, 1876, 514). "The organism, it is further urged, in its fundamental constituents, must be given, i.e. the soul must find it ready made in order to impress on it its own individuality, to govern it, or in any measure, however small, to modify or change it" (Fichte, 1860, 35). And if we see a necessity to "grant at all the reality of such a type [a real substantive existence], according to which the body is organised", we also have to accept that the soul "must be present [in its body], [...] with an *interpenetrating* [and at least form-giving capacity] from the very first commencement of the bodily existence, to the laying aside of the same in the hour of death" (ibid., 36). This means that the soul and spirit of the human being cannot merely be associated with the brain but must be thought of as connected to the *whole* bodily existence.

On the other hand, when asserting a pre-existing independent soul which is active in the body, this gives rise to the thought of its continued existence after death, independent of the body. This continued existence independently of the body can in fact be noticed during life because, from a physical viewpoint, the organism is constantly being created and dying: the body is constantly broken down in a "moulting process" and, in this respect, dying and death are an "integral product of the process of life" (Fichte, 1876, 324) in which the soul is constantly preserved. Actual death is only the final consequence of this process (ibid., 324):

> If we now ask what actual so-called death means, then we must follow the same analogy: because there is absolutely no *other* or *new* phenomenon present in it. This continuous dying, the repeated shedding of the sensory chemical substances, is completed in "death". The organic soul, the "inner body" lets the sensory media fall away *completely*, just as it does *incompletely* in every moment of its life. This *"natural death"* corresponds to the *beginning* of embodiment as its opposite and consequence.

This gives rise to Fichte's idea of immortality, the preservation of the human soul in death (ibid., 329–330):

> For the question of how the *human being himself* behaves in this process of death hardly needs to be asked. Even after the final act of the living process visible to us he remains just the same in terms of spirit and organising force ("inner body") as he was beforehand. His integrity is preserved: for he has not lost anything of what was his and belonged to his substance during his visible life. It is only in death that he returns to the invisible world, or rather, as he has never left this, as it is actually the *enduring element* in everything visible – he has only shed a particular form of visibility. "Death" means simply that the ordinary view of the senses is no longer perceptible in exactly the same way that the actually real element, the ultimate basis of the physical phenomena, are imperceptible to the senses.

Anthropology, which he describes as the "*scientific* examination of the human soul" (Fichte, 1876, XV), based on "the evidence of experience" (ibid., XXI) therefore actually provides the "ultimate reason to be permitted to speak of the immortality of the human personality" (ibid., 36).

But the method of this anthropology "could only take *one* course: that, namely, of starting from the facts which lie before us in actual consciousness, and from them of drawing our conclusions respecting those hidden conditions of the soul which are necessary to bring about such a result" (Fichte, 1860, 14). In terms of e.g. the action of the soul in the body, "we must admit, that in the present state of our psychological knowledge it is impossible, in the way of *direct* observation, to come even to an approximate certainty upon this question" (ibid., 29).

Anthropology has therefore reached its limits epistemologically. But Fichte nevertheless saw a prospect of crossing this boundary. For just as the human spirit has struggled upwards from unconscious physical action to conscious understanding of the "idea", a further advance from here seems conceivable. The idea would then be the "lowest stage" so to speak of life in the spirit (Fichte, 1876, 619):

> Just as this opens the entry inwards (or upwards) to the other world in this life, allowing him to enter step by step ever further into this world, it is equally true that this series of steps is not really a process of being raised beyond oneself, but a gradual shedding of the sheaths which darken the spirit in the sense world.

By taking this into account, Fichte saw before him a whole new perspective of scientific psychology which he described as follows (ibid., 620–621):

> Up until now scientific psychology has actually taken heed of only *one* half, of the physical and mental processes of consciousness and shown some understanding of this: the other equally important half, because it contains the basis of the former, does not yet exist for this psychology. However, if the *full* spiritual being of man is to be recognised, this can only happen by paying equal attention to both halves and by understanding their true relationships to each other. This is the future task of psychology which, it must be admitted, will obtain not merely significant new scope, but also a new, genuinely founded starting point. [...]
> This fundamental understanding of the human being now raises the final results of "anthropology" to *"anthroposophy"*, because it is quite justified to speak of the knowledge which at first appears as something sensory and transient, but justifies the character of inner eternity and transcendence as *"wisdom"*.

So the demand for a *spiritual scientific anthroposophy* as a consequence of scientifically-based anthropology also arises from the philosophy of Immanuel Hermann Fichte.

The history of European science and philosophy therefore shows that the need to cross the boundaries of ordinary cognition arises not only in a poetic form such as Goethe's *"Faust"*, but in the real academic work at universities in further developing the aims of German Idealism. This step is required in order to come from what has been *created* in man, nature and the universe to an empirical cognition of the *creative* which, as a result of the created, must be accepted by science and psychology but which is hidden from ordinary physical and psychological observation. This is the need to proceed from anthropology to anthroposophy, something for which Rudolf Steiner then laid the basis epistemologically, methodologically and in terms of content around the turn of the 19th and 20th centuries.

6.4 Summary

As a result of what has been lawfully created, objective empirical ontological idealism comes to the postulate of *active* laws (or lawful *action*) in the emergent phenomena of the physical, living, psychological and spiritual elements in nature and humanity. This gives rise to the need to distinguish not only *one* (the physical) but *four* classes of active forces and laws: physical, living, psychological and ego-related/spiritual, and to take account of these in such realms as medicine.

However, a problem exists in that empirically science and psychology have before them only what has been created and can at best discover its laws in an abstract form, i.e. as a pure idea, (in physics e.g. the laws of the magnetic force field, in the organic realm the organisational laws of the living world, in the bodily physiognomy the character given by the soul and spirit); but the *activity or energy itself* associated with the laws appears empirically to be *fundamentally imperceptible*.

A number of thinkers in the history of European science have seen this problem as a boundary to knowledge and looked for an epistemological solution. Following Schelling and Hegel, I. P. V. Troxler (1780–1866) demanded that pure *thinking* about active laws should proceed to an *empirical perception* of these i.e. that a higher form of science needed to be developed whose cognition in the supersensible realm should follow the model of scientific knowledge: perception and the lawful clarification through thinking of what has been perceived. While Troxler classifies the sciences of ordinary cognition as "anthropology", as early as 1828 he called the necessary higher science *"anthroposophy"*. On the basis of the psychological and scientific observations of the life of the soul available at the time, I. H. Fichte (1796–1879) concluded in his *Anthropologie* that the human soul is a self-contained entity which needs the body to develop ordinary consciousness, but before awakening to consciousness, it is unconsciously active in the building up of its body. Also, for the soul, an existence free of the body which is not empirically accessible to scientific anthropology can in principle be assumed before birth and after death. In this connection he therefore demanded that anthropology be expanded into an *anthroposophy* in order to extend human cognition to the higher realms of human existence and reality.

7 Anthroposophy as an empirical spiritual science

7.1 The limits to knowledge and their transcendence

The limit which "anthropology" comes up against epistemologically has been characterised above by the fact that, while – for science and psychology – the action of forces and laws in nature and human beings are necessary on account of what has been brought about, they cannot be observed empirically. In *"Riddles of the Soul"* Steiner describes how this boundary is often believed to be principally insurmountable, or theoretical hypotheses are developed which lack the perceptual correlate (Steiner, 1996, 14). In neither case can knowledge be achieved in terms of the concept of cognition supported here.

However, these are not the only options for action at this limit to knowledge. Steiner explains how abilities can be developed *at* this threshold which lead out beyond it from sense perception to a spiritual perception, faculties which in fact enable research into the non-sensory action of the forces in the human being, nature and the cosmos in the sense intended by Troxler and I. H. Fichte and how, in this way, a modern *anthroposophy* as spiritual science can be established, developed and used in practical life. This is expressed as follows in Steiner's *"Anthroposophical Leading Thoughts"* written in 1925 (Steiner, 1973b, 13–14):

1. Anthroposophy is a path of knowledge, to guide the Spiritual in the human being to the Spiritual in the universe. [...]
2. Anthroposophy communicates knowledge that is gained in a spiritual way. Yet it only does so because everyday life, and the science founded on sense-perception and intellectual activity, lead to a barrier along life's way – a limit where the life of the soul in man would die if it could go no farther. Everyday life and science do not lead to this limit in such a way as to compel man to stop short at it. For at the very frontier where the knowledge derived from sense perception ceases, there is opened through the human soul itself the further outlook into the spiritual world.
3. There are those who believe that with the limits of knowledge derived from sense perception the limits of *all* insight are given. Yet if they would carefully observe *how* they become conscious of these limits, they would find in the very consciousness of the limits the faculties to transcend them. The fish swims up to the limits of the water; it must return because it lacks the physical organs to live outside this element. Man reaches the limits of knowledge attainable by sense perception; but he can recognise that on the way to this point powers of soul have arisen in him – powers whereby the soul can live in an element that goes beyond the horizon of the senses.

To clarify exactly how this awareness of the threshold and the acquisition of soul forces which lead beyond this boundary are meant in terms of the above-mentioned scientific development of anthropology and anthroposophy since the 19th century, I shall first refer to the concept of the boundary in Hegel's *Logic* and to the striving of German Idealism to cross this threshold. For Hegel the *concept of the threshold* already contains the element which leads beyond this limit: because the boundary is determined by the fact that something comes up against its other, so that the element which stretches out beyond the boundary is already encountered at this limit (Hegel, 2010, 98–100):

> Something is therefore immediate, self-referring existence and at first it has a limit with respect to an other; limit is the non-being of the other, not of the something itself; in limit, something marks the boundary of its other. – But other is itself a something in general. The limit that something has with respect to an other is, therefore, also the limit of the other as a something; it is the limit of this something in virtue of which the something holds the first something as *its* other away from itself […].
> The limit is the *middle point* between the two at which they leave off. They have *existence beyond* each other, *beyond their limit*; the limit, as the non-being of each, is the other of both.
> [This means that] the something, which is now only in its limit, equally separates itself from itself, points beyond itself to its non-being and declares it to be its being, and so it passes over into it.

In terms of the boundary referred to here between knowledge of what has been effected and that of the effective force (cf. Section 6.1), this would mean that knowledge of what is effected must come up against that of the active force and be able to *pass over* into this.

In fact, in the everyday cognition of the anthropological sciences, not only what has been effected but also the element of *effective force* can be found empirically, and therefore that which leads beyond everyday cognition. And it is this everyday cognition which itself leads to this effective force, which in fact produces it.

This becomes clear if you reflect on the experience of the difference between thinking *activity* and the *content* of thinking. Anthropological knowledge deals initially with the *finished* objects of knowledge which have come about up to the moment of cognition and reveals the laws of these finished objects. This knowledge does not experience the *forces* which, connected to these laws in nature, create the becoming of these objects. Although the result of science is spirit, i.e. the lawful essence of the finished phenomena, it is a powerless and purely abstract spirit, in the end the "dead bones of logic", as Hegel put it (ibid., 32). "The system of logic is the realm of shadows, the world of simple essentialities, freed of all sensuous concretion" (ibid., 37). This is the nature of Faust's problem. For *the act of intellectual knowledge,* the action and life of reality have disappeared from this reality. And what Hegel formulated in this connection for philosophy, he meant in particular for what we have called the "anthropological" sciences here (Hegel, 2001a, 20):

> When philosophy paints its grey in grey, one form of life has become old, and by means of grey it cannot be rejuvenated, but only known. The owl of Minerva takes its flight only when the shades of night are gathering.

The question is therefore how knowledge could join *life* in the clear light of morning, as it were, or as Faust cries out after his striving to gain insight into the forces of life as such (cf. Section 6.1) has reached a degree of fulfilment (Goethe, 2005):

> *I feel a youthful, holy, vital bliss*
> *In every vein and fibre newly glowing. (430–432)*
> *In these pure features I behold*
> *Creative Nature to my soul unfold.*
> *What says the sage, now first I recognize:*
> *"The spirit-world no closures fasten;*
> *Thy sense is shut, thy heart is dead:*
> *Disciple, up! untiring, hasten*
> *To bathe thy breast in morning-red!" (440–446)*

Cognition *is able* to join life. Paradoxically, the more dead and abstract the science has become on the *one* hand, the more *life and activity* arise in it as real empirical *experience* on the *other*. This can become clear if you consider the difference for inner experience between the *content* of thought and the thinking *activity*. Although the *content* of thinking is abstract and lifeless, the thinking *activity* is living action, creation, active production of the human spirit, as Hegel also noted (Hegel, 2001b, 15):

> If we take our prima facie impression of thought, we find on examination first (a) that, in its usual subjective acceptation, thought is one out of many activities or faculties of the mind, *co-ordinate* with such others as sensation, perception, imagination, desire, volition, and the like. The *product* of this activity, the form or character peculiar to thought, is the *universal*, or, in general, the abstract. *Thought*, regarded as an *activity*, may be accordingly described as the *active* universal, and, since the deed, its product, is the universal once more, may be called the *self*-actualising universal. *Thought* conceived as a subject (agent) is a *thinker*, and the subject existing as a thinker is simply denoted by the term *I*.

And the thinking I not only lives in the dead objective side of the thought content, but *simultaneously* in the living *action of the thinking activity* (Hegel, 2010, 16):

> Thus, inasmuch as subjective thought is our own most intimately inner doing, and the objective concept of things constitutes what is essential to them, we cannot step away from this doing, cannot stand above it, and even less can we step beyond the nature of things.

And the more precise and abstract the thought content, the more the thinker must endeavour, to "take on oneself the strenuous effort of the Notion" (Hegel, 1977, 35) in order to be able to create and understand the lawful connection at all. It is precisely *because* the products of rational thinking, the logical lawful elements, only appear in normal consciousness in an abstract ineffective form that our *own* activity of thinking is essential for them to appear at all. (This is also the reason for the capacity of human freedom[158]). In contrast to mental images, memories, feelings, drives, etc., pure concepts never appear in consciousness *of themselves* but always and only through the agency of our thinking. And this is why the abstract thoughts of science and logic are the best training ground for the independent activity of the human mind. This enables our thinking to develop the *strength* to move with certainty *in the spiritual element* of the pure clear world of concepts and at the same time not to let itself be disturbed by any undesired influences from sensory perception or the inner world of feeling (Hegel, 2010, 37):

> To study this science, to dwell and to labor in this realm of shadows, is the absolute culture and discipline of consciousness. Its task is one which is remote from the intuitions and the goals of the senses, remote from feelings and from the world of merely fancied representation. Considered from its negative side, this task consists in holding off the accidentality of ratiocinative thought and the arbitrariness in the choice to accept one ground as valid rather than its opposite. But above all, thought thereby gains self-subsistence and independence. It will make itself at home in abstractions and in the ways of working with concepts without sensuous substrata [...].

Earlier eras were closer to the meaning and consciousness of the spiritual world than to the scientific one. In all pre- and non-scientific cultures

---

158 Human thinking activity therefore appears as the only place in nature and the human organisation in which the laws of the world do not realise themselves but are only brought into being through the (thinking) activity of the human being. This fact is also the basis of human *freedom* (cf. Section 5.8). This is why, in *"The Science of Knowing, Outline of an Epistemology Implicit in the Goethean World View"*, when speaking about the basis of the laws governing the world, Steiner states: "It does not live as will somewhere outside the human being; it has given up all will of its own in order to make everything dependent upon man's will. In order for the human being to be able to be his own lawgiver, he must give up all thoughts of such things as extra-human determining powers of the world, etc." (Steiner, 1988a, 111).

civilisation – and along with it medicine – were permeated by spiritual forms of experience and meaning. In the scientific age these experiences and meanings were increasingly suppressed by sensory perception and the purely intellectual treatment of what was observed. Science finally recognised only the material world of the senses as objective reality, leading to the materialistic outlook on life. Despite this, when compared to the preceding more spiritual eras, the *materialistic* age brought clear progress in terms of the human being's *spiritual* abilities, albeit in the form of intellectuality. This is far from being a paradox, because it is only the logical intellectual treatment of the sense observations performed by the individual ego's own thinking effort which leads to the power, independence and certainty in purely spiritual activity mentioned by Hegel which constitute the modern scientist and which were not present in this form in previous times. Steiner said of this (Steiner, 1973b, 57):

> He [the human being] fell into the materialistic outlook in the very epoch of time that brought his own spiritual being a stage higher in development. This is easily liable to misunderstanding. We may observe only the 'fall' into materialism and lament over it. Whilst, however, the *perception and vision* of this age had to be limited to the external physical world, there was unfolding within the soul, as *actual experience*, a *purified and self-subsisting spiritually* of the human being.

This pure spiritual *experience* is twofold: first the experience of the pure objective logical *content* of thinking and, *at the same time*, the experience of the pure thinking *activity* devoted to this thought content. Of course the experience of this activity often remains unnoticed, just *because* it devotes itself to the content as it were and in this devotion does not notice itself[159].

---

159 This is why, in his *"Philosophy of Freedom"* Steiner called the *observation* of thinking a kind of "exceptional state" (Steiner, 2011, 32): "The observation of a table, or a tree, occurs in me as soon as these objects appear upon the horizon of my experience. Yet I do not, at the same time, observe my thinking about these things. I observe the table, and I carry out the thinking about the table, but I do not at the same moment observe this. I must first take up a standpoint outside my own activity if, in addition to observing the table, I want also to observe my thinking about the table." The exceptional state therefore consists of the fact that we make the *thinking which we have already carried out* into the object of our inner observation. For thinking "must be there first, if we would observe it" (ibid., 35). "There are two things which are incompatible with one another: productive activity and the simultaneous contemplation of it" (ibid., 35). In relation to the observation of thinking, see also Section 2.1. Nevertheless, according to Steiner, thinking is experienced and perceived *while being carried out* because it is consciously produced by the thinker himself. Thinking "is a percept in which the perceiver is himself active, and a self-activity which is at the same time perceived" (ibid., 216). This does not contradict what has already been said because the spiritual activity of thinking is *experienced during* thinking, i.e. *in*

However, if we become aware of *this* experience, then we will also become aware that the *experience of the living* thinking activity, i.e. the *observation of an active force* has been developed *from* the experience of the dead thought content. This means that an element has been empirically found at the boundary of ordinary cognition of the kind which is sought on the other side of this boundary, and the ability has been developed by means of the dead abstract spirit (the thought content) to *live* in this element (by thinking activity), in other words in the real active living spirit. This is what Steiner means by the words already quoted above (Steiner, 1973b, 13ff.):

> Everyday life and science do not lead to this limit in such a way as to compel man to stop short at it. For at the very frontier where the knowledge derived from sense perception ceases, there is opened through the human soul itself the further outlook into the spiritual world. [...] Man reaches the limits of knowledge attainable by sense perception; but he can recognise that on the way to this point powers of soul have arisen in him – powers whereby the soul can live in an element that goes beyond the horizon of the senses.

Johann Gottlieb Fichte (1762–1814) made this spiritual perception in the performance of the act of thinking conscious and elevated it to a central element of his philosophy. He called it "intellektuelle Anschauung" which literally means "intellectual experience", often translated as "intellectual intuition". By this he meant the inner perception of the living thinking *activity* carried out by the human I itself, whereas Hegel used the same term to mean more the active observation of the objective thought *content*, something which was also the subject of his principal philosophical interest: "pure knowledge, also when defined as *intellectual intuition*" (Hegel, 2010, 54). In his *Wissenschaftslehre* from 1797, Fichte made a point of stressing this intuition of the living thinking activity in the actual process of its execution (Fichte, 1994, 46):

> "Intellectual intuition" is the name I give to the act required of the philosopher: an act of intuiting himself while simultaneously performing the act by means of which the I originates for him. Intellectual intuition is the immediate consciousness that I act and of what I do when I act. It is because of this that it is possible for me to know something because I do it. That we possess such a power of intellectual intuition is not something that can be demonstrated by means of concepts, nor can an understanding of what intellectual intuition is be produced

---

actu: but what has been experienced can only be *observed* post actum, because observation is a process of *facing* something. "I can never observe my present thinking; I can only subsequently take my experiences of my thinking process as the object of fresh thinking" (ibid., 35). The experience or perception of *active* thinking referred to in the text is not its observation post actum, but its experience in actu (cf. also Sections 6.1–6.3).

from concepts. This is something everyone has to discover immediately within himself; otherwise, he will never become acquainted with it at all.

And the ability for this spiritual intuition is like a new sense organ which gives access to content matter which is inaccessible to those who only know sensory perception. In his lectures on the Wissenschaftslehre in 1813, Fichte said (Fichte, 1834, 4):

> This science of knowing assumes a completely new inner sense instrument through which a new world opens up, one which is not present for ordinary people.

F. W. J. Schelling (1775–1854) also drew attention to this inner intuition of the spiritually active I which remains hidden from the outer life of the senses. He believed that the capacity for this kind of intuition was a requirement for being able to know anything of a supersensible world (Schelling, 1980, 180):

> We all have a secret and wondrous capacity of withdrawing from temporal change into our innermost self which we divest of every exterior accretion. There, in the form of immutability, we intuit the eternal in us. This intuition is the innermost and in the strictest sense our own experience, upon which depends everything we know and believe of a supersensuous world.

And Novalis (Friedrich von Hardenberg, 1772–1801), who studied J. G. Fichte in depth and even came to know him personally, was of the opinion that Fichte's "intellectual intuition" might be the first form of a higher, intuitive thinking cognition, which could bring yet more to light than one's own spiritual ego, but which needed to perceive everything in the way in which one's own thinking ego is experienced, i.e. as an *activity* (Novalis, 1997, 49):

> It might well be possible that Fichte is the inventor of an entirely new way of thinking – for which language has as yet no name. The inventor is perhaps not the most perfect and ingenious artist on his instrument [...]. [...] But it is probable that people exist and will exist – who are far better able to Fichtecize than Fichte himself!
> You should observe everything in the same way as you view your ego – as *your own activity*. But it is easiest with your own ego – this is the beginning, the principle of this practice (Novalis, 1960, 524).

This gives a perspective of a type of science which can empirically observe and therefore cognise the *active force in nature and the universe* in the same way as is initially done solely for the thinking *I* at the boundary of ordinary consciousness described above. A form of cognition and knowledge of this kind would equate to the *crossing* of the threshold which initially exists for the cognition of ordinary consciousness. Through first-hand experience of one's own spiritual activity, self-consciousness also finds the concept of the real – that is active – spirit. This reveals what to sensory object consciousness was previously a hidden world: real, active spirit. This is equivalent to Hegel's description of crossing the threshold in his *"Phenomenology of Spirit"* (Hegel, 1977, 110–111):

> It is in self-consciousness, in the Notion of Spirit, that consciousness first finds its turning point, where it leaves behind it the colourful show of the sensuous here-and-now and the nightlike void of the supersensible beyond, and steps out into the spiritual daylight of the present.

From this it can be seen how, through empirical cognition and the kind of science arising from ordinary consciousness, the most important poets and thinkers of the Central European history of philosophy and science in the 18th and 19th centuries endeavoured to develop a higher form of cognition and science dedicated to the spiritual and active force in the human being and nature, but which – in terms of its cognitive principle – was analogous to natural science and of equal status to it. Creating a science of this kind and introducing it as a complement to ordinary science in the conceptual and practical cultural sphere was the aim of Rudolf Steiner's life's work.

7.2 Rudolf Steiner's empirical spiritual science

Steiner wrote his basic philosophical and epistemological works from a completely *empirical* point of view. The full title of his main philosophical work, written in 1894, is therefore: *"The Philosophy of Freedom. The Basis for a Modern World Conception. Some results of introspective observation following the methods of Natural Science"* (Steiner, 2011). This also describes the principle of anthroposophical spiritual science as defined by Steiner: the application of the *cognitive principle based on perception and thinking* as developed *in science* to the realm of *emotional and spiritual* perception. For this reason, even as late as 1924 (a year before his death) and after developing most of his anthroposophy, Steiner wrote in the new edition of his early epistemological work *"The Science of Knowing, Outline of an Epistemology Implicit in the Goethean World View"* from 1886: "It is true to say that in none of my later [anthroposophical] books have I diverged from the idea of knowing activity that I developed in this one; rather I have only applied this idea to spiritual experience" (Steiner, 1988a, 125). As early as 1868 Franz Brentano postulated the theory that the true nature of philosophical research was none other than that practised in the sciences: "Vera philosophiae methodus nulla alia nisi scientiae naturalis est" (quoted in Steiner, 1996, 65–66) and he also made this method the basis of his *"Psychology from an Empirical Standpoint"* (Brentano, 1995). Steiner held the view that the method which begins at the above-mentioned threshold of ordinary consciousness could also be applied to higher spiritual experience. This would enable scientific and psychological anthropology to be carried over into anthroposophy.

Anthroposophy in our present era thus strives for *spiritual* cognition in a similar way to the beginning of the understanding of nature in the era of Copernicus and Galileo (Steiner, 1917):

> Now what natural science then sought to do for the interpretation and explanation of the mysteries of nature, spiritual science seeks to do for the spirit and soul at the present time. In its fundamental nature, spiritual science desires to be nothing else than something for the life of soul and spirit similar to what natural science then became for the life of external nature.

The ability to undertake *science* in a realm of spiritual perception is initially guided by natural science, as Steiner described in 1910 in his major anthroposophical work *"Occult Science – An Outline"* (Steiner, 1969b, 27):

> Occult science[160] seeks to free the scientific method and spirit of research, which in its own domain holds fast to the sequence and relationship of sense-perceptible events, from this restricted application, while maintaining the same essential attitude and mode of thought. Thus it would speak of the non-sensible in the same spirit in which Natural Science speaks of the sensible. While Natural Science, in the employment of scientific thought and method of research, stops short within the sense-perceptible, Occult Science would like to regard the work of the human soul on Nature as a form of self-education, and apply the faculties, thus educated in the soul, to the realms of the non-sensible. Such is its method and procedure. It does not speak of sense-phenomena as such, but of the non-sensible World-contents in the *same mood* as does the natural scientist of those accessible to sense-perception. It preserves the essential bearing which the soul maintains in scientific procedure – i.e. the very element whereby alone our knowledge of Nature becomes a science. Hence it may justly call itself a science.

The object of this spiritual science, as already stated, is the world of spiritual perception as an *active reality* where natural scientific knowledge comes up against the threshold described above. "In this respect anthroposophy believes it can start its investigation where anthropology stops" (Steiner, 1996, 7). And this threshold is, as already mentioned, the *experience of the thinking activity in pure thinking*, something every thinking person – in particular the scientist – can become aware of through their own inner

---

160 In this work Steiner usually uses the expression occult science in place of and with the same meaning as spiritual science. This does not mean something like "secret" in the sense of kept secret from other people (the publication of this *"Occult Science – An Outline"* would otherwise be totally pointless), but the – public – presentation of methods and material of spiritual scientific research whose subject, the world of spiritual perception in the above-mentioned sense, is not accessible to ordinary sensory awareness and is *from that point of view* "secret". (See for instance the chapter "The character of occult science" in Steiner, 1969b, 25ff.).

empiricism. In order to avoid the misunderstandings which frequently arise about the concept "spirit" in the sense used here, it must be expressly pointed out that by "spirit" Steiner initially means nothing more than that which every thinker can experience in themselves: "Indeed, we can even say that if we would grasp the essential nature of spirit in the form in which it presents itself *most immediately* to man, we need only look at the self-sustaining activity of thinking" (Steiner, 2011, 122).

What is experienced there is not only the pure abstract *content* of the concept, but – as emphasised by Johann Gottlieb Fichte – the *thinking activity* in the act of being produced by the ego. This is a spiritual *perceptual experience* which, by its nature, demonstrates the fundamental possibility that the human being can *perceive* the active spirit in their clear waking consciousness. This leads to the idea that it may be possible to perceive *other* spiritual realities. In the additions to the new edition of his *"Philosophy of Freedom"* from 1918, Steiner expressed this as follows (Steiner, 2011, 216):

> Would it be right to *expect*, from the point of view that this purely intuitively[161] experienced thinking gives us, that man could *perceive* spiritual things as well as those perceived with the senses? It would be right to expect this. For although, *on the one hand*, intuitively experienced thinking is an active process taking place in the human spirit, *on the other hand* it is also a spiritual percept grasped without a physical sense organ. It is a percept in which the perceiver is himself active, and a self-activity which is at the same time perceived. In intuitively experienced thinking man is carried into a spiritual world also as perceiver.

Of course this does not yet result in the experience of any *other* active spiritual element than that of our own thinking. But it is nevertheless the starting point for higher spiritual perception, as Steiner states in one of his main works on spiritual training, *"Knowledge of the Higher Worlds. How is it achieved?"* (Steiner, 1969a, 218):

---

161 In *"The Philosophy of Freedom"* Steiner does not mean the concept of *intuition* or of *intuitive experience* in the semi-conscious sense in which the word is normally used nowadays, but in the fully conscious sense of *pure thinking*. "Intuition" is both the thought *content* which must be brought to the perception in cognition and also the *form* in which it appears, in other words the thinking *activity* (cf. Section 2.3). And the *experience* of this thinking activity is also called intuition and this is what is meant in the above text as "intuitively experienced thinking". In general Steiner defines the concept of intuition, including both the experience of the content of thinking and the thinking activity, as follows: "Intuition is the conscious experience – in pure spirit – of a purely spiritual content. Only through an intuition can the essence of thinking be grasped" (Steiner, 2011, 123).

For the supersensible activity here meant, it is exceptionally important to have a clear understanding of the experience of pure thinking. Fundamentally speaking, this experience itself is already a supersensible activity of the soul, only it is one in which nothing supersensible is yet perceived. With pure thinking one lives in the supersensible, but one experiences *this* alone in a supersensible way; one does not, as yet, experience anything else of a supersensible nature. And supersensible experience must be a continuation of that experience in the life of soul which can be attained in union with pure thinking. Hence it is so important to understand this union rightly.

This experience gives rise to the idea of a *continuation* of this type of experience of spiritual reality to possible *further* experiences of the active spirit, as was postulated by Schelling and Novalis. Steiner expressed this as follows (Steiner, 2011, 216–217):

> In intuitively experienced thinking man is carried into a spiritual world also as perceiver. Within this spiritual world, whatever confronts him as percept in the same way that the spiritual world of his own thinking does will be recognized by him as a world of spiritual perception. *This* world of spiritual perception could be seen as having the same relationship to thinking that the world of sense perception has on the side of the senses. Once experienced, the world of spiritual perception cannot appear to man as something foreign to him, because in his intuitive thinking he already has an experience which is purely spiritual in character. Such a world of spiritual perception is discussed in a number of writings which I have published since this book first appeared. The *Philosophy of Freedom* forms the philosophical foundation for these later writings. For it tries to show that the experience of thinking, when rightly understood, *is* in fact an experience of spirit. Therefore it appears to the author that no one who can in all seriousness adopt the point of view of *The Philosophy of Freedom* will stop short before entering the world of spiritual perception. It is certainly not possible to deduce what is described in the author's later books by logical inference from the contents of this one. But a living [grasp] of what is meant in this book by intuitive thinking will lead quite naturally to a living entry into the world of spiritual perception.

The *"living grasp"* of the intuitive, actively experienced *thinking* "will lead quite naturally to a *living entry into the world of spiritual perception*". This points to the *method* which, according to Steiner, leads to an expansion of thinking into other living spiritual experience and about which Steiner's philosophical forerunners such as Troxler and I. H. Fichte had no clear ideas[162].

---

162 Troxler, for example, merely talks of the "self-development" which needs to be accomplished by the human soul with the help of the activity of thinking, because thinking is the element which links the human being with this object – with which he is confronted externally in the sense world – in an *inner spiritual way*. In order to reach a higher spiritual perception, the human being must therefore

Steiner explains how this method rests primarily on the *power of thinking* whose development since the advent of a form of cognition based on perception and thinking beginning from the time of Aristotle must be *strengthened* still further. With reference to this, Steiner states in the first chapter of the book written with the doctor Ita Wegman *"Extending Practical Medicine. Fundamental Principles based on the Science of the Spirit"* (Steiner & Wegman, 1996, 2–3):

> Before anything is said in anthroposophy about the spiritual aspect, methods are developed that entitle one to make such statements. To get some idea of these methods, readers are asked to consider the following. All findings made in established modern science are essentially based on impressions gained through the human senses. Human beings may extend their ability to perceive what the senses can provide by means of experiments or through observations made using instruments, but this adds nothing *essentially* new to knowledge gained in that world in which human beings live through their senses.
> Thinking, in so far as it is applied to investigating the physical world, also does not add anything to the evidence of our senses. In thinking we combine, analyse, etc. sensory impressions to arrive at laws (of nature); those who investigate the world of the senses must, however, say to themselves: the thinking which thus wells up in me does not add anything real to the reality of the world perceived by the senses.
> This will change as soon as human beings do not limit themselves to the level of thinking that they initially develop through life, upbringing and education. We can strengthen our thinking and increase its power.

What is therefore required is a *strengthening* of thinking, which can be acquired in particular by an academic training and the (natural) scientific activity which accompanies this. The manner in which this strengthening of thinking can be practised is explained in what follows (ibid., 3):

> We can focus the mind on simple, limited thoughts and then, excluding all other thoughts, concentrate the whole power of soul on such ideas. A muscle gains in strength if tensed repeatedly, the forces always being in the same direction. Inner powers of soul are strengthened in the sphere that normally governs thinking by doing exercises of the kind just mentioned. It has to be emphasized that the exercises must be based on simple, limited thoughts. For the soul should not

---

get "beyond the world of the senses" but "not in the direction of the outer world but towards ourselves". This would lead to a fundamental change or "inversion" of the direction of cognition: and this is connected to the development of thinking. Because thinking is "nothing other than a natural force subject to the power of our higher selves". But nowhere does Troxler explain how this force should be applied to reach a capacity for higher perception. Troxler only speaks of "combining all the powers of the soul" and in his *Logic* also refers to this activity as "meditation" and "contemplation" (cf. Heusser, 1984, 264–267).

be exposed to influences that are half or even fully unconscious during those exercises. (Only the principle of the exercises can be given here; for full details and directions on how to do such exercises, see Rudolf Steiner's *Knowledge of the Higher Worlds, Occult Science*, and other anthroposophical writings.) The most obvious objection to this is that if the whole power of soul is directed to a specific thought, focusing on it completely, all kinds of autosuggestion and the like may arise, and one simply begins to imagine things. It is, however, also shown in anthroposophy how the exercises should go, so that the objection is null and void. It is shown that in doing the exercises one proceeds in full presence of mind just as one does in solving a problem in arithmetic or geometry. The mind cannot lapse into unconscious spheres when solving such problems, nor can it do so if the directions given in anthroposophy are carefully followed.

The important point of these exercises which present the *principle* of *meditation* for acquiring higher knowledge as defined by Steiner is therefore a *strengthening* or *"reinforcing"* of the *activity* of thinking, such as is practised when doing geometrical or mathematical exercises. The "simple, easily encompassed ideas" on which thinking should concentrate, can be e.g. geometrical ideas (Steiner, unpublished) or simple symbolical pictures or "imaginations" (Steiner, 1969a), exercises to concentrate attention in nature (Steiner, 1969a), looking backwards on remembered events (Steiner, 1969b, 252ff.) and also mental efforts directed at so-called "boundaries of cognition", i.e. at scientific problems which at first appear to be insoluble at the frontiers of knowledge (Steiner, 1996). These are all repeated and sustained mental activities or exercises which can only be accomplished by *active concentration of the powers of thinking and imagination*. The point is always to strengthen the *power* of thinking and imagination in the same manner as muscle power can be strengthened by the regular practice of pull-ups on a bar. In Steiner's words: "The content of the thought-picture in the imaginative meditation is not the important thing; what is important is the faculty of soul that is thereby developed" (Steiner, 1969b, 236).

According to Steiner's description, as a *consequence* of this exercise it is not only the power of thinking or imagination which is *strengthened* but what has been strengthened is *perceived* more intensely in this process and the power of the *capacity of spiritual perception expands*, from an awareness of the power of thinking to further realms which were previously hidden from consciousness (Steiner & Wegman, 1996, 4):

> Doing the exercises strengthens the *powers of thought* to a previously undreamt-of degree. We feel powers of thought active in us like a new content in the essence of our being. And as our own being is given new content, the world, too, is perceived to have a content of which we may have had a vague idea before but which we have not known from experience. Considering our ordinary thinking

in moments of self-observation, we find our thoughts to be shadow-like and pale compared to the impressions gained through the senses.

Perceptions gained through enhanced powers of thinking are far from pale and shadowy; they are full of content, utterly real images; their reality is much more intense than is found in the content of our sensory impressions. A new world opens up for human beings when they have extended their powers of perception in the indicated way.

The *"new content in the essence of our being"* – as subsequently described – refers to that organisation of forces from which the powers of thought arise i.e. of which they form a part. According to Steiner these are the *etheric* forces which are responsible for growth and form in the organic development of the physical body and by which, towards the end of the first seven years of life, the maturation of the central nervous system is brought to a stage where a part can then separate from the physical organisation and be used for the powers of imagination and thinking in a form that makes the child ready to begin school[163].

The exercises provided thus have the aim of extending the faculty of spiritual perception from the person's own power of thinking to the living force of their own etheric organisation, in other words to make it possible to empirically experience the etheric body in the physical body (Steiner, unpublished). And with the newly perceived "world-content" Steiner means the objective etheric of the outer world which, according to his description, can be perceived by means of strengthened thinking, in other words our own etheric organisation, in a similar way to the objective outer world which we experience by means of our physical organisation. The faculty can thus be developed – as sought by Faust – in order to be able to *experience* empirically the active forces of the living world, what is active and formative in nature and the cosmos, the existence and characteristics of which could previously only be *inferred* intellectually from the sense-perceptible results.

By *strengthening* itself, scientific thinking would thus achieve the ability to think not only "shadow-like" abstract thought content, as is the case with the thinking of ordinary consciousness, but thoughts "full of inner content, vividly real and graphic" which, according to Steiner, are an even more intensive – because active and living – empirical experience than the (extensive) sense perceptions. This would equate to the Fichtecising, the "new way of thinking" which Novalis spoke of in connection with Johann Gottlieb Fichte and

---

163 Cf. Sections 4.11 and 5.3–5.4 and I. H. Fichte's hypothesis of a pre-conscious "imaginative activity of the soul" acting as an organic formative force in the body before being released for its conscious activity of thinking and imagination later in development.

which would permit people to "observe everything in the same way as you view your ego – as your *own activity*".

These viewpoints show why Steiner laid such importance on the exercises for spiritual perception being based on the *reinforcement of thinking* and therefore carried out in a completely sober-minded way "like finding the solution to an arithmetical or geometrical problem" and why the "healthy ordinary consciousness [based on thinking] is the necessary prerequisite for a seeing consciousness" (Steiner, 1996, 120). Because the freedom, self-consciousness and critical self-control of ordinary consciousness rest on the fact that the thinking we ourselves produce penetrates the clear thought content (cf. Section 5.8) and, through this, the world of experience and that this same thought content is abstract, powerless and "shadow-like" i.e. *has no power or compulsion over the thinker*. The thinker's own power gains strength through the powerless thought content of ordinary consciousness. The thinker therefore understands themselves and the world through their *own* power and insight. If *active* content is now perceived by thinking – strengthened by exercise – in terms of the above-mentioned intuitive consciousness, then this must be *retained* or even *increased* in the face of the human faculties of insight and freedom, in order to be able to withstand it consciously. This is why the thinking activity needs to have enhanced *its own* power in order to be able to face the *power of this content* with the same freedom and independence with which it faces powerless ideas. Without this prerequisite the healthy maintenance of self-consciousness would be called into question, as Steiner also points out in *"Riddles of the Soul"* (Steiner, 1996, 120–121):

> But if self-consciousness were not already something acquired by ordinary consciousness, self-consciousness could not be developed within a seeing consciousness[164]. One can understand from this that a healthy ordinary consciousness is the necessary prerequisite for a seeing consciousness. Someone who believes himself able to develop a seeing consciousness without an active and healthy ordinary consciousness is very much in error. In fact, ordinary normal consciousness must accompany seeing consciousness at every moment; otherwise the latter would bring disorder into human self-consciousness and therefore into man's relation to reality. Anthroposophy, with its seeing knowledge, can have to do only with this kind of consciousness, but not with any dimming down of ordinary consciousness.

And this sheds light on the fundamental importance possessed by *pure thinking* – especially that which scientists need to develop – as a requirement for all higher knowledge in the sense meant by Steiner (Steiner, 1965, 321–322):

---

164 Translator's comment: *Das schauende Bewusstsein* means i.e., a consciousness that not only thinks the spirit but sees it with spiritual organs as well.

> In my recently published book *The Riddle of Man* I described 'intuitive consciousness' – in line with the Goethean idea of 'the power to judge in beholding'. By this I mean the human ability to achieve direct intuition and observation of a spiritual world. My earlier works deal with pure thinking in such a way that it is clear that I count this as an essential element of 'intuitive consciousness'. In this pure thinking I see the first, still shadowy revelation of the stages of spiritual cognition. It is apparent from my later works that as higher spiritual forces of cognition I only consider those which man develops in a similar manner to pure thinking. For the realm of spiritual powers of cognition, I reject every human accomplishment which descends below pure thinking and recognise only that which leads beyond this pure thinking. A supposed cognition which does not recognise pure thinking as a kind of model and which does not move in the regions of the same careful consideration and inner clarity as the thinking of clear ideas, cannot lead into a real spiritual world.

The *western* spiritual path advocated by Steiner thus differs diametrically in this point from eastern and other meditation methods for expanding consciousness, if these latter are not based on the capacity for thinking and judgement. For example, *transcendental meditation* is based on the *suspension* of the *activity of thought* in order to allow free rein to the content of the meditation (mantric words from Sanskrit) or thoughts, feelings or even physical sensations arising from unconscious depths of the soul: "We live on the level of spontaneous speech, not on that of thinking and not even of complete understanding. [...] The experience of effortless thinking is the central point and the prerequisite for the experience of the absolute" (Mildenberger & Schöll, 1977, 83–84). The experience of the "absolute", "transcendental" would then not be conscious in actu, but "only afterwards" (ibid., 72). In contrast to this and in line with Fichte and Hegel, the exercises given by Steiner for meditation and contemplation are based on the *effort*, in fact the *intensification* of the thinking *activity* already practised in pure scientific thought, because this enters into the "absolute", as Hegel called the spiritual which forms the basis of the world (Hegel, 1977), in full consciousness actively and in actu. As already described this is initially only the case in the form of pure abstract thoughts, but in which the human spirit already finds itself in the "transcendental", in other words, in the inner being of the outer manifestation: "Becoming aware of the idea within reality is the true communion of man" (Steiner, 1988b, 91). With further intensification of the power of thinking the *living, active element*, i.e. the *power* with which the idea accomplishes its effects in the phenomenon is then *experienced in full consciousness* (Steiner, 1996). But the activation of the power of thinking is already the first point in which active living spirit is experienced in full consciousness. This is therefore the starting point from which to continue.

Steiner's attitude towards the role of active thinking when crossing the threshold between the experience of abstract and active spirit is of the greatest importance because, for the further development of consciousness, it links directly from the form of intellectual consciousness which began with the era of the Greek philosophers and on which the development of western *science* is based. In evolutionary terms this corresponds to a *further* development of human consciousness, whereas recourse to older often eastern methods involves a *return* to spiritual forms of consciousness corresponding to an earlier, pre-intellectual and pre-scientific era in a certain sense. The *developmental principle* in nature first discovered by western science, when applied to the spiritual level and consciousness could, however, lead to an important approach to these issues, as was stated by Troxler as far back as 1828 (Troxler, 1944, 53):

> However, all the more recent philosophy is not to be seen [...] as a fruitless aberration, a mere deterioration with no progress, giving rise to the need to score out whole centuries of ceaseless effort by the greatest thinkers from the diaries of the history of the human mind as vain and useless works and to doing nothing better than returning to Plato and Pythagoras or even to the wisdom of the Egyptians and Indians or the Chinese or even to the legends and myths of the ancient world.

This gives a concise summary of the aim and fundamentals of supersensible cognition as defined by western anthroposophy. This cognition rests on the transfer and application of the *scientific principle of knowledge* developed through the study of sensory objects *to supersensible perceptions.*

On the one hand this requires the *faculty of supersensible perception* developed through systematic practise[165]; and this in turn is an intensification and further development of the spiritual perception already present in the pure thinking of normal consciousness. On the other hand, mere *perception* is as inadequate for a spiritual "science" as it is for natural science: the *laws* of what has been perceived need to be *conceived* and clearly brought into harmony with what has been perceived. This is why, in *"The Philosophy of Freedom"*, when talking of the world of spiritual perception which is envisaged just like our own power of thinking, Steiner writes (as already quoted above) (Steiner, 2011): "*This* world of spiritual perception could be seen as having the same relationship to thinking that the world of sense perception has on the side of the senses". And this is why, in the penultimate year of his life, Steiner noted in the new 1924 edition of *"The Science of Knowing, Outline of an Epistemology Implicit in the Goethean World View"*: "It is true

---

[165] A detailed description of this path of knowledge is given in Steiner's basic anthroposophical works (Steiner, 1969b; Steiner, 1970a; Steiner, 1969a).

to say that in none of my later [anthroposophical] books[166] have I diverged from the idea of knowing activity that I developed in this one; rather I have only applied this idea to spiritual experience" (Steiner, 1988a, 125).

Anthroposophy is thus based on the *further development* of the spiritual faculty which is due to science and the thinking of ordinary consciousness, through which – while avoiding any irrationality and mysticism – a spiritual *science* becomes possible. Steiner expresses this as follows in his essay *"Philosophy and Anthroposophy"* (Steiner, 1929):

> Under Anthroposophy I denote a scientific investigation of the spiritual world which, while cognizant of the limitations of mere physical science and ordinary mysticism, and before attempting to penetrate into the spiritual world, first develops in the soul faculties not yet evident in ordinary consciousness and science. The development of these faculties renders this advance possible.

In this process Steiner distinguished *three stages* of spiritual perception and cognition (Steiner, 1969b) which can only be mentioned briefly in this context. The first is the one already described which arises due to a systematic intensification of *thinking* and which, according to Steiner's description, leads to perception of the *etheric forces* and thus the *processes of life* in man and nature. Steiner describes this stage as the *imaginative* stage of perception and cognition. Continuing from there, the second stage depends on a further systematic intensification of the powers of the soul, now specifically aimed at *feeling* and leading to direct empirical knowledge of the *emotional* powers and being in man and the cosmos, i.e. to a knowledge of the activity of the *astral body in the human and animal organisation*. Steiner calls this knowledge the *"inspirational"* or *"inspiration"*. The third stage of supersensible cognition described by Steiner is that of higher spiritual-scientific *"intuition"*, which rests on a systematic and fully conscious intensification of the *will* and which leads to knowledge of the purely spiritual being in man and the cosmos, to a knowledge of the action of the ego organisation in the human organism (Steiner, 1970a; Steiner, 1969b).

---

166 This excerpt and the previous one from *"The Philosophy of Freedom"* are not quoted here merely in order to repeat the explanation of Steiner's general concept of cognition which he applied to *all* forms of science, but in addition because, since Steiner's time, some critics have tried to see an inconsistency between Steiner's early epistemological and philosophical period of work and his later anthroposophical one. Take e.g. even Zander who, though he collated most earlier critics and despite his aim of proceeding in a "critical historical" manner, in his chapter "Science", did not even take the trouble to assimilate Steiner's concept of cognition from the fundamental epistemological works and to take note of its application to spiritual science, and therefore simply ignored the scientific basis for anthroposophy which Steiner presents there in detail (Zander, 2007).

However, each of these levels of cognition *pre*supposes the state of consciousness applicable to science and the principle of its mode of obtaining knowledge and *develops this further*. For, as Steiner described on the basis of the history of philosophy and consciousness in e.g. *"The Riddles of Philosophy"* (Steiner, 1973a), the principle of *development* does not apply only to the organic realm but also to *consciousness*: evolution has produced the modern *intellectual* consciousness and scientific knowledge from earlier forms of consciousness and cognition; and *this* can now form the basis for the following stages of consciousness and knowledge, as Steiner described e.g. in 1920 in his lectures on *"The Boundaries of Natural Science"* (Steiner, 1983a, 32–33):

> We must take fully seriously the concept of becoming as it applies to human life as well. We must begin by acquiring the discipline that modern science can teach us. We must school ourselves in this way and then, taking the strict methodology, the scientific discipline we have learned from modern natural science, transcend it, so that we use the same exacting approach to rise into higher regions, thereby extending this methodology to the investigation of entirely different realms as well. For this reason I believe – and I want this to be expressly stated – that nobody can attain true knowledge of the spirit who has not acquired scientific discipline, who has not learned to investigate and think in the laboratories according to the modern scientific method. Those who pursue spiritual science [Geisteswissenschaft] have less cause to undervalue modern science than anyone. On the contrary, they know how to value it at its full worth.

This applies especially to the relationship of anthroposophy to scientific medicine.

7.3 Summary

The above-mentioned boundary to knowledge has been noticed by other thinkers in the history of philosophy and science in Central Europe and a way of overcoming it sought. In his *Wissenschaftslehre*, J. G. Fichte (1762–1814), the father of I. H. Fichte, drew attention to the fact that in thinking it is possible to empirically experience not only abstract spirit (ideas, laws) but also *active, living spirit*, that being the thought *activity produced by the ego*. It is therefore *in thinking* that a higher spiritual form of experience begins. Schelling (1775–1854) and Novalis (1772–1801) believed that it was necessary to learn to observe – in the manner of Fichte – *other* spiritual facts in the same way as the ego in its activity of thinking: as active spiritual entities. Rudolf Steiner (1861–1925) described how the *activity* of thinking, as is normally used for mathematics or geometry, could be *intensified* by practice to such a degree that it would become able to perceive not only powerless abstract laws like everyday thinking but also their *effective forces*. Further, spiritual scientific *cognition* only comes about if the supersensible forces are not only *perceived* but their laws *conceived* at the same time. Steiner's anthroposophy is therefore based on an expansion of the faculty of perception into the spiritual and on the transfer of the principle of knowledge which has been developed in natural science to the realm of non-sensory reality. This is what Steiner, to a large extent in agreement with his predecessors Troxler and I. H. Fichte, described as *empirical spiritual science* and called *anthroposophy*. Besides the scientific cognition of sensory objects, Steiner describes three levels of spiritual-scientific cognition corresponding to the three higher organising elements of the human being: life, soul and spirit (spiritual "I"). The human being's four organising elements also relate him to the other realms of nature: the mineral realm (substances), plant realm (life), animal realm (soul) and the human being (spirit).

8 Anthroposophical spiritual science and natural scientific medicine

8.1 The fourfold image of man as a basis for medical anthropology, nosology and therapy

Although Steiner was asked for advice on personal matters including health by private individuals early on in his spiritual scientific work after the turn of the century, and also gave advice on medical matters from a spiritual scientific point of view to individual doctors and therapists, the beginning of anthroposophical medicine is rightly dated to 1920 (van Deventer, 1992; Zander, 2007, Vol. 2, 1455ff.) when, at the request of interested pharmacists and doctors, Steiner began to hold lecture cycles for experts on spiritual scientific aspects of medicine (Steiner, 1948). These lectures were continued up to 1924 (Steiner, 1991; Steiner, 1951b; Steiner, 1983b; Steiner, 1994b; Steiner, 2014; Steiner, 1987; Steiner, 1994a). Most were taken down in shorthand and are now available in the form of books as part of Steiner's complete works. In 1921 the first therapeutic clinical institutes were set up in Arlesheim near Basel in Switzerland and in Stuttgart (van Deventer, 1992) and, as a direct result, the first laboratories for manufacturing medicines were established, forming the basis for the company Weleda (Kugler, 1997). At the turn of the year from 1923/24, the Medical Section was set up as part of the newly established Free School of Spiritual Science at the Goetheanum in Dornach (Switzerland) headed by Ita Wegman (1876–1943), the Dutch doctor trained in Switzerland who was the founder of the first clinical therapeutic institute in Arlesheim. Steiner and Wegman jointly wrote the first book on anthroposophical medicine *"Extending Practical Medicine. Fundamental Principles based on the Science of the Spirit"* published in 1925 (Steiner & Wegman, 1996).

In this book and in many places in the medical lectures and other contexts, Steiner clearly points out that anthroposophical medicine is not opposed to conventional natural scientific medicine but is developed entirely from this and aims to *expand* the scientific findings on the human being through spiritual scientific findings in the sense described above (ibid., 1–2):

> This small book presents new approaches in medical knowledge and skills. A proper judgement of its contents will only be possible for those who are prepared to consider the points of view that were dominant when the medical views discussed in these pages evolved.
> It is not a matter of being in opposition to the school of medicine that is working with the accepted scientific methods of the present time. We fully acknowledge its principles. And in our view, the approach we present should only be used

by those who are fully able and entitled to practise medicine according to those principles.

We do, however, add further insights to such knowledge of the human being as is now available through accepted scientific methods. These are gained by different methods, and we therefore feel compelled to work for an extension of clinical medicine, based on these *wider* insights into the nature of the world and the human being.

Basically those who follow the established practice of medicine cannot object to what we are presenting because we do not go against that practice. The only people who can refuse to accept our attempt without further ado are those who not only demand that we accept their system of knowledge but also insist that no insights may be presented that go beyond their system.

Extended insight into the nature of the world and the human being is in our view offered in anthroposophy, an approach established by Rudolf Steiner. To our understanding of the *physical* human being, which can only be gained by the methods of natural science, it adds understanding of the *non-physical* or *spiritual* human being. Anthroposophy does not involve progressing from insight into the physical to insight into the spiritual aspect by merely thinking about it. This would only produce more or less well thought-out hypotheses, with no one able to prove that they are in accord with reality.

Before anything is said in anthroposophy about the spiritual aspect, methods are developed that entitle one to make such statements.

The "spiritual man" referred to here means the human being's etheric body, astral body and ego which, in terms of their *inner reality and action*, are only accessible empirically to the corresponding spiritual scientific forms of cognition (cf. Section 7.2), and in their outer appearance to *anthropological* natural science and psychology.

For knowledge of the whole physical and spiritual organisation of man and the beings of nature this equates to a differentiated, modulated cognition described in the previous chapter which can be presented schematically as follows (Tab. 2).

Tab. 2: Levels of being and cognition in nature and human beings

Mineral	Plant	Animal	Human being	Force organisation	Level of cognition
			Spirit	Ego ("I")	Spiritual scientific intuition
		Soul	Soul	Astral body	Spiritual scientific inspiration
	Life	Life	Life	Etheric body	Spiritual scientific imagination
Matter	Matter	Matter	Matter	Physical body	Natural scientific cognition

So anthropology and anthroposophy *complement* each other and anthroposophical medicine is the unity of the two, both in theory and practice. In terms of the view of science and reality presented here, both describe *the same thing*, though from two different sides. The findings made by the "anthropological" sciences from the sensory or psychological observation of ordinary consciousness in terms of physical, physiological, psychological and spiritual phenomena on the healthy or ill person, are the result of forces whose effects, based on what has been presented, can only be empirically observed by spiritual perception. There is therefore no contradiction between anthroposophy and conventional medical anthropology in terms of the actual phenomenologically observable *facts* but there is certainly a contradiction in terms of the *theories*.

The *reductionist* theory of medicine explains the phenomena of health and illness in the human being's life, soul and spirit as *causal results* of molecular interactions. The *empirical phenomenologically* based anthropology as defined by Goethe, Troxler, Franz Brentano, Steiner, Max Scheler, Nicolai Hartmann, Helmut Plessner, Gerhard Danzer and others recognises the *empirical emergent phenomena* of the body, life, soul and spirit as *ontologically independent* and acknowledges the action of their own specific laws in them. Anthroposophy examines *the action* of these laws empirically and, by doing so, arrives – as does anthropology in an external empirical way – at four classes of causally active organised forces.

However, in comparison to conventional medical anthropology, this gives rise to a modified *concept of health and illness*. Health and illness are causally no longer solely consequences of process cascades of molecular interactions, but the result of a complex harmonious or disharmonious interaction of the physical, etheric, astral and ego organisations in an organ, organ system or in the whole organism (Steiner & Wegman, 1996; Selg, 2004). Consequently, illnesses can have physical, etheric, emotional or spiritual causes, a view also held in modern medicine by psychosomatics for the realm of the soul and spirit without, however, having so far lead to the ensuing consequence for the issue of causality, i.e. to recognition of the soul and spirit as an *active* inner cause (Adler, 2005). Looked at like this, external physical causes such as toxins, microorganisms, etc. are not *directly* active causative agents, with the exception of physical acts of violence, but an *external inducement* for *individual* adequate or inadequate action (in the sense of self-organisation, cf. Chapters 3, 4 and 5) of these inner causes. How this concept of health, illness and therapy for individual organs, disease entities or medical specialisms is defined, has been described in detail in the anthroposophical medical literature and can be followed up there (cf. Section 8.4). Here I initially only wish to address the fundamentals of the anthroposophical medical concept and its connection to science.

Therapy is therefore not aimed solely at *physical* interactions at a molecular level based on the current models of active substance-receptor relationships, but at *differentiated* effects of laws and forces in the realms of life, soul and spirit. An introduced substance is therefore no longer responsible for an organic or psychological effect due to a *"mechanism"* but creates an *induction* of the body's own action due to forces of the physical, etheric, astral or ego organisation or their interaction in an organ, organ system or the whole organism. The therapeutic effect of substances and processes is therefore not determined merely by their physicochemical action which is produced in the inorganic realm but through the relationship it has to forces and laws in the physical, etheric, astral or ego organisation (Steiner & Wegman, 1996; Selg, 2004; Sieweke, 1982; Sieweke, 1994).

Apart from conventional, physicochemically based pharmaceuticals, anthroposophical medicine also uses specially prepared *substances from the mineral, plant and animal kingdoms* which influence the activation or inhibition of forces in the physical, organic, soul and spiritual organisation in a differentiated manner. The *special processes used in anthroposophical pharmacy* have the aim of focusing the action of these substances onto their targets (Glöckler, 2005). The same holds for the *different ways of applying substances* in anthroposophical medicine: *taken internally, injected or applied externally* in the form of ointments, compresses, baths or other external applications. According to anthroposophical concepts, the substances have a different systemic effect because the various, i.e. physical/etheric/astral/ego-related systems of forces display a different configuration in the metabolic processes of the digestive organs, in the rhythmical organisation of the circulatory system and in the skin which is strongly influenced by the processes in the senses.

Overall, apart from the *elimination* of microbes or *pathological* structures and the *replacement* of substances or structures which are lacking, it is a matter of *initiating the activity of the health-creating systems* in the whole human organisation. This is why anthroposophical medical thinking pays so much attention to the *forces of self-healing* (Heusser, 1999a; Hildebrandt, 1999; Heusser, 2002b). The ideas of orthodox medicine based on *heteronomy (outer control)* and *external pathogens* (pathogenesis) are *extended* by ideas of *autonomy and hygiogenesis* which take account of the above-mentioned *inner causal factors* of the whole human organisation (Bahrs & Matthiessen, 2007). *Deficits* are countered by the activation of *resources* arising from the patient's own forces, based not only on the physical organisation but on the *whole*, i.e. physical, living, emotional and spiritual aspects of the human being (Heusser, 2002c).

The consequence of this is *a holistic and multimodal integrative therapy concept* which, depending on the actual situation, applies *different types of therapy at different levels*. Even the use of *medicines* can – as shown – be applied to physical, etheric, astral and ego-related effects or to their inner relationships. In addition there are *artistic therapies* such as painting (Mees-Christeller et al., 2000), clay modelling (Golombek, 2000), music therapy (Felber et al., 2000) and therapeutic speech formation (Denjean-von-Stryk & von Bonin, 2000) which primarily act via the emotional realm, but due to its integration in the whole human being also have secondary effects on the physical/living and spiritual organisation. *Rhythmical massage*, on the other hand, acts primarily on the living/physical elements and, by harmonising these, also on the soul and spirit (Hauschka, 1971). A further therapeutic option is *curative eurythmy*, a new movement therapy developed by Steiner (Steiner, 1983b) which, in contrast to *physiotherapy* which is also used in anthroposophical medicine, does not work primarily on the structures and functions of the *physical body* but, in accordance with its basic concept, on the *etheric* and on the harmonious integration of body, soul and spirit. Curative eurythmy (also called therapeutic eurythmy) is therefore used to treat internal organic processes such as e.g. the different digestive functions of the gastrointestinal tract from the stomach to the large intestine, the kidney and bladder functions, etc. (Kirchner-Bockholt, 1997).

A further therapeutic field presents itself due to the fact that the human being – in contrast to the other creatures of nature – has a *spiritual ego*. This is where all *cognitive therapies and strategies for dealing with illness* are to be found; plus questions about the *meaning of life and illness, working with destiny* and *biography work* which deals with aspects of the development of the ego during life in relation to processes of disease and healing (Burkhard, 2005); and also *spirituality and religious beliefs* through which humans, as spiritual beings, relate to the divine spiritual foundation of the world (Steiner, 1987; cf. also Section 5.7).

This explains why anthroposophical medicine is diagnostically and therapeutically not only an integrative holistic medicine but also an *individualised one*. On the one hand individualisation arises purely quantitatively due to the *multifactorial causality* of the pathogenesis which includes not only the various physical and social conditions of the outside world but an individual configuration of the important physical, etheric, astral and ego-related *inner* causal factors for each patient. On the other hand, diagnosis and therapy are individualised for principal qualitative reasons, because the *spiritual individuality of the person* contains their completely individual essence, through which they are a unique being in their own right with their own development, biography, inner destiny and self-determination. This is what Goethe called the spiritual "entelechy", the "higher guide" (cf. Nager, 1991, 165) in the human being and what determines the "inner life plan" as it were (ibid., 119). A deviation from this innate "law of growth", for example due to an unbalanced lifestyle, can lead to illness as Goethe himself experienced with his own nearly fatal illness during his student days in Leipzig: "[...] and I irritated my happy organization to such a degree, that the particular systems contained within it necessarily broke out at last into a conspiracy and revolution, in order to save the whole" (Goethe, 1848, 282–283).

An illness based in the individual ego and destiny is naturally completely *unique* and must be assessed differently to an illness which is based on the *general* laws of the physical organisation which apply to everyone, or to illnesses with more *typological* characteristics, such as the constitutions and temperament tendencies based in the organic functions, or those caused by typical emotional reactions such as stress reactions and the like. Even with comparable external symptoms, illness is therefore not always the same thing and, when taking the patient's history, making a diagnosis and finding the right treatment, anthroposophical medicine always tries to understand the physical, etheric, astral and ego aspects and to apply these therapeutically. The outcome is the holistic, multimodal, integrative and individualised therapy concept of anthroposophical medicine (Soldner & Stellmann, 2011; Girke, 2012; Jachens, 2012).

8.2 Scientific examination of anthroposophical concepts and medical rationale

The concepts of anthroposophical medicine are derived from the spiritual scientific research which Steiner founded and subsequently developed both in terms of method and content. On the one hand – as per its definition – this form of research requires an expanded, i.e. supersensible faculty of perception and, on the other, the application of the cognitive principle – which also applies to natural science – to the world of spiritual perception. Further, it relates to the action of particular classes of forces and laws in the physical, living, emotional and spiritual elements of human beings and nature. These concepts, which in addition make use of their own appropriate nomenclature, vary considerably at first from the concepts and thinking habits of conventional medicine and it is therefore understandable from a psychological viewpoint that anthroposophical medicine often elicits considerable scepticism if not disapproval or even polemic (Stratmann, 1988; Bock, 1993; Burkhard, 2000; Zander, 2007; Ernst, 2008; Ernst & Schmacke, 2015).

How should we view these concepts from a *scientific* point of view, particularly since the supersensible knowledge on which they rely is certainly not a commonplace ability[167] but first has to be acquired via a specific spiritual

---

167 Zander even contests the existence of a supersensible faculty of cognition as defined by Steiner and claims that Steiner did not possess any faculty of this kind but derived his anthroposophical ideas from other authors and teachings

training? Must these concepts then be accepted as a matter of faith and is anthroposophical medicine a world view whose concepts are unverifiable and must therefore be placed outside rational and scientific medicine?[168]

Steiner expressed himself clearly on this matter in a wide variety of circumstances. For an understanding of his anthroposophical writings such as *"Occult Science – An Outline"* he wanted "readers who will not accept what is here presented on blind faith, but rather put it to the test of their own insight and experience of life. He desires careful readers – readers who will allow only what is sound and reasonable" (Steiner, 1969b, 24).

Such testing of the written material is therefore to be done using the reader's own *experience* and *logic*, i.e. both elements which are the basis for every cognition and every science (cf. Section 2.4). *Direct* examination would of course only be possible for someone who, due to their own supersensible faculty of cognition, could gain an *empirical experience* of supersensible facts such as the etheric body and its activities, which is clearly not an everyday occurrence. But Steiner refers "not *only* [...] to the spiritual test of supersensible research, but to the test – *unquestionably valid* – of open-minded thought, the test of healthy human intelligence and reflection" (Steiner, 1969b, 24).

Because empirical spiritual science as intended by Steiner must of course *understand* the spiritual perceptions in accordance with their laws, just as natural science does with the physical laws (cf. Section 7.2) and present the

---

and developed these into the philosophy of anthroposophy by the transfer and addition of his own constructions. However, these are Zander's *assumptions* for which he provides no factual proof but merely *supposes* and in the light of which he then *interprets* Steiner's work and its historical and contemporary connections in a more or less logical way (Ravagli, 2009). Even though Zander's type of rejection of supersensible faculties of perception and cognition is not *scientifically* tenable, it can still be understood *psychologically* because a faculty of this kind is no ordinary matter.

168 With reference to anthroposophical medicine, Barbara Burkhard therefore believed she had to construct a contrast of "convinced believers versus unbelieving researchers" (Burkhard, 2000, 160), and Franz Stratmann believed that anthroposophical ideas were "doctrines and mystical hypotheses which are inaccessible to examination and a scientific conception of medicine" (Stratmann, 1988, 70). Edzard Ernst misinterprets the fourfold nature of the human being and, like the other above-named authors, does not even go into Steiner's scientific explanation of anthroposophy and anthroposophical medicine, only reporting from secondary sources that anthroposophical medicine rests "on a wealth of conclusions through analogy which 'take up the magical, alchemical and astrological ideas of earlier centuries'" (Ernst, 2008, 1). However, he provides no evidence of this. In his most recent publication he makes similar claims on the basis of poorly researched literature, without paying any attention to the scientific basis of anthroposophy (Ernst & Schmacke, 2015).

results in the *form of thoughts*. However, the realities of the world of spirit will be cast into forms of thought "which the prevailing consciousness of our time – scientifically thoughtful and wide-awake, though unable yet to see into the spiritual world – can understand" (ibid., 8). And as thoughts are fundamentally subject to logic, spiritual scientific realities must and can – according to Steiner's account – be subject first and foremost to a test of *logic*.

But logic alone is not enough: for it would always be possible for someone to build a logical, internally coherent world view through purely intellectual constructions, independently of empirical reality, which was therefore of no relevance to this at all. This is why Steiner repeatedly pointed out the necessity for testing scientific statements not just for their *logic* but also for their *relevance to reality* or, as he succinctly put it in e.g. his course *"Fachwissenschaften und Anthroposophie"* from 1920: "A judgement must be logical and agree with reality" (Steiner, 2005, 42).

However, how is it possible to verify an anthroposophical spiritual scientific statement for its agreement with reality, i.e. by empirical perception, when the verifier has no access to a perception of this kind? An *indirect* test is both necessary and possible, in the manner customary in a field such as medicine, including in a *natural scientific* sense. Because only the tiniest part of what a medical student and later a doctor learns and makes use of in training or in practice has been *empirically* discovered by them. They study *in the form of thoughts* what specialists have discovered and recognised on account of their *perception* – potentially expanded by apparatus – through systematic research, for example how insulin regulates sugar mobilisation at a molecular level. Of course they have personally got to know *the nature* of such perceptions and knowledge acquisition from examples in practicals, they know from experience how the empirical material "feels" so to speak, and they can refer to demonstrations of the scientific methods which provide information on the details of this type of knowledge acquisition. They therefore know the principles of scientific empiricism and the formation of scientific judgement. This enables them to maintain a justified *trust* in the explanations of others which they acquire through thinking, but of course not as blind faith, but as a well-founded "leap of faith". For if the logical consistency of this explanation is tested in personal practice or via studies and its potential connection with the *accessible* phenomena makes rational sense, then this corresponds to an indirect empirical verification of this explanation's agreement with reality.

It is the same in principle with spiritual science, as understood by Steiner. Already in the *experience of our own thinking* it is possible to "feel" empirically the spiritual "material" to which Steiner refers, or it is possible to have further spiritual experiences based on spiritual exercises without having to be a systematic researcher or specialist in such matters. In this case, however, we are not subjected unquestioningly to the statements of a spiritual scientific specialist. We know *the nature* of actual spiritual perceptions and we know how perceptions of this kind can be subjected to intellectual judgement (cf. Section 6.1). We know, in fact, the spiritual scientific process as such from the experiences of our own ordinary consciousness[169] and, in addition, we can take note of Steiner's related descriptions of the methodological development of the higher spiritual scientific faculty of cognition from this ordinary one (Steiner, 1970a; Steiner, 1969b; Steiner, 1969a). This then justifies approaching the spiritual scientific information with an open mind, initially with a similar scientific attitude, thinking through what has been described *rationally*, relating this mentally to our own world of experience and then testing it *indirectly on the experience itself*. As far as this goes, the relationship of the practising doctor or scientist to scientific anthropology and to spiritual scientific anthroposophy is basically the same.

It should also be noted that *scientific rationality* is not limited to the natural sciences, as some seem to assume – otherwise pure mathematics or the science of logic which are based on pure inner experiences would not count as sciences. As shown, *every* level of the complex and hierarchically arranged world has *its own* laws and therefore *its own rational content* which cannot be derived from that of the other levels, but is only compatible with it and can therefore be understood along with it in a unified but internally differentiated *rational overall view* of reality (cf. Section 5.9). *The reason, the law* is therefore the element which links all the phenomena of the complete but internally differentiated world.

The idea, the law, the logical or rational content is therefore also the element which brings together anthropology and anthroposophy "at the same point" as it were (Steiner, 1996, 25). This is particularly important for the *verifiability of spiritual scientific statements in medicine* because the laws and forces of the physical, living, soul and spiritual organisation of the human being are not levels which are abstractly stacked one on top of the other, but work *into each other*. For example, the physical body is not something mechanical which is unconnected to the life organisation, but *is filled with*

---

169 Anyone who wishes to acquire a personal empirical insight into the nature of this spiritual scientific process in the realm of pure spiritual experience, as can already be had within ordinary consciousness, can do so using Steiner's *"Philosophy of Freedom"*. In this book everything is aimed at the activation *of the empirical spiritual experience* within our own thinking and at the development of a scientific *judgement* in this experience. The same applies to Steiner's other basic philosophical anthroposophical works. He himself called his *"Philosophy of Freedom"* and *"Truth and Knowledge"* a "means of self-education for the soul and spirit", through which the reader – aside from the intellectual absorption of a scientific content – could "train" themselves in the specified direction (Steiner, 1983c, lecture given on 15.12.1910).

life. According to the basic anthroposophical concepts, the structures and functions of the physical body are *internally completely interwoven and determined* by the laws and forces of life i.e. by the etheric organisation which, although superior to it, nevertheless acts *within it* (cf. Section 4.11). This etheric activity influences the physical so that *empirically an impression of the etheric* must be present *in the physical*. Anthropology looks for the *laws* of this action in what has been created, anthroposophy deals empirically *with this action itself*. The *law* of the organic action which can be expressed in the form of an idea is identical in anthropology and anthroposophy, thus forming the bridge between the two kinds of science. And, in accordance with this principle, this applies to all active laws of the physical, etheric, astral and ego-organisation to the extent that they manifest in the physical or psychological phenomena (cf. Section 8.1).

What anthroposophy therefore wishes to add to medical anthropology for the understanding of health and illness and for therapeutic practice is an empirically obtained, rational, internally coherent *knowledge* of the active forces in nature and the inner human forces as a differentiated physical, living, soul and spiritual whole: in other words, of the inner activity of that which anthropology also postulates as the action of forces using external sensory or conventional psychological empiricism, and can recognise as a differentiated physical, living, psychological and spiritual whole. Steiner's spiritual scientific statements on physiology, pathology and therapy aimed in this sense for the presentation of a "rationale" (Steiner, 1989, 138) as the basis of a spiritual scientific expansion of medicine so that, in an overall medical anthropological sense, the inner relationship between a "rational diagnosis" and a "rational therapy" can be found (Steiner, 1994a, 122). Both scientific disciplines, the anthroposophical and the anthropological, thus come empirically to a rational picture of the human being in their own fields and "those who look at both pictures will be able to find a harmony between their mental pictures similar to that between the negative of a photograph and the corresponding positive print" (Steiner, 1996, 25).

However, such an agreement can only be expected if the relevant spiritual scientific, psychological and natural scientific pictures do not consist of *models* but contain only the *laws* which can actually be found in the corresponding *empirical phenomena*. "The spiritual science intended here is therefore at pains to develop within the inorganic and organic sciences a pure phenomenalism and to simply present the processes themselves as they appear, without applying any atomistic or other hypotheses" (Steiner, 2005, 318). In fact it must not be forgotten that, with its models, science *goes beyond* the actual phenomena and their laws *and makes assumptions* which are meant to explain the phenomena. These explanations are then usually of a reductionist nature. In this sense it is "actually a world view which the sciences express, even though this is usually denied" (Steiner, 2005, 250). The incompatibility of anthroposophical concepts with science which is sometimes claimed (Stratmann, 1988; Bock, 1993, 64ff.; Burkhard, 2000; Zander, 2007; Ernst & Schmacke, 2015) is usually based on its incompatibility with certain scientific *models*, whereas internally it can accord very well with the scientific *facts* and their laws, something which Steiner repeatedly pointed out.

This inner agreement also settles the question of the scientific competence of anthroposophical medicine in the context of modern medical research. The a priori rejection of this basic possibility of agreement can only be made by those who want to restrict the real and rational in medicine to the physical in the human being and who cannot accept an empirical spiritual science in the sense presented here.

8.3 Anthroposophical medicine and modern scientific medical research

Steiner's own view of the relationship between anthroposophy and natural science *in medical research* is completely in accord with the general relationship between anthropology and anthroposophy outlined above. Even the first course of twenty lectures on Spiritual Science and Medicine which Steiner held in 1920 at the request of doctors, contained more than fifty references to and suggestions on scientific and psychophysiological studies, preclinical, experimental and clinical studies, literature reviews on current or historical medical issues, on doctoral and habilitation theses, etc., all on topics for which Steiner presented views from an anthroposophical and spiritual scientific standpoint (Steiner, 1948). Due to the extent of the complex subject matter and the time available for the lectures, the spiritual scientific viewpoints themselves were not developed as a structured system of theories "that begins with the axioms and ascends to more and more complex ideas" (ibid., 144), but were presented like a "rough guiding thread" (ibid., 265) which approaches the subject by "a circuitous route" (ibid., 144) from the most varied perspectives in order to "acquire a general view and conception of man" (ibid., 254), namely "in a certain relation to the outside world" (ibid., 254), through which we can better understand the connection of man to the processes and substances of nature and can therefore work in a practical way with the spiritual scientific findings (ibid., 12). However, this is all presented very aphoristically, like a sketch[170] which, amongst other things, is intended to prompt the listener "to

---

170 Steiner himself was aware that this approach made his lectures on medicine "among the most difficult to comprehend of all lectures presenting the anthroposophical point of view" (Steiner, 1948, 193) but he regarded this as a difficulty which, in view of the objective pursued "can hardly be otherwise" (ibid., 193).

observe nature independently" which "may indicate a pathway in a certain sense, to a particular realm" (ibid., 131), whether this be for medical practice or for the gradual development of a medical system expanded through spiritual science.

Steiner was perfectly aware that "all this can only be a preliminary outline," (ibid., 217), because he considered the important point of these and other lectures on medical topics to be the willingness of doctors to "participate" (ibid., 11) in a general reform of medical thinking *altogether*. In the debate between "allopathic medicine" and the various attempts at medical reform, Steiner did not wish "to 'take sides', but [...] simply to put before you the facts as they really are" (ibid., 265), as far as they appear from spiritual scientific research. However, this does not mean the primacy of spiritual scientific viewpoints over those of natural science, as Zander claims[171] but, as presented in *"Riddles of the Soul"* in 1917 (cf. also Section 8.2), a *complementary* relationship between anthropology and anthroposophy or, as described in lectures on the academic scientific disciplines and anthroposophy (Steiner, 1950, 24):

> It looks like this. Natural science applies on one side, spiritual science, spiritual research on the other side. Just as when you dig a tunnel and when everything has been calculated properly, tunnelling from two sides you meet in the middle, so spiritual *research and natural scientific research* meet and produce the whole of knowledge which people seek.

According to Steiner's repeated statement in his medical lecture courses, in practical terms this means that, for the relationship of anthroposophy to medical research, *both* sides have to make their – equally justified and necessary – contributions. Although anthroposophical viewpoints supply the rationale for the meaning and effect of immaterial factors in health, disease and recovery, this rationale has no authoritative value for natural scientific medical research as Zander believes (Zander, 2007, Vol. 2, 569) but only one of "guiding thoughts" or "regulatory principles" or, as we would say nowadays *working hypotheses* which must *first be verified by natural scientific empirical research* i.e. confirmed or verified *indirectly* in the above-mentioned sense (Steiner, 1951b, 59):

---

171 Zander claimed: "In whatever way the importance of conventional and alternative medicine will be apportioned in anthroposophical medicine through future research, in Steiner's 'expansion' of empirical medicine, 'conventional medicine' was not put on a par with anthroposophical medicine, but the latter interpreted as dominant" (Zander, 2007, Vol. 2, 1569). This is absurd, not least from the point of view of the possibility of spiritual scientific error mentioned by Steiner himself (cf. also Note 173).

I wanted to tell you this as a principle in order to make you understand that these things depend upon a [rationale]; but the [rationale] is merely a regulating principle. You will find that the statements based on this principle can be verified, as all such facts are verified by the methods of modern medicine. There is no question of asking you to accept these things before they have been tested, but it is really true that anyone who enters into them can make remarkable discoveries.

"Verification"[172] here does not mean merely a simple scientific "confirmation" of the spiritual scientific fact, but usually an *elaboration* of the subject using scientific methods. After all, the spiritual scientific idea is often very general and requires the reification of the physical level of its actual *manifestation* which can be understood by natural scientific methods. In addition, Steiner never claimed that spiritual scientific research was free of errors[173]. So it is generally true to say that (Steiner, 1951b, 7):

> Indeed, all that comes from anthroposophical investigation in regard to medicine and, for instance, physiology, can be nothing more than a stimulus which must then be worked out empirically. Only on the basis of this empirical study can there arise valid and convincing judgements of the matters in question – and this is the kind of judgement that is needed in the domain of therapy.

When it comes to medical research questions in terms of an integrative medical anthropology, the point in question is the inner *connection* between the factors and effects of the physical, living, soul and ego-related spiritual levels of the complete human organisation which are emergent to each other but which act upon each other. From the anthroposophical perspective of an expanded medicine, it is therefore not only possible but necessary to describe a physiological or psychological process at the level of its physical material conditions and, in addition, at the levels of its etheric, astral and ego-related events. The things which critics like Zander or Burkhard view as problematic[174] such as Steiner's multiperspective view and description of the reciprocal

---

172 With reference to the current often – under the influence of Popper – disputed possibility of scientific *verification*, see also the statements in Section 2.4, on the relationship between verification and falsification.

173 His comment on this in e.g. the course *"Fachwissenschaften und Anthroposophie"* from 1920 was: "I openly admit that spiritual science can be mistaken in certain matters. It is only at the beginning. But this is not the point. The main point is the direction in which we aim to go" (Steiner, 2005, 400); or: "Mistakes may be made in some details, but the main aim is to show a new direction" (ibid., 412).

174 For example, both Burkhard and Zander see a contradiction in the fact that the anthroposophical view holds etheric and other immaterial forces put forward by anthroposophy responsible for the effects of the mistletoe therapy recommended by Steiner for treating cancer on the one hand, and substances such as lectins found by modern science (with a significant contribution by researchers

dependencies of such perspectives, is simply in the nature of the matter i.e. of the multiperspective reality of the human being: and *no perspective can replace the other*, something which already follows on epistemological grounds alone (cf. Chapter 2). Steiner himself pointed out the importance and necessity – also from the *spiritual scientific* standpoint – of the autonomous *natural scientific* side of research and its technical measures and he expected anthroposophical medical research to take proper account of this. In October 1922 he told the doctors interested in anthroposophy (Steiner, 1989, 88ff.):

> I must stress how important it is that there should be no misunderstanding, specifically as regards these basic matters. [...] If, for example, someone were to think that it is pointless to engage in sensory empiricism in physiology or biology, believing there to be no need for any specialised branch of science because all it takes is to develop spiritual faculties, look into the spiritual world and then arrive at a view of the human being in health and sickness and then more or less found a spiritual medicine – this would be a great error. In fact some people do this, but it leads nowhere. The most that happens is that they complain loudly about empirical medicine, but they are only complaining about something that they do not understand. It is definitely not a matter of writing off conventional empirical science and founding a spiritual science from a spiritual cloud-cuckoo-land. This is a completely wrong approach to the empirical sciences [...] For example, if you undertake spiritual scientific research you will not find the same things as can be examined under the microscope. If anyone

---

in anthroposophical institutes, as is well known) on the other; and that, furthermore, these lectins are interpreted in the context of anthroposophical concepts (Burkhard, 2000, 118ff.; Zander, 2007, Vol. 2, 1569). Zander even believes that in Steiner's "holistic body, soul and spirit" view of the human being a "crass materialism repeatedly [enters] into specific issues besides the dominant spiritualism" (ibid., 1495). Materialism does not arise because we acknowledge the material side of existence but if we accept it as the *only* side and reduce the non-material phenomena to it ontologically and/or epistemologically. Given that Zander appears to accept neither the existence nor the cognoscibility of the spiritual side of existence claimed by Steiner to which of course matter also belongs, then this materialism must be ascribed to *him*. Steiner's achievements in fact include having raised materialism through the epistemological explanation of the *spiritual essence of material phenomena, without* denying the *reality* of the material phenomena or having underestimated the justification, or rather necessity, of material research. Steiner's basic scientific attitude in concurrence with Goethe even enables a unified, i.e. *monistic overall scientific view* to develop which epistemologically and ontologically also acknowledges the living, soul and spiritual sides of existence and therefore the human being *as a human being* in addition to the material one, whereas *material monism* from which even authors such as Burkhard and Zander cannot be completely exonerated, has scientifically lost sight of the human being as such due to its restriction to matter devoid of spirit.

tries to get you to believe that they can find the same things through spiritual science as can be found under the microscope, you can safely regard them as a charlatan. It is not the case. The findings of present-day empirical research are valid. What is more, in order to have a complete science in any field – including that of spiritual scientific anthroposophy – it is not permissible to do away with sensory empiricism, it is essential to take account of this sensory empiricism. Anyone who is, if I may use the expression, an expert in anthroposophical spiritual science, will find that, by engaging in spiritual science, they will first need to seriously concern themselves with the phenomena of the world in the sense of sensory empiricism.

However, this is a lengthy and *far from complete process even today*, never mind what might have been possible in Steiner's lifetime. For this is not simply a matter of applying some different or additional "expanding" views to medicine or to another region of life, in order to establish an "alternative" view of life and a special ideological group: Steiner's aim was actually to contribute to the development of a *complete overall view of the natural and spiritual science of man, nature and the cosmos* based on a sound epistemological and methodological foundation. This was the point not only of his teachings on medicine but also in other realms such as pedagogy, agriculture, the arts, and social and economic issues. In 1920 when lecturing to teachers on physics he argued (Steiner, 1988c, 176):

> You will certainly find many things in these lectures that are unsatisfying, for I could only offer initial suggestions at best. One thing is clearly shown in these lectures, however, and that is the necessity to build anew our whole physical, chemical, physiological, and biological approach to the world. It must be rebuilt from the ground up. We will accomplish this [...] only when we have developed further not just the schools but the nature of science itself. And until we have succeeded in the Waldorf schools to the extent that the subjects for instruction have been renewed [...], only then will we achieve what should and must be achieved [...].

In medicine – which relies on all these sciences – it is a matter of developing *sound academic scientific research*, besides the practical side to which Steiner attached great value from the start. This requires us to aim for "the same exactitude" which is customary in science (Steiner, 1994a, 88). An interdisciplinary approach is needed ("We are specialized enough. What is important now is to bring together the individual specialties again", Steiner, 1986); and research results which have already been published in the literature must be viewed together in a synthetic manner. This applies both to the results of natural science ("if you will set yourselves to an exhaustive study of the results of the most orthodox empirical science, if you will relate the most obvious with the most remote", Steiner, 1951b, 39) and to the spiritual science which he founded ("it is possible now to expand the short sketch that has been given by many details contained in my lecture courses and writings", Steiner, 1994b, 42). "The cursory way of dealing with these matters which is necessary here may make a great deal appear fantastic. Everything can, nevertheless, be verified in detail" (Steiner, 1951b, 55). What is important is "to follow up the things which I have now mentioned only briefly and then bring the results into a proper scientific system" (Steiner, 1989, 131). This scientific system must be pictured as integrating the natural and spiritual scientific aspects into a unified overall view. "And I believe that it would be beneficial if we could have a literature which bridges the gap between spiritual science and natural science at the earliest opportunity" (ibid., 162).

The longer term goal of these efforts is the gradual development of a *"system of medicine"* founded on natural and spiritual science (Steiner, 1970b, 182) of which Steiner, by his own account, was able to provide only a rough outline and suggestions in his medical and other scientific lecture courses. This is the real reason for the aphoristic and therefore difficult style of these lectures that caused problems even for the listeners of the day and with which critics like Zander like to find fault (Zander, 2007, Vol. 2, 1489ff.). Steiner did not intend these lectures to be a medical "course" as such, although he also uses this term at times ("It is my wish that this attempt should not be confused with an actual medical course, which it nevertheless will be in a sense", Steiner, 1948, 11); but first and foremost wanted to point the thinking and research of his medical or otherwise scientifically engaged audience in a new *direction* (ibid., 11); and furthermore to lay the responsibility for everything else on *their own research* ("These considerations must now be brought to a close, and their further progress will depend on your own work", Steiner, 1980b, 177). And it has become even clearer in modern medical research than at Steiner's time that: "The proper end can be attained satisfactorily only through people working in an orderly way together" (ibid., 177). Research by definition must be set up with a *long-term view* and must, as I have described and as is always the case for medical research, apply the usual *scientific research methods* for the verification and actual application of spiritual scientific points of view (Steiner, 1951b, 28):

> Everything I say will be aphoristic, merely hinting at ultimate conclusions. Our starting point, however, must be the objective and empirical investigations of modern times, and the intermediate stages will have to be mastered by the work of our doctors. This intermediate path is exceedingly long [...].

Even the first and only *systematic* work on medicine by Steiner and Wegman (Steiner & Wegman, 1996) is unable to meet the aim of a medical system in the above-mentioned form in any way whatsoever and the same would have applied to the planned continuation of this work which did not come about due to Steiner's death in March 1925 (cf. Postscript by Ita Wegman in: Steiner &

Wegman, 1996, 123). In the last year of his life when announcing the publication of the first volume, Steiner remarked: "The medical book of which I spoke yesterday can only offer a first, elementary introduction to what will [only] become a fully developed science in the distant future" (Steiner, 1985c, 202–203). The development since then and the current situation of anthroposophical medicine make it clear that this was a very realistic assessment. This must take account on the one hand of the development of a medical system expanded by spiritual science as such, in other words *a systematic medical anthropology comprising natural and spiritual science* and, on the other, of the *natural scientific and psychological verification of anthroposophical concepts on physiology, pathology and therapy* i.e. experimental, clinical and health services research in terms of conventional medicine.

To conclude, these considerations show that there is not only no contradiction between anthroposophical medicine and modern scientific research as defined by Steiner but that the latter is an essential component and basis of the former.

8.4 Concerning the extension of medical anthropology through anthroposophy

There is now a wealth of material on an *anthroposophically expanded view of nature and medical anthropology* which has been developed by doctors, pharmacists and scientists since the founding of anthroposophical medicine at the beginning of the 20th century. This provides a very different, considerably expanded basis for the "medical system" referred to by Steiner in comparison to the situation in his lifetime, but this system still cannot be viewed as complete by any means. The *experimental and scientific clinical research* initiated by Steiner has developed to an even lesser degree for a variety of reasons, so that all in all anthroposophically extended medicine must be assumed to have a very provisional status. Only the main topics of this development and the current situation can be described here[175] in order to give an idea of the directions and ways in which the efforts to realise the medical development initiated by Steiner have progressed.

The development of an anthroposophically expanded view of nature and medical anthropology began even with Steiner's introductions and commentary on Goethe's natural scientific writings (Steiner, 1988b) and was later continued especially in his science courses (Steiner, 2001; Steiner, 1980b).

---

175 The scope of this work does not permit a *historical* appraisal of this development, something also not addressed nor achieved by Zander in the most comprehensive (though also problematical, cf. Introduction and Sections 8.2, 8.3 and 8.5) work on the history of anthroposophy in Germany to date (Zander, 2007).

Although these are not medical works they have numerous references to medicine.

The same applies to the scientific works of other anthroposophical authors on the nature of the mineral realm and the *concept of substance*. These include the scientifically-based comments by Walther Cloos and Wolfgang Wimmenauer on minerals and their processes of formation (Cloos, 2015; Wimmenauer, 1992), the studies by Eugen Kolisko on hypothesis-free chemistry mentioned above (Kolisko, 1922a) and the debates of his time on the problem of atomism (Kolisko & Rozumek, 2012), Manfred von Mackensen's books on process chemistry (von Mackensen & Schoppmann, 2001), Martin Rozumek's phenomenological and conceptual writings (Rozumek & Buck, 2008; Rozumek, 2008) and Otto Wolff's *"Grundlagen einer geisteswissenschaftlich erweiterten Biochemie"* (Wolff, 2013) which discusses the relationship of substances to the physical, etheric, astral and ego forces in the human organism. A similar angle is evident in Wilhelm Pelikan's book on the metals (Pelikan, 2006) and Gisbert Husemann's work on tin (Husemann, 1962). Following Goethe's scientific method, these writings are based on the natural phenomenology of the substances and make the connection between these and the human organisation whereas, in her book on metals and their therapeutic properties, Hilma Walter primarily takes an anthroposophical and spiritual scientific viewpoint (Walter, 1966). The same applies to Rudolf Hauschka's books on substance and therapeutic teachings (Hauschka, 2002; Hauschka, 1974). Alla Selawry, on the other hand, has related the phenomenological natural scientific and spiritual scientific aspects of the effects of metals to homoeopathic phenomenology (Selawry, 1985).

There are also numerous goetheanistic works in *botany* which contain some important references to the human organisation (Schad, 1982c; Schad, 1982d; Suchantke, 2012). This is taken up again in the actual medicinal plant books, such as Pelikan's famous Heilpflanzenkunde (medicinal botany) (Pelikan, 1997–2012), in the books by Christian Simonis (Simonis, 1955; Simonis, 1981), and in Jochen Bockemühl's introductions to medicinal plants based primarily on qualitative phenomenology (Bockemühl, 1996–2003; Bockemühl, 2010). These are supplemented by Hilma Walter's medicinal plants book written from a completely anthroposophical spiritual scientific viewpoint (Walter, 1971).

A particular group of works is formed by those dealing with the nature and activity of the *etheric*, partly from a phenomenological natural scientific side and partly from an anthroposophical spiritual scientific one. These include the experimental work on sensitive flow forms in liquids by Theodor Schwenk (Schwenk, 2014), Johanna Zinke's related study on the flow forms in air caused by spoken sounds (Zinke, 2003) and Hans Jenny's experiments on sound forms in various materials (Jenny, 2009). Although based on spiritual science, Guenther Wachsmuth's studies on the etheric formative forces in the cosmos, earth and man try to gain an external empirical verification by reference to numerous scientific studies (Wachsmuth, 1932; Wachsmuth, 1965), whereas Ernst Marti confines himself to anthroposophical aspects (Marti, 1994). Numerous studies deal with the physical – e.g. geometrical – forms of expression of the etheric in plants (Bockemühl, 1985; Edwards, 2006), with some including the effects of cosmic etheric forces on plants (Adams & Whicher, 1980), animals (Endres & Schad, 1997) and in the geometry of the human form (Schüpbach, 1947–1948). Hermann Poppelbaum has addressed the question of the scientific basis of the concept of the etheric body (Poppelbaum, 1924), and Jochen Bockemühl (Bockemühl, 1996–2003; Bockemühl, 2010) and Andreas Suchantke (Suchantke, 2012) describe training methods for attaining experiential knowledge of the etheric in nature.

There are also studies on *zoological and human morphology*, amongst which mention must be made in particular of those which compare the human and animal forms. These include studies which deal with the special ontogenetic (Kipp, 1948) or morphological positions of the human being (Poppelbaum, 1981; Verhulst, 1999; McKeen, 1996) who, e.g. with his upright stance, is an expression of the human ego. These can be complemented by studies on the human inner organ systems (Kranich, 2003).

Attention is also drawn to Wolfgang Schad's classic work which, using comparative morphology, demonstrates how the three main groups of mammals – the rodents, carnivores and ungulates – show a relative one-sided dominance of what appears in balance in the human being and how the overall form of the human being and the various groups of animals is related to their way of life: the sense and nerve system in the rodents, the rhythmical system in the carnivores and the metabolic system in the ungulates (Schad, 1983; Schad, 2012). Armin Husemann demonstrates the same principle by comparing the dentition and its function in relation to the way of life in rodents, carnivores and ungulates on the one hand and human beings on the other (Husemann, 1996). Husemann also pointed out the sculptural and musical construction of the human being (Husemann, 1994). Based on a recommendation by Steiner, he uses clay modelling and musical exercises in the medical training institution, the Eugen Kolisko Akademie in Filderstadt near Stuttgart, of which he is the head, as a systematic training tool for medical students and doctors in order to develop their sensitivity for the formative forces in the human organisation (Husemann, 2007). Husemann's newest book demonstrates the results of such a training, namely the dynamic understanding of the relationships between organic form, function and consciousness in human physiology and pathology (Husemann, 2015).

Steiner first mentioned the existence and importance of the *functional threefold nature* of the human organisation in his book *"Riddles of the Soul"*, published in 1917 (Steiner, 1996, 131–144). He drew attention to the fact that, between the nerve-senses system in which breakdown is caused by the functions of consciousness and the metabolic system which serves the process of synthesis, there is a functional polarity which is balanced by the rhythmical system with its functions of breathing and circulation (Heusser, 1999b). These viewpoints were developed further by a variety of authors, particularly from a morphological angle (Schad, 1977; Schad, 1985a; Vogel, 1979) and are of fundamental importance for anthroposophical medicine. In his much-used textbooks on anatomy and histology and later on embryology and morphology, the renowned Nuremberg anatomist Johannes Rohen confirmed Steiner's idea of functional threefoldness as an organising principle in these fields (Rohen et al., 1998; Rohen & Lütjen-Drecoll, 2000; Rohen & Lütjen-Drecoll, 2006a; Rohen & Lütjen-Drecoll, 2006b; Rohen, 2008). Based on Steiner's concept, he introduced the general classification of all organs, both anatomically and functionally, in the three main systems of metabolism, respiration and circulation (the rhythmic system) and "exchange of information and control" served by the nervous system. In German-speaking Europe, Rohen's textbooks are a well-known basis for mainstream medical education.

Physiology has also been studied from the viewpoint of threefoldness, especially for the heart and circulation (Schad, 1985a; Schad, 1998). In terms of the functioning of the heart, attention was drawn, for example, to the justification of Steiner's criticism of the mechanical pump model of the heart, based on relevant findings in the literature on the circulation (Bavastro & Kümmell, 1999). Recently Branko Furst has published detailed accounts of Steiner's basic concepts of cardiovascular physiology with their critique of the mechanistic model of the heart as a pump and demonstrated their validity by reviewing the accumulated anatomical, physiological, evolutionary, embryological, experimental and clinical evidence (Furst, 2014; Furst, 2015).

Much work has been done in the realm of *rhythmology and chronobiology*. Based on a synopsis of a large number of studies on rhythm from the scientific literature, Wachsmuth described rhythm as an expression of life processes in the earth and in man (Wachsmuth, 1965); Hoerner wrote about the importance of rhythms for the process of psychological development (Hoerner, 1991); and Bernd Rosslenbroich described the rhythmical organisation of the human being in relation to functional threefoldness (Rosslenbroich, 1994). A very important contribution was made by the studies of the researchers working with the Marburg occupational physiologist, Gunther Hildebrandt (1924–1999), on therapeutical physiology and hygiogenesis (Hildebrandt, 1998) and on chronobiology and chronomedicine (Gutenbrunner et al., 1987; Gutenbrunner et al., 1993). Hildebrandt also confirmed Steiner's fundamental views on rhythm research, in particular the significance of the frequency ratio between breathing and the pulse rate as an indicator of the health status of individuals (cf. the overview in: Hildebrandt, 1999; also: Heusser, 1999b). In our own research, the working group connected to the author was able to show experimentally that the pulse respiration quotient which can be calculated by means of the heart frequency variability (HRV) represents an independent variable compared to other HRV rhythm parameters (Cysarz et al., 2008), and that this quotient correlates to subjectively perceived health, better mental well-being and less severe depressive moods (von Bonin et al., 2014).

The laws of the rhythmic organisation are put to practical use in various types of applications in anthroposophical medicine. A number of *new therapies* have been developed on this basis: *rhythmical massage* in physiotherapy (Hauschka, 1971), *rhythmical embrocation* in nursing care (Layer, 2003), special forms of art therapy such as *therapeutic clay modelling* (Golombek, 2000), *painting* (Mees-Christeller et al., 2000), *interactive music therapy* (Felber et al., 2000), *therapeutic speech formation* (Denjean-von-Stryk & von Bonin, 2000), *curative eurythmy* (Hachtel & Gäch, 2007; Kirchner-Bockholt, 1997), and the application of rhythm in the *manufacture* of *medicines*[176]. In the author's working group, research has been able to show an effect of curative eurythmy on children with attention deficit disorder syndrome (Majorek et al., 2004): the recitation of hexameters using therapeutic speech formation results in a synchronisation of the breathing and pulse frequencies, in a similar way as happens in regenerative sleep during the night (Cysarz et al., 2004). This shows that these kind of artistic therapies can have demonstrable physiological effects which provide a basis for recovery. The physiological and subjective psychological effects of curative eurythmy against stress have also been substantiated in more recent studies (Kanitz et al., 2011; Kanitz et al., 2013; Seifert et al., 2012; Seifert et al., 2013; Edelhäuser et al., 2015; Berger et al., 2015).

An interesting series *"A New Image of Man"* has been compiled by professors including Karl E. Schaefer (Brown University), Herbert Hensel (University of Marburg), Ronald Brady (Ramapo College, New Jersey), Gunther Hildebrandt (University of Marburg), Uwe Stave (University of Miami) and Wolfgang Blankenburg (University of Bremen) and contains approaches to an *individualised* physiology, psychology and medicine

---

[176] For example the company WALA which produces anthroposophical medicines and cosmetics from natural substances was originally established from a suggestion by Rudolf Steiner. In 1924 Dr. Rudolf Hauschka (1891–1969), a chemist from Vienna, travelled to Holland and attended an anthroposophical summer school in Arnheim where he met Rudolf Steiner with whom he had a number of conversations. When Hauschka asked Steiner what *life* was, in his opinion, the latter answered: "Study rhythm, rhythm carries life." (www.wala.de/unternehmen/gruendungsidee/. Viewed 26/12/2015).

(Schaefer et al., 1977; Schaefer et al., 1979a; Schaefer et al., 1979b). These contributions were the result of a symposium at the Herdecke Community Hospital in 1973 which brought together the above-mentioned researchers with a view to the founding of the Witten/Herdecke University (1982).

Anthroposophical aspects of the human *nerves and senses system* have been compiled by Wolfgang Schad (Schad, 2014), and a start has been made on an interpretation of the physiology of the senses which take psychological aspects of the sensory-motor neuronal system into account (Heusser, 2000a) and which interpret the occurrence of pain in a similar perspective (Heusser, 2007). Attention is drawn to the dissertation by Walter Johannes Stein from 1919 which presents the entire physical/living/emotional/spiritual process in the sensory experience. Steiner provided Stein with important suggestions for this (Stein & Steiner, 1985). In addition, attention is drawn particularly to the physiology textbook *"Allgemeine Sinnesphysiologie. Hautsinne, Geschmack, Geruch"* (a general physiology of the senses, skin, taste and smell) by Herbert Hensel which points out such things as the irreducibility of the sensory experience in the sense meant by Steiner (Hensel, 1966), as well as the habilitation thesis by Gerhard Kienle on *"Die optischen Wahrnehmungsstörungen und die nichteuklidische Struktur des Sehraums"* (optical perceptual disorders and the non-Euclidian structure of visual space) (Kienle, 1965). Mention should also be made of two treatises which have developed Rudolf Steiner's concept of the "sense of speech" from the viewpoints of developmental psychology, moto-neuro-biology and social perception (Lutzker, 1996; Peveling, 2015).

To the above-mentioned selection of monographs on particular topics in a goetheanistically and anthroposophically expanded natural science and medical anthropology can be added the extensive literature which has been published in *journals* established specifically for that purpose. In the first instance, this includes the public periodical *Natura*, started by Ita Wegman and published in eight volumes between 1926 and 1940, containing very varied contributions on morphological, physiological, psychological, pathological, material, therapeutic, cosmological, history of medicine and other topics of relevance to medicine (Wegman, 1981). In addition to this are the *Beiblätter* (supplements) to the journal *"Natura"* published by Wegman for the doctors working in the Medical Section at the Goetheanum, the first of which she wrote herself and which, in terms of content, initially corresponded roughly to the second volume of the book planned jointly with Steiner (Wegman, 2000).

There are also numerous articles in the periodicals published by the Natural Science Section of the Goetheanum *"Gäa Sophia. Jahrbuch der Naturwissenschaftlichen Sektion der Freien Hochschule am Goetheanum"* (Wachsmuth, 1926–1932) and, since 1964, the *"Elemente der Naturwissenschaft"* (http://science.goetheanum.org/index.php?id=348). Since 1984 there is also the *"Tycho de Brahe Jahrbuch für Goetheanismus"* published by the Carl

Gustav Carus-Institut in Öschelbronn with goetheanistic contributions from the spectrum of the natural sciences, medicine and pharmacy (www.carus-institut.de/tycho). The *"Persephone"* series of books published by the Medical Section at the Goetheanum in 1992 on the whole medical field and the journal for anthroposophical medicine *"Der Merkurstab"* published by the Medical Section at the Goetheanum and the Gesellschaft Anthroposophischer Ärzte in Germany which has been in existence since 1988 and which forms a continuation of the *"Ärzte-Journals"* (1946–1949) and *"Beiträge zur Erweiterung der Heilkunst"* (1950–1987) (www.merkurstab.de) are more directly focussed on medical topics. Both the anthroposophical medical journals and the books dating from the beginning of the anthroposophical medical movement around 1920 have been indexed in clearly arranged documentation volumes and provided with comprehensive indices of key words and authors by Angelika Overstolz (Overstolz, 1995; Overstolz, 2001). Recently, the Medical Section at the Goetheanum has started the project *"Anthromedics"*, an online portal in English and German offering information on and literature access to anthroposophical medicine by providing a digital archive to all the issues of the journal *"Der Merkurstab"* and digital access to textbooks on anthroposophical medicine (German Anthroposophical Medical Association & Medical Section, 2014).

Systematic works giving an overview of the fundamentals of anthroposophical medicine as a whole are to be found in the spiritual scientific descriptions by Herbert Sieweke (Sieweke, 1982; Sieweke, 1994) and Volker Fintelmann (Fintelmann, 2007). Particular mention should be made of the three-volume complete overview of anthroposophical medicine *"The Anthroposophical Approach to Medicine"* first published by Friedrich Husemann and later by Otto Wolff in several repeatedly expanded editions, which was the standard work on anthroposophical medicine for a long time. This work in three volumes, now in need of revision, ranges from anatomical and physiological principles via general and specific pathology to the description of individual specialisms with practical recommendations for treatment based on the specific indications (Husemann & Wolff, 1982, 1987, 1989). This book can now be replaced to a certain extent or at least complemented by Matthias Girke's excellent new book on internal medicine (Girke, 2012) and in the field of metabolism additionally by Andreas Goyert's book on processes of the metabolic system (Goyert, 2015).

In addition there is a series of *textbooks or compendia on individual medical disciplines, disease entities and types of treatment which are important for the conceptual basis of anthroposophical medicine and treatments.* Apart from the *internal medicine* mentioned above (Girke, 2012), this covers *geriatrics* (Girke, 2014), *dermatology* (Jachens, 2012), *paediatrics* (Soldner & Stellmann, 2011); *psycho-physical development and psychopathology* (Walter,

1955; Treichler, 1981b), *psychiatry* (Treichler, 1981a; Bissegger et al., 1998) and *psychotherapy* (Vandercruysse, 1999; Priever, 1999); *curative education* (Bort et al., 1956; von Zabern, 2002; Weihs, 1971; Grimm & Kaschubowski, 2008); *nursing care* (Heine & Bay, 1995); the field of *external applications such as compresses* (Fingado, 2001); the art therapies and curative eurythmy mentioned above; and also *pharmacy* (Glöckler, 2005), *oncology* and anthroposophical mistletoe therapy (for reviews see Fintelmann, 2002; Kienle & Kiene, 2003; Ostermann et al., 2009; Kienle & Kiene, 2010; Büssing et al., 2011c; Ostermann & Büssing, 2012; Heusser & Kienle, 2014a); *intensive care* (Bavastro, 1994); *nutrition* (Schmidt, 1975–1979); the relationship of *medicine and pedagogy* (Kolisko, 2002; Glöckler, 1998; Glöckler & Goebel, 2007; Marti & Heusser, 2009; Fischer et al., 2013)[177]; and *individual anthroposophical medicines* (Wolff, 1990; Schramm, 1983; Soldner, 2008; Basold et al., 1995), etc. A number of fields are missing from the range of topics in these works which, while being practised in anthroposophical medicine and taught on courses and in practices or clinics, are at best only documented to a certain degree in individual contributions in books or journal articles. These fields include gynaecology, ophthalmology, diseases of the ear nose and throat, and others.

Mention should also be made of the literature on the *history of anthroposophical medicine* (van Deventer, 1992; Kugler, 1997; Selg, 2000a; Selg, 2000b; Selg, 2003; Selg, 2015; Zander, 2007), the *history of curative education and social therapy* (Frielingsdorf et al., 2013) and the *relationship of anthroposophical medicine to the history of medicine*, in particular to Paracelsus (Daems, 2001); further, its relationship to Christianity (Selg, 2005; Husemann, 2009), to *spirituality* (Heusser, 2006b) and to *medical ethics* (Heusser & Riggenbach, 2003; Glöckler, 2002). Works on the epistemological and attitudinal inner *ethical and moral path of learning and development of the doctor* (Selg, 2006; Selg, 2013) also belong in this context. This path has always played a special part in anthroposophical medicine following on from Steiner's medical course for young doctors given in 1924 (Steiner, 1994b), and has also been important for the relationship of anthroposophical medicine to scientific methodology (cf. Section 8.5).

Despite having increased enormously in extent since Steiner and Wegman and despite several partial synopses, the sum total of this literature provides only an *initial foundation* for an anthroposophical medical system in the

---

177 This *"Guide to Child Health"* by Michaela Glöckler und Wolfgang Goebel has been very widely read by parents (including 3 editions of the English translation and 20 editions of the German original). It has been translated into various languages and found global dissemination, underlining the great need for spiritually expanded medical points of view amongst the general public.

sense intended by Steiner. The material is still too heterogeneous, selective and topically diffuse to be able to provide a system of this kind. The concepts often remain quite general and are of very varied quality. Many articles are based primarily on the formation of anthroposophical ideas and often show a great lack of conceptual anthroposophical scientific development and an even greater lack of the corresponding empirical validation. Due to the many separate studies, the outlines of a systematic natural and spiritual scientific synthesis have so far only emerged in parts, such as for the functional threefold nature of man and its morphological, anatomical and physiological basis and the therapeutic physiology and hygiogenesis research which rests on this. In these fields, the material which has arisen from Steiner's anthroposophical suggestions, particularly through the conceptual and empirical studies of the working groups lead by Wolfgang Schad, Johannes Rohen, Gunther Hildebrandt and Herbert Hensel, and also through salient reviews of scientific evidence such as that of Branko Furst on the heart and circulation, anthroposophical concepts have been integrated seamlessly and on a scientifically sound basis into the academic system of theory and research and can be viewed as a solid contribution by anthroposophy to science and medical anthropology.

However, there is still a long way to go to conceptually reach a compelling unified natural and spiritual scientific synthesis in the sense of the "medical system" intended by Steiner as described in Section 8.3 both in terms of breadth and depth. Four lines of endeavour will be necessary for working towards this goal. This book is intended to make a preliminary contribution to this from the angle of the scientific fundamentals of anthroposophy.

To begin with, a sound epistemological basis has to be laid for building a solid bridge between the natural and spiritual perspectives of reality in scientific terms. This comprises a) the justification of the spiritual as an integral and even essential part of reality itself (cf. Chapter 2); b) the transformation of the natural sciences into a new non-reductionist form compatible with spirituality (cf. Chapters 3, 4 and 5); and c) the justification of spiritual cognition as an expansion of scientific cognition from the materiality of nature to the realm of spirituality (cf. Chapters 6 and 7).

Second, the multifaceted spiritual elements and perspectives which are widely dispersed in the enormous body of work by Steiner and in the secondary anthroposophical literature will need to be synthesised into a coherent anthroposophical conception of the human being in relation to health, disease and treatment. The same applies to further or new spiritual insights obtained by others. This synthetic work has not been reached satisfactorily by any means. Yet an important step in this direction has been achieved by Peter Selg with the first systematic description of the development of an anthroposophical human physiology in Rudolf Steiner's complete works (Selg, 2000c). Selg has also produced the first systematic compilation of source texts on physiology, pathology and therapy from Rudolf Steiner's Complete Works (Steiner, 2004a; Steiner, 2004b). This is very valuable for establishing a coherent conception of this kind.

The third and very important line of work will be the systematic examination and elaboration of this conception with the appropriate natural scientific methodology in the sense described above, the "verification" in Steiner's own words (see Sections 8.2 and 8.3 above).

The fourth angle is the personal efforts of anthroposophical doctors, pharmacists and scientists to gradually develop their own faculties for spiritual perception and spiritual scientific cognition (cf. Chapter 7), alongside the sound natural scientific training and systematic qualification of a larger number of anthroposophically oriented scientists in medical specialisms.

If thought is given to how much work is required to accomplish individual factual results in any of these lines of work, then it is clear that the development of a comprehensive systematic spiritual and natural scientific medical anthropology and medical theory will be a gradual process and, by its very nature, require the work of many individuals and research groups over several generations. The research groups and infrastructure available at the moment are too small to be able to entertain optimistic expectations. Having said that, the same applies to this new anthroposophical anthropological research as to the approx. 250 years of natural science: sound research and real progress only come about one step at a time, as a *development* over the long-term and through the cooperation of many and varied scientists.

8.5 The status of clinical scientific research in anthroposophical medicine

This development also includes the scientific testing of the efficacy, effectiveness, safety and suitability of anthroposophical medical treatment methods through experimental and clinical studies, as Steiner himself demanded and expected (cf. Section 8.3). There has also been development in this field since Steiner's lifetime, but significantly less than the basic anthropological medical literature. This is largely due to structural reasons, but it also has intrinsic causes inherent in the particular holistic, multimodal and individualised approach to treatment in anthroposophical medicine.

The structural reasons are immediately obvious when you see how few institutes and researchers are available for this work. For instance, the total of 16 anthroposophical clinics and five scientific institutes for anthroposophical medicine, mainly based in Central Europe, have fewer than 50 research posts between them. Most of the research departments which have been set up in the larger hospitals were established after 1980. In addition, the anthroposophical medicine manufacturers like Weleda and Wala can never match the conventional pharmaceutical giants in terms of research funding.

Lili Kolisko was the first to undertake *experimental scientific work* when she took up Steiner's suggestions on this in his first medical course in 1920. She started by carrying out a nutritional study with rabbits and human test subjects using natural scientific methods in order to follow up Steiner's spiritual science-based theory that the spleen has a functional importance in integrating the food ingested at various times into the body's regular rhythms (Steiner, 1948, 203). Kolisko's work *"Milzfunktion und Plättchenfrage"* (the function of the spleen and platelets) (Kolisko, 1922b) published in 1922 provided some indirect empirical evidence for the validity of this theory which Steiner accepted, while pointing out the existence of some methodological objections in certain details (Steiner, 1994a, 87–89). In fact, from the perspective of present-day requirements for experimental design, the research would need to be regarded as a pilot study both in quantitative and qualitative terms which, despite its positive result, cannot lay claim to being a proof. It was criticised for this even at the time (Zander, 2007, 1468–1472). Despite this, Steiner praised Kolisko because her work was the first time that research of an anthropological/anthroposophical nature in the above-mentioned sense had been carried out and because this study, despite being only a beginning, provided the first example "of how we aim for the same exactitude which is aimed for in the scientific basis of medicine today" (Steiner, 1994a, 87–88); and he was therefore annoyed at the attempt by an anthroposophical doctor to suppress this report.

Zander took this event and particularly Steiner's support for the study as grounds for accusing anthroposophical medicine, – and in this he concurred with other authors – of having the result to be proven "already in the bag" from a spiritual scientific angle, which "to this day [is] a point of contention between university and anthroposophical medicine in terms of the empirically inaccessible assumptions of anthroposophy" (Zander, 2007, 1477). Zander implied with this that the natural scientific research advocated by Steiner was only used to justify his medical philosophy and claimed: "In the final instance anthroposophical medicine is not based on the experiment but on the [spiritual] 'view'" (ibid., 1469). Zander's claim overlooks the fact that, first, experiments are always carried out on the grounds of ideas and from that point of view, it is never the experiment but always *the idea* which forms the basis of medicine (Martin, 2000); and, in this context, the experiment only serves to confirm or reject the idea in a scientific manner. The fact that, in the case of anthroposophical medicine, the idea can come from a spiritual scientific source whose *field of perception* is not a priori accessible to everyone (so at first forms an "empirically inaccessible assumption" for many) is correct. But the same thing applies to many fields of natural science, as shown in Section 8.2. And because the idea – as an idea – has to be accessible at least to *logic* and thinking, it is basically accessible to science as a *hypothesis*. The scientific experiment then serves to test the idea in a field which is accessible to everyone and, what is more, the validity of such ideas can always be tested in their applicability to actual practice. This is how Steiner intended his spiritual scientific "guiding thoughts" mentioned in Section 8.3. Second, Zander does not realise that, in the case of research to support a particular theory, it makes no sense to want to test an idea experimentally *in the first place*, because without *justifiable* prior knowledge it cannot be assumed in advance that the experiment has a realistic probability of having a positive result. Third, this experiment *did* have a positive result, even though it had acknowledged methodological deficits. The positive outcome could of course be coincidence, but nobody conducting a mere piece of justification research would wish to risk to expose their ideas to experiment if they had no basis for thinking that the idea had a *potential for according with reality*. Finally, it should be mentioned that Steiner's theory about the spleen is currently being reappraised in terms of both natural and spiritual science in our department, as part of a larger research project (Scheffers & Weinzirl, 2015), and that we have recently been able to verify the enlargement of the spleen after food intake using ultrasound in a controlled experimental study (Garnitschnig et al., 2015).

The same applies to the experimental proof of the *effect of homoeopathically potentised substances* first suggested by Steiner and successfully carried out by Lili Kolisko. Zander, however, does not discuss this although this field of work was the one to which Kolisko devoted her life, whereas *"Milzfunktion und Plättchenfrage"* remained a one-off piece of research. In his course for doctors given in 1920 Steiner requested that the effect of potentised substances be tested experimentally and described "by means of curves" (Steiner, 1948, 1). Lili Kolisko, who was a laboratory technician rather than a doctor, took up this suggestion and investigated the effect of potentised substances on plant growth in controlled experiments. Publication of her first study *"Physiologischer und physikalischer Nachweis der Wirksamkeit kleinster Entitäten"* followed in 1923 (Kolisko, 1997); and in 1961 Kolisko compiled the results of her almost thirty years of experimental work in this field which form *the* pioneering work on the experimental assessment of the effects of potentised substances (Righetti, 1988, 77) in book form (Kolisko, 1961). In the 1960s Wilhelm Pelikan and Georg Unger continued this work using an elaborate experimental design and were able to confirm her findings by means of a full statistical analysis (Pelikan & Unger, 1965). Since 1995, the Department for Anthroposophical Medicine built up by the author of this book has developed a leading international research group on homoeopathically potentised substances under the leadership of Dr. Stephan Baumgartner at the Institute of Complementary Medicine at the University of Bern (www.ikom.unibe.ch). The group has established modern evaluation standards (Baumgartner et al., 1998) and conducted the corresponding rigorously controlled studies in physical systems, plant models and microorganisms, fundamentally confirming the existence of the effect of potentised substances as described by Lilli Kolisko (Baumgartner et al., 2004; Guggisberg et al., 2005; Wälchli et al., 2006; Scherr et al., 2006; Scherr et al., 2007; Baumgartner et al., 2008; Baumgartner et al., 2009; Jäger et al., 2011a; Klein et al., 2013; Majewsky et al., 2014). Results are also compiled in reviews (Witt et al., 2007; Majewsky et al., 2009; Betti et al., 2009; Jäger et al., 2011b) and new methodological aspects of this research using potentised substances (Baumgartner, 2005; Jäger et al., 2011c).

According to Steiner's own demand for empirical scientific verification discussed above, the *clinical effects and effectiveness* of medicaments and non-pharmacological interventions developed from an anthroposophical rationale must also be demonstrated using acknowledged natural scientific methods. However, due to the reasons indicated above, the implementation of a systematic research infrastructure could only be developed gradually and at an academic level only over the last two decades. Most anthroposophical therapy methods were first developed and introduced into direct medical practice according to the anthroposophical rationale and in the context of immediate therapeutic needs (Steiner & Wegman, 1996; van Deventer, 1992), and the established use is now gradually being examined for its efficacy, effectiveness and mode of action.

An example of this is the use of the homoeopathically potentised semi-metal *antimony* to improve blood coagulation in the case of bleeding disorders. Steiner was the first to propose antimony for this indication as early as 1920 in his first medical course (Steiner, 1948, 241ff.). In their description of the anthroposophical rationale, Steiner and Wegman summarise the well-known chemical and physical (especially crystallographic) properties of antimony and describe how, through these properties, antimony can interact easily with those etheric formative forces in the organism which are normally under the control of the ego-organisation and are active in the process of blood coagulation (Steiner & Wegman, 1996, Chapter 16). According to the rationale, if the activity of the ego-organisation is weak, with ensuing insufficient blood coagulation, antimony can be used to improve coagulation. From an epistemic viewpoint it is clear that the activity and interaction of physical, etheric, ego and other forces[178] cannot be observed directly by ordinary

---

178 As explained in Section 8.1, from an anthroposophical perspective the processes of health and disease cannot be explained solely as a matter of molecular interactions, but of interactions of specific and emergent lawful forces on the

consciousness (cf. Section 6.1), but it is certainly possible to observe the *effect* of these interactions indirectly in the processes of the physical body, be it in the modification of clinical symptoms (in this case bleeding) or in appropriate laboratory tests.

As in many other cases of medications proposed by Steiner on the basis of spiritual research (which deals with the forces as such, see Section 7.2), antimony was directly introduced into medical practice and established through practical experience in the 1920s and 30s by anthroposophical doctors and, besides other rationally deduced indications, it has always been used for bleeding disorders in a homoeopathic potentised form, usually as D6 which is a toxicologically harmless dose (Heusser et al., 2004). But for many years this preparation was not tested experimentally in preclinical or clinical trials. This was not only true of the new anthroposophical preparations introduced directly into practice based on a spiritual scientific rationale, but also of other substances which were basically imported into anthroposophical medicine from homoeopathic or phytotherapeutic traditions and which were likewise interpreted according to a spiritual rationale on the basis of their observable effects on the levels of the physical body, organic processes and/or emotional and cognitive processes. In addition, in the small clinical establishments of those days with their modest infrastructure, there was no question of clinical research in the present-day sense, something that naturally applied to most areas of orthodox medicine as well.

For this reason, after establishing the first department for anthroposophical medicine at the University of Bern in 1995, we conducted a survey amongst the anthroposophical doctors in Switzerland on the retrospective judgement of the benefits and side effects of this treatment (Meier & Heusser, 1999) and, following this in a joint project with the thrombosis laboratory of the Department for Clinical Research at the University of Bern, started a systematic investigation of the effects of antimony. We found that antimony in vitro significantly increased the clotting firmness in the blood of healthy donors and tended to do so in patients with blood clotting disorders, plus slightly reducing the clotting time (Heusser et al., 2004). An exactly comparable result was then found in vivo in a randomised, placebo-controlled double-blind crossover study with 30 healthy test subjects: significantly reduced clotting time and significantly increased clotting capacity 30 minutes after intravenous injection of antimony D6 compared to the placebo with perfect tolerance (Heusser et al., 2009a). Based on these results, a randomised double-blinded placebo-controlled study is now being performed in cooperation with the university clinic for urology at the University of Bern in order

---

emergent levels of the physical body (the physical organisation), life (the etheric organisation), soul (the astral organisation), and spirit (the ego-organisation).

to examine whether intra- and postoperative bleeding in patients undergoing a trans-urethral prostate resection can be reduced using antimony. This may serve as an example of the basic relationship between a rationale obtained from anthroposophy and the empirical scientific research of our times, much in the spirit of Steiner's statement already cited (Steiner, 1951b, 7):

> Indeed, all that comes from anthroposophical investigation in regard to medicine and, for instance, physiology, can be nothing more than a stimulus which must then be worked out empirically. Only on the basis of this empirical study can there arise valid and convincing judgements of the matters in question – and this is the kind of judgement that is needed in the domain of therapy.

This current example from the author's own research may show that anthroposophical kinds of therapy are certainly capable of being tested experimentally, i.e. "as all such facts are verified by the methods of modern medicine" (Steiner, 1951b, 59). Additional examples for further empirical investigation of ideas which initially come from spiritual science can be found in the rhythm and hygiogenesis research mentioned in Section 8.4. This shows that anthroposophy does not only intend to contribute *ideas* to a more profound and encompassing understanding of the human being in medical anthropology, but also to make *practical, evidence-based* therapeutic contributions to the extension of the range of treatments.

All in all, however, the development of experimental and clinical research in anthroposophical medicine has fallen behind that of anthropological and medical theories. A comprehensive analysis of effectiveness, appropriateness, safety and costs of the entire body of anthroposophical medicine was undertaken between 1999 and 2005 as part of the Swiss Complementary Medicine Evaluation Programme (PEK), on whose steering committee the author of this book acted as a university representative for complementary medicine (Melchart et al., 2005). On the one hand the PEK programme undertook a comprehensive evaluation of the existing international clinical literature and documented the individual therapy fields (homoeopathy, anthroposophical medicine, traditional Chinese medicine, neural therapy, phytotherapy) as part of Health Technology Assessments; on the other hand a large health service research study was carried out in Switzerland to compare doctors' practices in complementary medicine and orthodox medicine in terms of the practice and patient structures, patient care, patient satisfaction and the subjective feeling of patient benefit. Development of the HTAs was undertaken by Dr. Gunver Kienle and Dr. Helmut Kiene from the Institute for Applied Epistemology and Medical Methodology (IAEMM) in Freiburg and Dr. Hans Ulrich Albonico from the Swiss Anthroposophic Medical Association (VAOAS), in cooperation with Prof. Peter Matthiessen from the Gerhard Kienle Chair of Medical Theory, Integrative Medicine and Anthroposophical Medicine at the University of Witten/Herdecke. The HTA for anthroposophical medicine was published in 2006 and gave an overview of the entire clinical literature on anthroposophical medicine (AM) (Kienle et al., 2006). Its findings were summarised as follows:

> *Background and Objective:* The aim of this Health Technology Assessment Report was to analyse the current situation, efficacy, effectiveness, safety, utilization, and costs of Anthroposophic Medicine (AM) with special emphasis on everyday practice.
> *Design:* Systematic review.
> *Material and Methods:* Search of 20 databases, reference lists and expert consultations. Criteria-based analysis was performed to assess methodological quality and external validity of the studies.
> *Results:* AM is a complementary medical system that extends conventional medicine and provides specific pharmacological and non-pharmacological treatments. It covers all areas of medicine. 178 clinical trials on efficacy and effectiveness were identified: 17 RCTs, 21 prospective and 43 retrospective NRCTs, 50 prospective and 47 retrospective cohort studies/case-series without control groups. They investigated a wide range of AM treatments in a variety of diseases, 90 × mistletoe in cancer. 170 trials had a positive result for AM. Methodological quality differed substantially; some studies showed major limitations, others were reasonably well conducted. Trials of better quality still showed a positive result. External validity was usually high. Side effects or other risks are rare. AM patients are well educated, often female, aged 30–50 years, or children. The few economic investigations found less or equal costs in AM because of reduced hospital admissions and fewer prescriptions of medications.
> *Conclusion:* Trials of varying design and quality in a variety of diseases predominantly describe good clinical outcomes for AM, few side effects, high patient satisfaction and presumably slightly lower costs. More research and more methodological expertise and infrastructure are desirable.

The studies covered indications such as depression, chronic fatigue syndrome, anxiety disorders, burnout, attention deficit hyperactivity disorder (ADHD), anorexia nervosa, migraine, trigeminal neuralgia, occipital pain and other pain syndromes, lumbar and cervical spine syndromes, acute ischialgia, hip and knee joint arthritis, intervertebral disc conditions, chronic inflammatory rheumatic diseases, wound healing after carpal tunnel syndrome, care of the umbilical cord, acute infections of the upper respiratory tract and ears, sinusitis, pseudocroup, fever in children, childhood asthma, lung disease (incl. sarcoidosis), cancer, burns, cardiovascular diseases, chronic hepatitis C or B, colitis ulcerosa, infections of the gastrointestinal tract, hyperlipidemia, hyperuricemia, indications in gynaecology and obstetrics and diseases of the thyroid.

The patient care studies carried out by PEK showed that patients treated in anthroposophical practices had a significantly higher level of patient satisfaction than those treated in conventional practices although on average they had significantly more chronic and serious illnesses than the latter. Their expectations of the treatment were also fulfilled to a higher degree (Esch et al., 2008). The PEK's final report summarised all these findings as follows (Melchart et al., 2005):

> Anthroposophical medicine does not have an adequate number of controlled randomised studies in comparison to conventional medicine. When including the wider findings from well-documented observational studies and case histories, however, there is satisfactory evidence for effectiveness and benefit to the patients. Safety can be considered as being largely verified.

Five years later an update of this HTA report was published which gave the following overall summary (Kienle et al., 2011b):

> *Results:* 70 new clinical studies were found. Altogether, 265 clinical studies investigated efficacy and effectiveness of AM: 38 randomized controlled trials, 36 prospective and 49 retrospective non-randomized controlled trials as well as 90 prospective and 52 retrospective trials without control groups. They investigated a wide spectrum of AM treatments in a multitude of diseases; the whole AM system in 38 trials, non-pharmacological therapies in 10 trials, AM mistletoe products in cancer therapy in 133 trials, and other AM medication treatments in 84 trials. Most studies showed a positive result for AM. Methodological quality differed substantially; some studies showed major limitations, others were reasonably well conducted. Trials with better quality
> still showed a positive result. External validity was usually high. Side effects or other risks were rare and usually described to be mild or moderate. Studies regarding safety showed a good tolerability altogether.
> *Conclusion:* Trials of varying design and quality in a variety of diseases predominantly describe good clinical outcomes for AM, only marginal side effects, high satisfaction of patients with regard to results and safety and presumably slightly lower costs. Further high-quality evaluations are desirable.

The main point of criticism of the proof of efficacy of anthroposophical medicine is the *lack of sufficient randomised studies*. There are a variety of reasons for this. As we have seen from the example of the antimony study, it is basically possible to carry out these kind of studies in an experimental setting, however this is not always possible, even where they are explicitly planned. For example, the author of this book designed four prospective, randomised, partially blinded and placebo-controlled studies on mistletoe treatment of neoplasms at the University of Bern in compliance with the standards of good clinical practice (GCP), had these approved by the ethics commission and started with the work. The studies were run in interdisciplinary cooperation with different clinical departments at the Bern University hospital and one study was multicentric with a total of seven centres in four Swiss cantons. However, all four projects had to be terminated early either because too few patients were recruited by the doctors in the various departments or because patients did not want to participate in a randomised, blinded placebo-controlled study. In addition, patients who want to receive anthroposophical treatment, have a verifiable preference for this type of treatment, as our own research shows (von Rohr et al., 2000a). This makes it practically impossible to randomise studies at anthroposophical hospitals (von Rohr et al., 2000b).

What is more, the anthroposophical treatment approach itself causes difficulties for a randomised evaluation as it is a *complex, integrative, multimodal* process where treatment is also *individualised*. It is difficult or impossible to study individual aspects of the therapy or to produce groups which are comparable in terms of treatment, particularly as large treatment groups are not available due to the overall small clinical facilities to date. This leaves *evaluation of the treatment system as a whole*, whether in a randomised or non-randomised design. One example comes from our own research. As co-chair of a large Swiss National Science Foundation project to research quality of life in patients with advanced cancers, the author was involved in a randomised study carried out in the Bern University hospital in which patients were to be given either a complex individualised anthroposophical treatment (patient group 1) or a psycho-oncological group therapy (patient group 2) or no additional treatment (patient group 3, control) in addition to conventional palliative tumour therapy (Hürny et al., 1994; Cerny & Heusser, 1999). Parallel to this a similar patient cohort was studied (patient group 4) by the same method in an anthroposophical tumour hospital where, due to explicit patient preference, randomisation was not possible, after the patient populations of both hospitals had been compared with each other (Pampallona et al., 2002). The overall anthroposophical medical system as such was therefore evaluated in groups 1 and 4. Lack of recruitment to the randomised study prevented its evaluation (this was one of the studies which was terminated prematurely). However it was possible to evaluate the patient cohort from the anthroposophical hospital (group 4), though only using a pre-post design and by comparing the results with the international literature. The quality of life of the 144 patients in the study increased during the three week average in-patient treatment in this hospital in all 20 parameters investigated, in 12 of these significantly and, specifically, in all the main quality of life domains: global, physical, emotional, cognitive, spiritual and social. The results are some of the best in the international literature (Heusser et al., 2006b; Heusser et al., 2006a); and in the follow-up after 4 months with the patients whom it was still possible to study, there were clear indications of a sustained improvement in quality of life despite a reduction in the increase recorded after discharge from hospital: all quality of life scores were still over baseline values four months after hospitalisation, and the qualitatively evaluated retrospective patient judgements of their treatment during hospitalisation also given four months later showed that patients, who in about 98% of cases did not have an anthroposophical background, had clearly experienced a differentiated and in part lasting effect of the comprehensive anthroposophical treatment on body, soul, spirit and social relations, both in quantitative and qualitative parts of the study (Heusser et al., 2006a; Heusser et al., 2009b).

In this therapeutic concept individualisation not only means selecting an individual combination of treatment elements from a multimodal range of therapies but that, depending on the system levels involved, individualisation is *essential to the method*. For if the human being really is a whole in which physical, living, emotional and spiritual-ego forces are causative agents (cf. Sections 5.9 and 8.1), then this has a major effect at each level involved, both for the disease and for the degree to which treatment can be generalised. The laws of the physical organisation lend themselves best to generalisation and, at the lowest level of elementary particles and inorganic substances, it is in fact possible to generalise across all realms of nature, but in the realm of organic substances this can only be done within the world of organisms. The laws of the living realm, e.g. of growth, defence etc. have significantly greater variation in responsiveness compared to the non-living world, e.g. of a crystal. In the human being this expresses itself for example through differences in "vitality" which are generally attributed to the "constitution". Compared to the living realm, the emotional one is even more responsive in a much more subtle way and, with its inwardness, displays individual "characteristics" even in animals which the merely living realm does not have. The emotional life nevertheless follows typical patterns familiar in animal and human psychology. All this likewise applies to human beings but they also have a spiritual individuality, the person's innermost "I", which makes them a unique, unmistakeable individual or "personality" whose decisions – based on insight and taken in a state of freedom – cannot be explained by the disposition of their character or "psychologically" predicted (cf. Sections 5.8 and 8.1). In this realm everything is individual and unique and can no longer be categorised by types as in mere psychology (the emotional realm) or constitutionally (the realm of organic vitality) or from the even more general generic nature of the physical body and its parts.

The conscious inclusion of these kind of points of view in diagnosis and therapy is one of the main reasons why relatively few randomised studies are carried out in the anthroposophical field – in contrast for instance to homoeopathy, traditional Chinese medicine or phytotherapy (Melchart et al., 2005). This kind of conception of the human being produces both ethical and methodological problems for randomised study[179].

---

[179] A more detailed discussion of these and other ethical, methodological and practical problems associated with randomised studies by anthroposophical authors

The *ethical problems* arise from the fact that, from the perspective of the anthroposophical conception of the human being, not only the patient but also the doctor plays a role in the therapeutic process as a being of soul and spirit. The doctor cannot therefore be seen merely as an agency for diagnosis and the handing-out of pills according to clinical guidelines and the hierarchy of evidence, but also as a personality in their own unique way who interacts with the equally unique individual patient, thus creating a doctor-patient relationship and a therapeutic process which are unique in *this* way. Everyone who has treated patients for many years knows that every doctor-patient relationship has a unique character, something which should not be overvalued but is simply a fact. In anthroposophical medicine *the individual doctor-patient relationship is an integral factor in the therapeutic situation* because listening, encouraging, reassuring, a friendly smile, advice on health and lifestyle, psychotherapy or logotherapy in the broadest sense, help in coping with the illness and with questions about meaning and destiny as well as in cases of existential or spiritual distress, are actions which accompany almost every consultation in the widest possible variety of ways, even if, in addition to the actual medical treatment, this is expressed only in a handshake or the general attitude with which you address the patient in a five-minute consultation. In this respect the doctor has a *responsibility* towards this individual patient, because they are actually not an illness but an *individual person* who is ill, who comes to the doctor and it is *they* who need to be helped with their illness and, what is more, in the "best way that knowledge and conscience" permit. If a process of randomisation is now introduced into this situation, then the doctor removes the opportunity of working in the best way possible in terms of his remit which is always an *individual* one, including in terms of the law. In the case of randomisation this has a particular effect on the therapeutic decision, more or less the most important part of the medical treatment. For if chance i.e. the computer generating random numbers makes the treatment decision instead of the "best knowledge and conscience" of the medical ethic, then the doctor cannot do justice to *his individual patient* not only in a legal sense but also personally. This is why the great German haematologist Begemann considered randomisation as "principally contrary to the doctor's remit" and also claimed that these objections should not be ignored by the ethics commission (Begemann, 1988).

---

can be found in: Kienle, 1974; Kienle & Burkhardt, 1983; Heusser, 1999d; Kiene, 2001; Kiene, 1994; Kienle et al., 2006; Kienle, 2009; Matthiessen, 2009. Here I can only touch on a few of the main points of this topic which explain why many anthroposophical doctors have considerable reservations about randomised studies due to their basic anthropological and medical ideas.

For obvious statistical reasons, the explanatory randomised clinical trial has *methodically* been the "golden standard" in clinical evaluation for a long time. However, the differentiated approach to the various system levels of the medical view of man sketched out here qualifies this view to a degree. If both the causes and the effects of disease and healing can lie not only on the physical level but also on the living, emotional or spiritual ones, then more importance may be attached to the completely individual human ego, the admittedly typological emotional level or the vital constitution, or indeed the most generalisable physical findings, depending on the role played by the individual causal factors. Because the laws and forces of the body, life, soul and spirit always act together in the whole, something which happens in the whole human being can never be generalised completely but – perhaps with the exception of the condition of freedom – is also never completely individual. However, depending on the circumstances, more weight will attach to what is general, more to the typological or more to the individual, unique. As a result, the focus of evaluation not only for diagnosis and therapy but *also for therapy research* must be of either a more *nomothetic, generalising*, or a *typological*, or an *idiographic* nature, addressing what is unique and distinctive in the human individual, as has been repeatedly demanded by Peter Matthiessen (Matthiessen, 2009; Ostermann & Matthiessen, 2003).

Group[180] randomised studies, however, only permit statements about what is general or at best what is typical: their purpose and method consist precisely in disregarding everything individual. The "best evidence" from the point of view of evidence-based medicine is therefore in fact that from randomised studies which are concerned neither with the individual patient nor with the judgement of the individual doctor, but only with the "biometric judgement" (Matthiessen, 2009). But randomised studies are not the "best" evidence for the evaluation of ideographic, individual aspects of patients and diseases, but the *worst* in that they block out everything individual. The "best evidence" for ideographic research is that in which the *facts* can only be discovered through *individual* observation and listening (*"narrative based medicine"*, ibid., 51), and whose method can only be the *evaluation* of these facts by *individuals*, e.g. by doctors or scientists (Kienle, 2005). From this point of view it is necessary to complement the "evidence-based medicine" with a "cognition-based medicine" (Kiene, 2001). Cognition-based medicine is indispensable in idiographic research, but it is also important for typological research, such as e.g. in typological gestalt recognition (ibid.). Treatment

---

180 A distinction must be made between this and randomised individual studies (n = 1) in which e.g. a range of treatments or a placebo are given to *one* patient for a period in a randomised sequence. This is only possible in a few therapy situations, however.

evaluation is therefore not a matter of a one-sided evidence-based medicine whose "best evidence" can only come from randomised studies, but is first and foremost about *"medicine-based evidence"* (Heusser, 1999d). This means that the *method* of obtaining evidence needs to adapt to the *type* of question and not the other way round. *Evidence-based medicine presupposes medicine-based evidence* and what then constitutes the "best" evidence, depends on the nature of the question. In practical terms this means that, for evaluating a particular therapeutic approach or method, it is necessary to have a meaningful synthesis of information from all available types of evidence, from the individual medical judgement based on experience all the way to randomised double-blind studies – with the necessary awareness of which types of results these kinds of evidence allow (Heusser, 2001b).

This view was methodologically substantiated by the author (Heusser, 1999d) under contract to the Swiss Federal Social Insurance Office to establish criteria for evaluating the benefit of complementary medical methods (Heusser, 2001b) and these were then adopted by the Swiss Federal Department of Home Affairs and incorporated in the *"Handbuch zur Standardisierung der medizinischen und wirtschaftlichen Leistungen"* (handbook for the standardisation of medical and economic services) published by the Federal Social Insurance Office. This was also the basis for the evaluation of the anthroposophical treatment studies for the HTA report for the Swiss national Complementary Medicine Evaluation Programme PEK (Kienle et al., 2006). In Germany the "special therapy approaches" including anthroposophical medicine, homeopathy and phytotherapy were recognised in the German Medicines Act of 1976 due to the efforts of Gerhard Kienle. According to the German Medicines Act, these special therapy approaches can – on the basis of their intrinsic clinical methodology and practical contexts – also use "other types of evidence" for establishing effectiveness, apart from randomised studies. As specified in Section 5 of the German guidelines for the testing of medicinal products (Bundesanzeiger, 1994) these include: uncontrolled studies, non-interventional studies, collections of reports on individual cases which enable a scientific evaluation, material from experience and expert assessment reports by specialist associations prepared according to scientific methods. The assessment procedure described by Kiene for his complementary methodology ("cognition-based medicine") must logically also be included here (Kiene, 2001).

This also corresponds to the gradually increasing methodological insight in conventional medicine that, depending on the medical question and the context to be evaluated, *different* evaluation procedures or trial designs are feasible and methodologically the most appropriate: explanatory or pragmatic randomised trials (Witt, 2009) or randomised or non-randomised studies (Verde & Ohmann, 2015), whereby observational studies may have effect measures comparable to those of randomised trials (Anglemyer et al., 2014). Currently, even the use of single case studies is experiencing an astonishing revival internationally, and the corresponding methodological guidelines have recently been developed (Gagnier et al., 2013). This means that setting up *one* "golden standard" to which all clinical research has to comply is a one-sided undertaking. Different questions demand different methodological standards for their answers. This applies to all forms of medicine, including naturally to anthroposophical medicine. If, for example, a whole system approach with multimodal treatment packages, a context or relation-dependent and an individualised treatment approach have to be evaluated such as in anthroposophical medicine, then there will be a need for pragmatic, non-randomised trials (Heusser et al., 2006a) or even of single case studies (Werthmann et al., 2014). If however, it is wished to establish the benefit of a defined mistletoe preparation for the average overall survival of patients with pancreatic carcinoma, irrespective of context and other individualising factors, then a randomised controlled trial may be appropriate (Tröger et al., 2013).

For this reason, various forms of clinical trials are and have been used in anthroposophical medicine to evaluate efficacy and effectiveness. This also includes randomised controlled trials, despite the ethical limits described above. Over the last few decades the quantity and quality of clinical studies in anthroposophical medicine has visibly improved; and nowadays almost all clinical studies have an adequate methodology and are published in international peer-reviewed journals. As the focus of this chapter is to give a general overview of the status of clinical scientific research in anthroposophical medicine, we cannot go further into the methodological discussion as such. This has been done elsewhere and also by others (Kienle, 1974; Kienle, 1980; Kienle & Burkhardt, 1983; Kiene, 1994; Heusser, 1999d; Heusser, 2001b; Kiene, 2001; Kienle, 2005; Walach et al., 2006; Boon et al., 2007; Fønnebø et al., 2007; Walach, 2011). Here we wished to draw attention to the multiperspective nature of the human being as a physical, organic, emotional, spiritual and social entity, and point out that this can have consequences for scientific methodology. The multiperspectivity of human nature is linked to a multiperspectivity in the evaluation process (Matthiessen, 2009).

This has to be kept in mind when judging the effectiveness and efficacy of anthroposophical medicine. In 2005 the final report of the Swiss national Complementary Medicine Evaluation Programme PEK came to the conclusion that "there is satisfactory evidence for the effectiveness and benefit" to patients of anthroposophical medicine, even if criticism can be levelled at the lack of quality (especially in older studies) and at the rather low number of randomised controlled trials to prove efficacy (Melchart et al., 2005, 94; Kienle et al., 2006). Five years later the 2011 update to the HTA reported a considerable increase in this evidence (Kienle et al., 2011b). Since then, not even five years later (mid-July 2015), the number and quality of basic, clinical and health service research studies in anthroposophical medicine has grown further, and a new branch of research has formed, i.e. that of medical education and professionalism in anthroposophical medicine. To give a real insight into the content of this diverse anthroposophical research landscape, we report here all the topics in the relevant publications in international peer reviewed journals listed in the meta-database PubMed from 2011 to 12 July 2015, without further comment[181]. This currently comprises a total of 76 studies: 14 studies in basic research, 27 in clinical research (5 single case studies and case series, 12 observational studies, 4 non-randomised controlled and 6 randomised trials), 23 studies in health care service evaluation and health economics, 4 reviews and 8 studies on medical training and professionalism.

Basic research: Immunomodulatory properties of a lemon-quince preparation as an indicator of anti-allergic potency (Gründemann et al., 2011); homeopathic preparations of quartz, sulfur and copper sulfate assessed by UV-spectroscopy (Wolf et al., 2011); inner correspondence and peacefulness among participants in eurythmy therapy and yoga: a validation study (Büssing et al., 2011b); comparative in vitro study of the effects of separate and combined products of Citrus e fructibus and Cydonia e fructibus on immunological parameters of seasonal allergic rhinitis (Baars et al., 2012); impact of coloured light on cardiorespiratory coordination (Edelhäuser et al., 2013); two new flavonol glycosides and a metabolite profile of Bryophyllum pinnatum (Fürer et al., 2013); preclinical assays to investigate an anthroposophic pharmaceutical process applied to mistletoe (Baumgartner et al., 2014); effects of chronic Bryophyllum pinnatum administration on Wistar rat pregnancy (Hosomi et al., 2014); cytotoxic properties of Helleborus niger L. on tumour and immunocompetent cells (Schink et al., 2015); cardio-respiratory balance during day-rest compared to deep sleep, a possible indicator for quality of life (von Bonin et al., 2014); improvement of circadian rhythm of heart rate variability by eurythmy therapy (Seifert et al., 2013); training effect of eurythmy therapy on specific oscillations of heart rate variability (Edelhäuser et al., 2015); identification of predictive psychometric measures for therapy responsiveness for a multimodal therapy concept from anthroposophical medicine in chronic cancer-related fatigue (Kröz et al., 2015); immunomodulatory effects of preparations from anthroposophical medicine for parenteral use (Gründemann et al., 2015).

Clinical research: Single case studies and case series: Eurythmy therapy in anxiety (Kienle et al., 2011a); inpatient treatment of community-acquired pneumonias with integrative medicine (Geyer et al., 2013); facilitating

---

181 A thorough appraisal would again be a task in itself, something which lies beyond the scope of this work.

self-healing in anthroposophic psychotherapy (Lees, 2013); durable response of cutaneous squamous cell carcinoma following high-dose peri-lesional injections of Viscum album extracts (Werthmann et al., 2013); tumour response following high-dose intratumoural application of Viscum album on a patient with adenoid cystic carcinoma (Werthmann et al., 2014).

Clinical research: Observational studies: Quality of life in breast cancer patients during chemotherapy and concurrent therapy with a mistletoe extract (Eisenbraun et al., 2011); influence of self-regulation and autonomic regulation on survival in breast and colon carcinoma patients (Kröz et al., 2011); pulpa dentis D30 for acute reversible pulpitis: a prospective cohort study in routine dental practice (Hamre et al., 2011); improvement of heart rate variability by eurythmy therapy after a 6-week eurythmy therapy training (Seifert et al., 2012); long-term outcomes of anthroposophic treatment for chronic disease: a four-year follow-up analysis of 1510 patients from a prospective observational study in routine outpatient settings (Hamre et al., 2013); effects of eurythmy therapy in the treatment of essential arterial hypertension: a pilot study (Zerm et al., 2013); intratumoural mistletoe therapy in patients with unresectable pancreas carcinoma: a retrospective analysis (Schad et al., 2013b); eurythmy therapy in the aftercare of pediatric posterior fossa tumour survivors – a pilot study (Kanitz et al., 2013); Mesembryanthemum crystallinum L. as a dermatologically effective medicinal plant (Raak et al., 2014); sleep quality in pregnancy during treatment with Bryophyllum (Lambrigger-Steiner et al., 2014); improvement of sleep quality during treatment with Bryophyllum (Simões-Wüst et al., 2015); effect of Colchicum autumnale in patients with goitre and euthyroidism or mild hyperthyroidism (Scheffer et al., 2016).

Clinical research: Non-randomised controlled trials: Anthroposophic supportive treatment in children with medulloblastoma receiving first-line therapy (Seifert et al., 2011); the impact of eurythmy therapy on stress coping strategies and health-related quality of life (Kanitz et al., 2011); antibiotic use in children with acute respiratory or ear infections in comparison of anthroposophic and conventional treatment under routine primary care conditions (Hamre et al., 2014b); targeting inflammation in cancer-related-fatigue: a rationale for mistletoe therapy as supportive care in colorectal cancer patients (Bock et al., 2014).

Clinical research: Randomized controlled trials. Citrus/Cydonia Compositum subcutaneous injections versus nasal spray for seasonal allergic rhinitis: a randomized controlled trial on efficacy and safety (Baars et al., 2011); treating menopausal symptoms with a complex anthroposophic remedy or placebo: a randomized controlled trial (von Hagens et al., 2012); five-year follow-up of patients with early stage breast cancer after a randomized study comparing additional treatment with a mistletoe extract to chemotherapy alone (Tröger et al., 2012); mistletoe therapy in patients with locally advanced or metastatic pancreatic cancer: a randomised clinical trial on overall survival (Tröger et al., 2013); quality of life of patients with advanced pancreatic cancer during treatment with mistletoe: a randomized controlled trial (Tröger et al., 2014b); additional therapy with a mistletoe product during adjuvant chemotherapy of breast cancer patients improves quality of life (Tröger et al., 2014a).

Health care service evaluation and health economics: Anthroposophic medicine in pediatric primary care (Jeschke et al., 2011); usage of alternative medical systems and anthroposophic medicine by older German adults (Büssing et al., 2011a); adverse drug reactions in a complementary medicine hospital: a prospective, intensified surveillance study (Süsskind et al., 2012); adverse drug reactions for anthroposophical and conventional drugs detected in the network of anthroposophical physicians (Tabali et al., 2012); adverse drug reactions to anthroposophic and homeopathic solutions for injection: a systematic evaluation of German pharmacovigilance databases (Jong et al., 2012); depression, comorbidities, and prescriptions of antidepressants in a German network of GPs and specialists with subspecialisation in anthroposophic medicine: a longitudinal observational study (Jeschke et al., 2012); prescribing pattern of Bryophyllum preparations among a network of anthroposophic physicians (Simões-Wüst et al., 2012); patients whose GP knows anthroposophic and other complementary medicine tend to have lower costs and live longer (Kooreman & Baars, 2012); 6-year comparative economic evaluation of healthcare costs and mortality rates of Dutch patients from conventional and CAM GPs (Baars & Kooremann, 2014); cost analysis of anthroposophical inpatient treatment based on DRG data (Heinz et al., 2013); the consumer quality index anthroposophic healthcare: a construction and validation study (Koster et al., 2014); parental attitudes and decision-making regarding MMR vaccination in an anthroposophic community (Byström et al., 2014); influence of alternative lifestyles on self-reported body weight and health characteristics in women (Simões-Wüst et al., 2014); human breast milk in relation to allergic sensitization and anthroposophic lifestyle (Torregrosa Paredes et al., 2014); influence of anthroposophic lifestyle on the concentration of metals in placenta and cord blood (Fagerstedt et al., 2015); Network Oncology (NO) – a clinical cancer register for health services research and the evaluation of integrative therapeutic interventions in anthroposophic medicine (Schad et al., 2013a); use of complementary and alternative medicine in healthy children and children with chronic medical conditions in Germany (Gottschling et al., 2013); complementary and alternative medicine in paediatrics in daily practice in Europe (Längler & Zuzak, 2013); integration of anthroposophic medicine in supportive breast cancer care (Ben-Arye et al., 2013); adverse drug reactions and expected effects to therapy with subcutaneous mistletoe extracts in cancer patients (Steele et al.,

2014a); safety of intravenous application of mistletoe preparations in oncology (Steele et al., 2014b); the use and safety of intratumoral application of mistletoe preparations in oncology (Steele et al., 2015); the use of anthroposophical and other complementary approaches in integrative oncology centres in Europe (Rossi et al., 2015).

Reviews: Quality of life and related dimensions in cancer patients treated with mistletoe extract, a meta-analysis (Büssing et al., 2011c); music therapy as part of integrative neonatology: 20 years of experience (Thiel et al., 2011); anthroposophic medicine: an integrative medical system originating in Europe (Kienle et al., 2013); overview of the publications from the Anthroposophic Medicine Outcomes Study (AMOS): a whole system evaluation study (Hamre et al., 2014a); systematic review on the effectiveness of eurythmy therapy (Lötzke et al., 2015).

Medical education and professionalism: Non-reductionistic medical anthropology, medical education and practitioner-patient-interaction in anthroposophic medicine (Heusser et al., 2012); integrative medical education: educational strategies and preliminary evaluation of the Integrated Curriculum for Anthroposophic Medicine (Scheffer et al., 2012); active student participation may enhance patient centeredness: patients' assessments of the clinical education ward for integrative and anthroposophical medicine (Scheffer et al., 2013); quality of postgraduate medical training in anthroposophic hospitals in Germany and Switzerland (Heusser et al., 2014b); problems and problem solving in integrative postgraduate medical training at anthroposophic hospitals (Heusser et al., 2014c; Eberhard et al., 2014); spirituality in medical doctors and their relation to specific views of illness and dealing with their patients' individual situation (Büssing et al., 2013b); influence of spirituality on cool down reactions, work engagement, and life satisfaction in anthroposophic health care professionals (Büssing et al., 2015).

When looking back at the beginning of anthroposophical medicine in the 1920s and on its development since then, it is apparent how right Steiner was when he demanded the development of empirical natural scientific research to develop the indications from spiritual science provided by anthroposophy (Steiner, 1951b, 7):

> Indeed, all that comes from anthroposophical investigation in regard to medicine and, for instance, physiology, can be nothing more than a stimulus which must then be worked out empirically. Only on the basis of this empirical study can there arise valid and convincing judgements of the matters in question – and this is the kind of judgement that is needed in the domain of therapy.

The overview presented of the current empirical research fields in anthroposophical medicine shows how diversified and fruitful this empirical work has been and still is, and how it continues to expand. It might actually be said that it has only been in the last few decades that it has been possible to create conditions which allow the thorough and qualitatively high standard of research necessary to work academically in the way in which Steiner demanded. These conditions comprise the availability of academic research facilities and professional scientists. Both have developed since the 1970s through the efforts of anthroposophical hospitals, the manufacturers of anthroposophical pharmaceuticals and physicians with scientific interests in Switzerland and Germany. These last-named started to develop collaborative research with academic scientists, perform clinical studies in conventional hospitals and establish their own research institutes which also attracted scientists. Although all of this was possible on a small scale, a number of anthroposophical scientists with academic connections evolved, some working as professors at various universities but in non-anthroposophical scientific fields. This, however, as well as the increasing public demand for complementary medicine and the debates on the necessity of generating a scientific evidence base for complementary medicine, gave rise to the situation where anthroposophical medicine as such was ready to spread into the universities. The first academic institutions explicitly dedicated to anthroposophical medicine are the Department for Anthroposophical Medicine at the Institute for Complementary Medicine (KIKOM) at the University of Bern, Switzerland, set up in 1995[182], and the Chair for Theory of Medicine, Integrative and Anthroposophical Medicine at the University of Witten/Herdecke[183], Germany, established in 2009. The author of this book has been fortunate to be appointed to and made responsible for both of these institutions, in Bern from 1995–2008 and in Witten/Herdecke from 2009–2016. Over the past five years further anthroposophical scientists have achieved postdoctoral lectureship qualifications or professorships at other universities such as Charité Berlin, Freiburg, Tübingen, Witten/Herdecke, Leiden (the Netherlands), Bern (Switzerland), Ann Arbor (Michigan, USA) and Sao Paolo (Brazil). At present (January 2016), at least twelve anthroposophical scientists hold such qualifications and are faculty members of these universities, and this number is growing. Anthroposophy and anthroposophical medicine have therefore clearly found their place in mainstream science and medicine, even though for many conventional scientists their aims and principles are still unknown, foreign or even suspect.

But this is the very reason, apart from the production of evidence for the efficacy, safety and effectiveness of anthroposophical therapies, why it is nec-

---

182 http://www.ikom.unibe.ch/. The Bern Institute for Complementary Medicine (Institut für Komplementärmedizin, IKOM) was initially called KIKOM: "Kollegiale Instanz für Komplementärmedizin", i.e. cooperative institution for complementary medicine.
183 http://www.uni-wh.de/gesundheit/lehrstuhl-medizintheorie/.

essary to develop the scientific basis of anthroposophy itself and its concepts. This is the purpose for which this book was written. In learning from and endorsing the ideas of Rudolf Steiner, it has always been my conviction that the relationship between anthroposophy and conventional science needs to be clarified through the intrinsic scientific foundation of both: epistemology. Anthroposophical medicine must likewise be built up on a solid epistemological basis and its concepts justified on an empirical basis. It requires to be developed in congruence with mainstream science and its therapeutic options must be tested on the basis of conventional methodology, albeit in accord with the specific features of the human being as a physical, organic, emotional, spiritual and social being and the corresponding medical questions.

This book has been specifically written to provide a demonstration of this span from epistemology to clinical research in order to give an introduction to the scientific foundation of anthroposophy and anthroposophical medicine in the context of academic science. By its very nature, anthroposophy can contribute to solving "the great problem of the next 100 years", as John Martin formulated it at the onset of this new millennium: "to understand what makes the human being a human being" (Martin, 2000). What has been achieved is certainly preliminary, but the potential is enormous.

8.6 Summary

Anthropology based on natural science and psychology and anthroposophy based on spiritual science as described by Steiner complement each other. Based on what nature constantly generates, anthropology can infer the existence of forces at different emergent levels apparent to phenomenological observation: matter, life, soul and spirit. However, in practice and due to habituation, scientific knowledge and the notion of reality nowadays are confined almost exclusively to matter, and in the organism to its physical composition though in great detail. Relatively little is said about the properties of soul and spirit as such, and they are scarcely distinguished from one another. Life is neither known nor recognised as an independent active principle; reductionism ascribes forces and reality only to matter, not to life, soul or spirit. However, on the basis of the apparent phenomena of life, soul and spirit, the existence of forces belonging to these realms can be inferred as causative agents in the same way as the existence of forces is inferred in the realm of matter on the basis of their phenomena. By virtue of its epistemology and self-conception as an empirical spiritual science, anthroposophy attempts to provide direct descriptions and concepts of the inner activity of the forces of matter, life, the soul and the spirit as well as their interactions in the human organism in real terms, whereas anthropology *infers* these forces logically. The aim is to make practical use of the relevant knowledge for areas of life such as medicine and to provide theoretical and practical contributions to medical anthropology and medical practice, amongst these a deeper understanding of the processes of health, illness, treatment and healing. Health and illness do not need to be seen as simply a consequence of the correct or incorrect interactions of molecules, but occur as a harmonious or disharmonious interaction of forces of the physical, living, emotional and spiritual organisation of the human being in an organ, organ system or in the whole organism. Anthroposophy can also make conceptual and practical contributions to therapy and the understanding of healing processes where it can expand the reparative and eliminative measures of conventional medicine – which operates predominantly via physical (and some emotional) means – by pharmaceutical or other therapy measures whose aim is to support the regenerative capacity of the forces of life, soul and spirit and their interactions with the physical forces and substances in the organism.

But what kind of attitude should we adopt in the scientific sense about "spiritual scientific" statements which we might not accept or at least may not be able to verify empirically ourselves? As Steiner himself described, the principle is to take these statements – which have of course to be provided in the form of thoughts and to this extent must be logically comprehensible, even though a perceptible correlate is lacking – as "guiding thoughts" or working hypotheses which can then be tested using empirical natural scientific methods as understood by modern medicine, just like any other hypotheses. For if etheric processes or those of the soul and spirit interact with those of the physical body as postulated by anthroposophical theory, then they must be expressed in the physical organism, and in this form they must be empirically indirectly detectable there by natural scientific methods. Therefore anthroposophical medical research aims to test and elaborate anthroposophical concepts obtained from spiritual science using the anthropological research methods of natural science and psychology. This applies to the theoretical side of medical anthropology and to practical therapy research.

The final parts of the chapter give a brief overview of the status of the development of anthroposophical medical theories since Steiner and of the evidence in terms of efficacy and effectiveness of anthroposophical therapy methods. It is shown that to date only a relatively small – but in part very important – portion of the anthroposophical conceptions of medical anthropology has been worked out or verified or even become part of conventional science and medicine, and a similar statement can be made about the evidence for the effectiveness of anthroposophical medicine. However, due to the gradual establishment of academic research facilities within the last two decades and the increasing availability of professional academic scientists with an anthroposophical background, the quantity and quality of research projects and high quality research publications are growing visibly. In this context, some methodological issues in clinical research are pointed out. A justification is provided based on the anthroposophical conception of the human being and current discussions of methodology as to why other types of trial designs apart from randomised studies may be necessary for therapy research. While this fact is well known, here it is linked to the nature of the human being as a physical, living, emotional and spiritual reality, a conception which has been justified in the previous chapters on the basis of epistemology, basic natural sciences, psychology, philosophy of mind, and spiritual science in the sense of anthroposophy. Over approximately the last 20 years anthroposophy has become increasingly established in the universities, a fact which has made it necessary to clarify the scientific basis of anthroposophy in an academic context. This has been my purpose in writing this book.

Bibliography

Adams, G.; Whicher, O. 1980. The Plant between Sun and Earth and the Science of Physical and Ethereal Spaces. Rudolf Steiner Press, London.

Adler, R.H. 2005. Einführung in die biopsychosoziale Medizin. Schattauer, Stuttgart, New York.

Adolphs, R. 2009. Who are we? Science 323: 585.

an der Heiden, U.; Schneider, H. 2007. Hat der Mensch einen freien Willen? Die Antworten der großen Philosophen. Philipp Reclam jun., Stuttgart.

Andres, K. 1992. Konzepte der anthroposophischen Psychiatrie und deren Beitrag zum Verständnis der Schizophrenie. Psychother Psychosom Med Psychol 42: 362–369.

Anglemyer, A.; Horvath, H.T.; Bero, L. 2014. Healthcare outcomes assessed with observational study designs compared with those assessed in randomized trials. Cochrane Database Syst Rev 2014; 4: MR000034. doi:10.1002/14651858.MR000034.pub2.

Antonovsky, A. 1979. Health, stress, and coping. New perspectives on mental and physical well-being. Jossey-Bass, San Francisco.

Aquin, T.v. 1977. Die Philosophie des Thomas von Aquin. 2nd ed. Felix Meiner, Hamburg.

Aspect, A.; Dalibard, J.; Roger, G. 1982. Experimental test of Bell's inequalities using time-varying analyzers. Phys. Rev. Lett. 49: 1804–1807.

Australian Government. 2014. What is genetic information? [Internet] 22/07/2014 [Viewed 12/08/2014]. URL: http://www.alrc.gov.au/publications/3-coming-terms-genetic-information/what-'genetic-information'.

Azpiroz, A.; De Miguel, Z.; Fano, E.; Vegas, O. 2008. Relations between different coping strategies for social stress, tumor development and neuroendocrine and immune activity in male mice. Brain Behav. Immun. 22: 690–698.

Baars, E.W.; Jong, M.; Nierop, A.F.; Boers, I.; Savelkoul, H.F. 2011. Citrus/Cydonia Compositum Subcutaneous Injections versus Nasal Spray for Seasonal Allergic Rhinitis: A Randomized Controlled Trial on Efficacy and Safety. ISRN Allergy 2011: 836051. doi:10.5402/2011/836051.

Baars, E.W.; Jong, M.C.; Boers, I.; Nierop, A.F.; Savelkoul, H.F. 2012. A comparative in vitro study of the effects of separate and combined products of Citrus e fructibus and Cydonia e fructibus on immunological parameters of seasonal allergic rhinitis. Mediators Inflamm. 2012: 109829. doi:10.1155/2012/109829.

Baars, E.W.; Kooreman, P. 2014. A 6-year comparative economic evaluation of healthcare costs and mortality rates of Dutch patients from conventional and CAM GPs. BMJ Open 2014; 4(8): e005332. doi:10.1136/bmjopen-2014-005332. Erratum in: BMJ Open 2014; 4(9): e005332corr1.

Bacon, F. 1990a. Neues Organon. Teilband 1. Felix Meiner, Hamburg.

Bacon, F. 1990b. Neues Organon. Teilband 2. Felix Meiner, Hamburg.

Bahrs, O.; Matthiessen, P.F. (Editors). 2007. Gesundheitsfördernde Praxen. Die Chancen einer salutogenetischen Orientierung in der hausärztlichen Praxis. Hans Huber, Bern.

Basfeld, M. 1998. Wärme: Ur-Materie und Ich-Leib. Verlag Freies Geistesleben, Stuttgart.

Basold, A.; Cloos, W.; Daems, W.F.; Krüger, H.; Pelikan, W.; et al. 1995. Heilmittel für typische Krankheiten nach Angaben von Rudolf Steiner. Aufbauprinzipien – Kiesel – Antimon – Eisen – Blei und Silber – Merkur. Edition Persephone. Medizinische Sektion am Goetheanum, Dornach.

Baumann, P.H.; von Berlepsch, K. 1999. Komplementärmedizin aus der Sicht der Wissenschaft. Bericht der Expertengruppe zum Nationalen Forschungsprogramm 34, Komplementärmedizin, 1992–1998. Forsch Komplementärmed 6 (Supplementum).

Baumgartner, S. 2005. Reproductions and Reproducibility in Homeopathy: Dogma or Tool? J Altern Complement Med 11: 771–772.

Baumgartner, S.; Heusser, P.; Thurneysen, A. 1998. Methodological Standards and Problems in Preclinical Homoeopathic Potency Research. Forsch Komplementärmed 5: 27–32.

Baumgartner, S.; Thurneysen, A.; Heusser, P. 2004. Growth Stimulation of Dwarf Peas (Pisum sativum L.) through Homeopathic Potencies of Plant Growth Substances. Forsch Komplementärmed 11: 281–292.

Baumgartner, S.; Shah, D.; Schaller, J.; Kämpfer, U.; Thurneysen, A.; Heusser, P. 2008. Reproducibility of dwarf pea shoot growth stimulation by homeopathic potencies of gibberellic acid. Complement Ther Med 16(4): 183–191. doi:10.1016/j.ctim.2008.03.001.

Baumgartner, S.; Wolf, M.; Skrabal, P.; Bangerter, F.; Heusser, P.; Thurneysen, A.; Wolf, U. 2009. Highfield 1H T1 and T2 NMR relaxation time measurements of H2O in homeopathic preparations of quartz, sulfur, and copper sulfate. Naturwissenschaften 96(9): 1079-1089. doi:10.1007/s00114-009-0569-y.

Baumgartner, S.; Flückiger, H.; Kunz, M.; Scherr, C.; Urech, K. 2014. Evaluation of Preclinical Assays to Investigate an Anthroposophic Pharmaceutical Process Applied to Mistletoe (Viscum album L.) Extracts. Evid Based Complement Alternat Med 2014: 620974. doi:10.1155/2014/620974.

Bausewein, C.; Roller, S.; Voltz, R. 2007. Leitfaden Palliativmedizin Palliative Care. 3 ed. Elsevier, München.

Bavastro, P. 1994. Anthroposophische Medizin auf der Intensivstation. Edition Persephone. Medizinische Sektion am Goetheanum, Dornach.

Bavastro, P.; Kümmell, H.C. (Editors). 1999. Das Herz des Menschen. Verlag Freies Geisteleben, Stuttgart.

Beckermann, A. 2001. Analytische Einführung in die Philosophie des Geistes. Walter de Gruyter, Berlin.

Begemann, H. 1988. Therapie als Wissenschaft. Bemerkungen zum Problem der vergleichenden Therapiebewertung am Beispiel der Zytostatika. Dtsch. Med. Wochenschr. 113: 1198–1203.

Belousov, B.P. 1981. Eine periodische Reaktion und ihr Mechanismus. In: Kuhnert, L.; Niedersen, U. (Editors), Selbstorganisation chemischer Strukturen. Harri Klein, Frankfurt a.M., pp. 73–82.

Beloussov, L.V.; Opitz, J.M.; Gilbert, S.F. 1997. Life of Alexander G. Gurwitsch and his relevant contribution to the theory of morphogenetic fields. Int. J. Dev. Biol. 41: 771–779.

Ben-Arye, E.; Schiff, E.; Levy, M.; Raz, O.G.; Barak, Y.; Bar-Sela, G. 2013. Barriers and challenges in integration of anthroposophic medicine in supportive breast cancer care. Springerplus 2013; 2: 364. doi:10.1186/2193-1801-2-364.

Berger, B.; Bertram, M.; Kanitz, J.; Pretzer, K.; Seifert, G. 2015. "Like walking into an empty room": Effects of Eurythmy therapy on Stress perception in comparison with a sports intervention from the subjects' perspective – a qualitative study. Evid Based Complement Alternat Med 2015: 856107. doi:10.1155/2015/856107.

Bergmann, L.; Schaefer, C. 1981. Lehrbuch der Experimentalphysik. Vol. IV: Aufbau der Materie. Edited by H. Gobrecht. Walter de Gruyter, Berlin.

Bermúdez, J.L. 1997. What is at stake in the debate on nonconceptual content? Phil. Perspect. 21(1): 55–72.

Betti, L.; Trebbi, G.; Majewsky, V.; Scherr, C.; Shah-Rossi, D.; Jäger, T.; Baumgartner, S. 2009. Use of homeopathic preparations in phytopathological models and in field trials: a critical review. Homeopathy 98(4): 244–266.

Bieri, P. 2001. Das Handwerk der Freiheit. Über die Entdeckung des eigenen Willens. Hanser, München.

Bissegger, M.; Bräuner-Gülow, G.; Brongs, S.; Klitzke-Pettener, S.; Reinhold, S.; Ruckgaber, K.-H.; Schäfer, M.; Weik, J. 1998. Die Behandlung von Magersucht – ein integrativer Therapieansatz. Verlag Freies Geisteleben, Stuttgart.

Björkman, J.; Nagaev, I.; Berg, O.G.; Hughes, D.; Andersson, D.I. 2000. Effects of environment on compensatory mutations to ameliorate costs of antibiotic resistance. Science 287: 1779–1782.

Bleuler, E. 1976. Das autistisch-undisziplinierte Denken in der Medizin und seine Überwindung. Springer, Berlin, Heidelberg, New York.

Bock, K.D. 1993. Wissenschaftliche und alternative Medizin. Paradigmen – Praxis – Perspektiven. Springer, Berlin, Heidelberg.

Bock, P.R.; Hanisch, J.; Matthes, H.; Zänker, K.S. 2014. Targeting inflammation in cancer-related-fatigue: a rationale for mistletoe therapy as supportive care in colorectal cancer patients. Inflamm Allergy Drug Targets 13(2): 105–111.

Bockemühl, J. (Editor). 1985. Toward a Phenomenology of the Etheric World: Investigations into the Life of Nature and Man. Anthroposophic Press, Spring Valley, NY.

Bockemühl, J. 1996–2003. Ein Leitfaden zur Heilpflanzenerkenntnis. 3 Vols. Verlag am Goetheanum, Dornach.

Bockemühl, J. 2010. A Guide to Understanding Healing Plants. Vol. 1. Mercury Press, Spring Valley, NY.

Bohm, D. 1986. Fragmentierung und Ganzheit. In: Dürr, H.-P. (Editor), Physik und Transzendenz. Scherz, Bern, München, Wien, pp. 263–293.

Bohm, D.; Peat, F.D. 1990. Das neue Weltbild. Naturwissenschaft, Ordnung und Kreativität. Goldmann, München.

Boogerd, F.C.; Bruggeman, F.J.; Hofmeyr, J.-H.S.; Westerhoff, H.V. 2007. Towards philosophical foundations of Systems Biology: introduction. In: Boogerd, F.C.; Bruggeman, F.J.; Hofmeyr, J.-H.S.; Westerhoff, H.V. (Editors), Systems Biology. Philosophical Foundations. Elsevier, Amsterdam, pp. 3–19.

Boon, H.; MacPherson, H.; Fleishman, S.; Grimsgaard, S.; Koithan, M.; Norheim, A.J.; Walach, H. 2007. Evaluating complex healthcare systems: a critique of four approaches. Evid Based Complement Alternat Med 4(3): 279–285.

Bort, J.; Holtzapfel, W.; Kirchner, H.; Löffler, F.; Maikowski, R.; Pache, W.; Pracht, E. 1956. Heilende Erziehung. Vom Wesen seelenpflege-bedürftiger Kinder und deren heilpädagogische Förderung. Natura-Verlag, Arlesheim.

Breaker, R.R. 2008. Complex Riboswitches. Science 319: 1795–1797.

Breidbach, O. 2001. Hirn und Bewusstsein – Überlegungen zu einer Geschichte der Neurowissenschaften. In: Pauen, M.; Roth, G. (Editors), Neurowissenschaften und Philosophie. Wilhelm Fink, München, pp. 11–58.

Breitbart, W. 2002. Spirituality and meaning in supportive care: spirituality and meaning-centered group psychotherapy interventions in advanced cancer. Support Care Cancer 10: 272–280.

Brendza, R.P.; Serbus, L.R.; Duffy, J.B.; Saxton, W.M. 2000. A function for kinesin I in the posterior transport of oskar mRNA and Staufen protein. Science 289: 2120–2122.

Brentano, F. 1907. Untersuchungen zur Sinnespsychologie. Duncker & Humblot, Leipzig.

Brentano, F. 1911. Von der Klassifikation der psychischen Phänomene. 2nd ed. Duncker & Humblot, Leipzig.

Brentano, F. 1995. Psychology from an Empirical Standpoint. Routledge, London, New York.

Bundesanzeiger. 1994. No. 244, 22.12.1994: Allgemeine Verwaltungsvorschrift zur Änderung von Arzneimittelprüfrichtlinien.

Burkhard, B. 2000. Anthroposophische Arzneimittel. Eine kritische Betrachtung. Govi-Verlag, Eschborn.

Burkhard, G. 2005. Das Leben in die Hand nehmen: Arbeit an der eigenen Biographie. 10th ed. Verlag Freies Geistesleben, Stuttgart.

Büssing, A.; Matthiessen, P.F.; Ostermann, T. 2005. Engagement of patients in religious and spiritual practices: confirmatory results with the SpREUK-P 1.1 questionnaire as a tool of quality of life research. Health Qual Life Outcomes 3: 53.

Büssing, A.; Keller, N.; Michalsen, A.; Moebus, S.; Dobos, G.; Ostermann, T.; Matthiessen, P.F. 2006. Spirituality and Adaptive Coping Styles in German Patients with Chronic Diseases in a CAM Health Care Setting. J Complement Integr Med 3: 1–24.

Büssing, A.; Michalsen, A.; Balzat, H.J.; Grünther, R.A.; Ostermann, T.; Neugebauer, E.A.M.; Matthiessen, P.F. 2009. Are Spirituality and Religiosity Resources for Patients with Chronic Pain Conditions? Pain Med 10: 327–339.

Büssing, A.; Ostermann, T.; Heusser, P.; Matthiessen, P.F. 2011a. Usage of alternative medical systems, acupuncture, homeopathy and anthroposophic medicine, by older German adults. Zhong Xi Yi Jie He Xue Bao 2011; 9(8): 847–856.

Büssing, A.; Edelhäuser, F.; Weisskircher, A.; Fouladbakhsh, J.M.; Heusser, P. 2011b. Inner Correspondence and Peacefulness with Practices among Participants in Eurythmy Therapy and Yoga: A Validation Study. Evid Based Complement Alternat Med 2011; pii:329023. doi:10.1155/2011/329023.

Büssing, A.; Raak, C.; Ostermann, T. 2011c. Quality of life and related dimensions in cancer patients treated with mistletoe extract (Iscador): a meta-analysis. Evid Based Complement Alternat Med 2012: 219402.

Büssing, A.; Janko, A.; Baumann, K.; Hvidt, N.C.; Kopf, A. 2013a. Spiritual needs among patients with chronic pain diseases and cancer living in a secular society. Pain Med 14(9): 1362–1373.

Büssing, A.; Hirdes, A.T.; Baumann, K.; Hvidt, N.C.; Heusser, P. 2013b. Aspects of spirituality in medical doctors and their relation to specific views of illness and dealing with their patients' individual situation. Evid Based Complement Alternat Med 2013: 734392. doi:10.1155/2013/734392.

Büssing, A.; Baumann, K.; Hvidt, N.C.; Koenig, H.G.; Puchalski, C.M.; Swinton, J. 2014. Spirituality and health. Evid Based Complement Alternat Med 2014: 682817. doi:10.1155/2014/682817.

Büssing, A.; Lötzke, D.; Glöckler, M.; Heusser, P. 2015. Influence of spirituality on cool down reactions, work engagement, and life satisfaction in anthroposophic health care professionals. Evid Based Complement Alternat Med 2015: 754814. doi:10.1155/2015/754814.

Byström, E.; Lindstrand, A.; Likhite, N.; Butler, R.; Emmelin, M. 2014. Parental attitudes and decision-making regarding MMR vaccination in an anthroposophic community in Sweden – a qualitative study. Vaccine 32(50): 6752–6757. doi:10.1016/j.vaccine.2014.10.011.

Cairns, J.; Overbaugh, J.; Miller, S. 1988. The origin of mutants. Nature 335: 142–145.

Cerny, T.; Heusser, P. 1999. Untersuchungen der Lebensqualität von Patienten mit metastasierendem Brust- oder Darmkrebs, behandelt in der Anthroposophischen Medizin oder in der Schulmedizin, letztere mit oder ohne psychoonkologische oder anthroposophische Zusatztherapie. Programm Bericht NFP 34. Forsch Komplementärmed 6 (Suppl. 1): 35–37.

Chakalova, L.; Debrand, E.; Mitchell, J.A.; Osborne, C.S.; Fraser, P. 2005. Replication and Transcription: Shaping the landscape of the genome. Nat. Rev. Genet. 6: 669–677.

Chalmers, D. 1996. The conscious mind. Oxford University Press, Oxford.

Christiaen, L.; Davidson, B.; Kawashima, T.; Powell, W.; Nolla, H.; Vranizan, K.; Levine, M. 2008. The transcription/migration interface in heart precursors of Ciona intestinalis. Science 320: 1349–1352.

Cleeremans, A. 2005. Computational correlates of consciousness. Prog. Brain Res. 150: 81–98.

Cloos, W. 2015. The Living Origin of Rocks and Minerals. 2nd ed. Floris Books, Edinburgh.

Commoner, B. 1968. Failure of the Watson-Crick theory of the chemical explanation of inheritance. Nature 220: 335.

Cotman, C.W.; Berchtold, N.C.; Christie, L.A. 2007. Exercise builds brain health: key roles of growth factor cascades and inflammation. Trends Neurosci. 30: 464–472.

Couzin, J. 2008. MicroRNAs make big impression in disease after disease. Science 319: 1782–1784.

Crick, F.H.C. 1966. The genetic code. Scientific American 215(4): 55–62.

Crick, F.H.C. 1970. Central dogma of molecular biology. Nature 227(5258): 561–563.

Cruse, H. 2004. Ich bin mein Gehirn. Nichts spricht gegen den materialistischen Monismus. In: Geyer, C. (Editor), Hirnforschung und Willensfreiheit. Suhrkamp, Frankfurt a. M., pp. 223–228.

Cysarz, D.; von Bonin, D.; Lackner, H.; Heusser, P.; Moser, M.; Bettermann, H. 2004. Oscillation of heart rate and respiration synchronize during poetry recitation. Am. J. Physiol. Heart Circ. Physiol. 287: H579–H587.

Cysarz, D.; von Bonin, D.; Brachmann, P.; Buetler, S.; Edelhäuser, F.; Laederach-Hofmann, K.; Heusser, P. 2008. Day-to-night time differences in the relationship between cardiorespiratory coordination and heart rate variability. Physiol Meas 29: 1281–1291.

Daems, W.F. 2001. Streifzüge durch die Medizin- und Pharmaziegeschichte. Edition Persephone. Medizinische Sektion am Goetheanum, Dornach.

Damásio, A.R. 1994. Descartes' Irrtum – Fühlen, Denken und das menschliche Gehirn. List, München.

Danzer, G. 2013. Personale Medizin. Huber, Bern.

Darlington, A.S.; Dippel, D.W.; Ribbers, G.M.; van Balen, R.; Passchier, J.; Busschbach, J.J. 2009. A prospective study on coping strategies and quality of life in patients after stroke, assessing prognostic relationships and estimates of cost-effectiveness. J Rehabil Med 41: 237–241.

Darwin, C. 1871. The descent of man, and selection in relation to sex. John Murray, London.

Darwin, C. 1998. The Expression of the Emotions in Man and Animals. With an Introduction, Afterword and Commentaries by Paul Ekman. 3rd ed. Oxford University Press, New York, Oxford.

Davidson, D. 1993. Der Mythos des Subjektiven. Philosophische Essays. Translated and with an afterword by J. Schulte. Reclam, Stuttgart.

De Robertis, E.M.; Kuroda, H. 2004. Dorsal-ventral patterning and neural induction in Xenopus embryos. Annu. Rev. Cell Dev. Biol. 20: 285–308.

de Vries, J. 1983. Grundbegriffe der Scholastik. Wissenschaftliche Buchgesellschaft, Darmstadt.

Dekker, J. 2008. Gene regulation in the third dimension. Science 319: 1793–1797.

Denjean-von-Stryk, B.; von Bonin, D. 2000. Therapeutische Sprachgestaltung. Anthroposophische Kunsttherapie. Vol. 4. Urachhaus, Stuttgart.

Descartes, R. 1954. Rules for the Direction of the Mind. Translated by Elizabeth Anscombe and Peter Thomas Geach for "Descartes Philosophical Writings". https://en.wikisource.org/wiki/Rules_for_the_Direction_of_the_Mind (Viewed 13/10/2015).

Deschamps, J. 2004. Hox genes in the limb: a play in two acts. Science 304: 1610–1611.

Dinger, T.J. 1996. Homöopathie und Anthroposophische Medizin: Ursprünge, Abhängigkeiten und Kontroversen von 1910–1990. Thesis. Universität Münster.

Donlea, J.M.; Ramanan, N.; Shaw, P.J. 2009. Use-dependent plasticity in clock neurons regulates sleep in Drosophila. Science 324: 105–112.

Driesch, H. 1921. Philosophie des Organischen. 2nd ed. Wilhelm Engelmann, Köln.

Driesch, H. 1929. Zur vitalistischen Begriffsbildung. Wilhelm Rouxs Arch. Dev. Biol. 116: 1–6.

Driesch, H. 1931. Entelechie und Materie. A. Klein, Leipzig.

Du Bois-Reymond, E. 1882. Über die Grenzen des Naturerkennens. Veit & Comp, Leipzig.

Du Bois-Reymond, E. 1918. Jugendbriefe von Emil Du Bois-Reymond an Eduard Hallmann. Dietrich Reimer, Berlin.

Dürken, B. 1936. Entwicklungsbiologie und Ganzheit. B. G. Teubner, Leipzig, Berlin.

Eberhard, S.; Weinzirl, J.; Orlow, P.; Berger, B.; Heusser, P. 2014. Recommendations for problem solving in integrative postgraduate medical training of physicians at anthroposophic hospitals in Germany and Switzerland. Forsch Komplementärmed 21(5): 284–293. doi:10.1159/000366186.

Eccles, J.C. 1994. Wie das Gehirn sein Selbst steuert. Piper, Berlin, Heidelberg.

Edelhäuser, F.; Hak, F.; Kleinrath, U.; Lühr, B.;, Matthiessen, P.F.; Weinzirl, J.; Cysarz, D. 2013. Impact of colored light on cardiorespiratory coordination. Evid Based Complement Alternat Med 2013: 810876. doi:10.1155/2013/810876.

Edelhäuser, F.; Minnerop, A.; Trapp, B.; Büssing, A.; Cysarz, D. 2015. Eurythmy therapy increases specific oscillations of heart rate variability. BMC Complement Altern Med 2015: 167. doi:10.1186/s12906-015-0684-6.

Edwards, L. 2006. The Vortex of Life: Nature's Patterns in Space and Time. 2nd revised ed. Floris Books, Edinburgh.

Efficace, F.; Marrone, R. 2002. Spiritual issues and quality of life assessment in cancer care. Death Stud 26: 743–756.

Einstein, A. 1989. The Collected Papers of Albert Einstein. Vol. 2. The Swiss years: Writings, 1900–1909 [E-book]. Princetown University Press, Princetown, NJ. [Viewed 13/10/2015]. URL: http://tinyurl.com/p4y7r48.

Eisberg, R.; Resnick, R. 1985. Quantum Physics of Atoms, Molecules, Solids, Nuclei, and Particles. John Wiley & Sons, New York.

Eisenbraun, J.; Scheer, R.; Kröz, M.; Schad, F.; Huber, R. 2011. Quality of life in breast cancer patients during chemotherapy and concurrent therapy with a mistletoe extract. Phytomedicine 18(2–3): 151–157. doi:10.1016/j.phymed.2010.06.013.

Ellis, G.F.R. 2005. Physics and the real world. Phys Today 58(7): 49–54.

Endres, K.-P.; Schad, W. 2002. Moon Rhythms in Nature. How Lunar Cycles Affect Living Organisms. Floris Books, Edinburgh.

Engel, G.L. 1977. The need for a new medical model: a challenge for biomedicine. Science 196: 535–544.

Ernst, E. 2008. Anthroposophische Medizin. Eine kritische Analyse. Fortschr Med Orig I; 150: 1–6.

Ernst, E.; Schmacke, N. 2015. Anthroposophische Medizin: eine kritische Analyse mit besonderer Berücksichtigung der Misteltherapie. In: Schmacke, N. (Editor), Der Glaube an die Globuli. Die Verheißungen der Homöopathie. Suhrkamp, Berlin, pp. 148–162.

Esch, B.M.; Marian, F.; Busato, A.; Heusser, P. 2008. Patient satisfaction with primary care: an observational study comparing anthroposophic and conventional care. Health Qual Life Outcomes 6: 74. doi:10.1186/1477-7525-6-74.

Fagerstedt, S.; Kippler, M.; Scheynius, A.; Gutzeit, C.; Mie, A.; Alm, J.; Vahter, M. 2015. Anthroposophic lifestyle influences the concentration of metals in placenta and cord blood. Environ. Res. 136: 88–96. doi:10.1016/j.envres.2014.08.044.

Feigl, W.; Bonet, E.M. 1989. Systemtheorie in der Medizin und Biologie. Wien Med Wochenschr 139: 87–91.

Felber, R.; Reinhold, S.; Stückert, A. 2000. Musiktherapie und Gesang. Anthroposophische Kunsttherapie. Vol. 3. Urachhaus, Stuttgart.

Fichte, I.H. 1856. Anthropologie. Die Lehre von der menschlichen Seele. Begründet auf naturwissenschaftlichem Wege für Naturforscher, Seelenärzte und wissenschaftlich Gebildete überhaupt. 1st ed. Brockhaus, Leipzig.

Fichte, I.H. 1859. Zur Seelenfrage. Eine philosophische Konfession. Brockhaus, Leipzig.

Fichte, I.H. 1860. Contributions to Mental Philosophy [E-book]. Longman, Green, Longman and Roberts, London. [Viewed 14/10/2015]. URL: https://archive.org/stream/contributionsto00moregoog#page/n60/mode/2up.

Fichte, I.H. 1876. Anthropologie. Die Lehre von der menschlichen Seele. Begründet auf naturwissenschaftlichem Wege für Naturforscher, Seelenärzte und wissenschaftlich Gebildete überhaupt. 3rd ed. Brockhaus, Leipzig.

Fichte, J.G. 1834. Johann Gottlieb Fichte's Einleitungsvorlesungen in die Wissenschaftslehre, die transcendentale Logik und die Tatsachen des Bewusstseins. Vorgetragen an der Universität zu Berlin in den Jahren 1812 u. 13. Edited from the estate by I.H. Fichte. Marcus, Bonn.

Fichte, J.G. 1994. Introductions to the Wissenschaftslehre and other writings 1797–1800. Translated by Daniel Breazeale. Hackett Publishing Company, Inc., Indianapolis, IN.

Fingado, M. 2001. Therapeutische Wickel und Kompressen. Handbuch aus der Ita Wegman Klinik. Natura-Verlag, Arlesheim.

Fintelmann, V. (Editor). 2002. Onkologie auf anthroposophischer Grundlage. Loseblattwerk mit Register. Johannes M. Mayer, Stuttgart.

Fintelmann, V. 2007. Intuitive Medizin – Anthroposophische Medizin in der Praxis. 5th ed. Hippokrates, Stuttgart.

Fischer, H.F.; Binting, S.; Bockelbrink, A.; Heusser, P.; Hueck, C.; Keil, T.; Roll, S.; Witt, C. 2013. The effect of attending Steiner schools during childhood on health in adulthood: a multicentre cross-sectional study. PLoS ONE 2013; 8(9): e73135. doi:10.1371.

Fischer, G.S. 1993. Faltung der Proteine: Der Weg vom Baustein zur Funktion. Verhandlungen der Gesellschaft Deutscher Naturforscher und Ärzte. Wissenschaftliche Verlagsgesellschaft, Stuttgart.

Fønnebø, V.; Grimsgaard, S.; Walach, H.; Ritenbaugh, C.; Norheim, A.J.; MacPherson, H.; Lewith, G.; Launsø, L.; Koithan, M.; Falkenberg, T.; Boon, H.; Aickin, M. 2007. Researching complementary and alternative treatments – the gatekeepers are not at home. BMC Med Res Methodol 2007; 7: 7.

Foote, J. 2003. Isomeric antibodies. Science 299: 1327–1328.

Fortlage, K. 1869. Acht psychologische Vorträge. Mauke, Jena.

Foster, R.; Kreitzman, L. 2004. Rhythms of Life. The biological clocks that control the daily lives of every living thing. Profile Books, London.

Fox Keller, E. 2000. The Century of the Gene. Harvard University Press, Cambridge, MA.

Frankl, V.E. 2005. Ärztliche Seelsorge. Grundlagen der Logotherapie und Existenzanalyse (1946). 11th ed. Franz Deuticke, Wien.

Frielingsdorf, V.; Grimm, R.; Kaldenberg, B. 2013. Geschichte der anthroposophischen Heilpädagogik und Sozialtherapie: Entwicklungslinien und Aufgabenfelder 1920–1980. Verlag am Goetheanum, Dornach.

Fuchs, T. 2005. Ökologie des Gehirns. Eine systemische Sichtweise für Psychiatrie und Psychotherapie. Nervenarzt 76: 1–10.

Fuller, S. 2007. Science vs. religion. Intelligent design and the problem of evolution. Polity Press, Malden, MA.

Fürer, K.; Raith, M.; Brenneisen, R.; Mennet, M.; Simões-Wüst, A.P.; von Mandach, U.; Hamburger, M.; Potterat, O. 2013. Two new flavonol glycosides and a metabolite profile of Bryophyllum pinnatum, a phytotherapeutic used in obstetrics and gynaecology. Planta Med. 79(16): 1565–1571. doi:10.1055/s-0033-1350808.

Furst, B. 2014. The Heart and Circulation. An Integrative Model. Springer, London.

Furst, B. 2015. The Heart: Pressure-Propulsion Pump or Organ of Impedance? J. Cardiothorac. Vasc. Anesth. 29(6): 1688–1701. doi.org/10.1053/jvca.2015.02.022.

Gagnier, J.J.; Kienle, G.S.; Altman, D.G.; Moher, D.; Sox, H.; Riley, D.; CARE Group. 2013. The CARE Guidelines: Consensus-based Clinical Case Reporting Guideline Development. Glob Adv Health Med 2013; 2(5): 38–43. doi:10.7453/gahmj.2013.008.

Gallagher, S.; Zahavi, D. 2012. The Phenomenological Mind. 2nd ed. Routledge, London, New York.

Garnitschnig, L.; Weinzirl, J.; Andrae, L.; Scheffers, T.; Ostermann, T.; Heusser, P. Postprandiale Dynamik des Milzvolumens bei gesunden Probanden. Tag der Forschung. Universität Witten/Herdecke. Gesundheitsforschung 2015. Abstract-Band, p. 38.

Gazzaniga, M.S. 2008. Human. The Science Behind What Makes Us Unique. Ecco (Harper Collins), New York.

Gazzaniga, M. 2011. Who's in Charge? Free Will and the Science of the Brain. Ecco (HarperCollins), New York.

German Anthroposophical Medical Association (GAÄD); Medical Section, School of Spiritual Science, Dornach. 2014. Anthromedics. Newsletter No.1/2014. [Internet] 18/03/2014 [Viewed 26/12/2015]. URL: http://www.medsektion-goetheanum.org/EYED2/files/file/Research%202010/ANTHROMEDICS%20Newsletter%20No%201%202014-03-18_EN_V03.pdf.

Geyer, C. 2004. Vorwort. In: Geyer, C. (Editor), Hirnforschung und Willensfreiheit. Suhrkamp, Frankfurt a. M., pp. 2–19.

Geyer, U.; Diederich, K.; Kusserow, M.; Laubersheimer, A.; Kramer, K. 2013. Inpatient treatment of community-acquired pneumonias with integrative medicine. Evid Based Complement Alternat Med 2013: 578274. doi: 10.1155/2013/578274.

Girke, M. 2012. Innere Medizin: Grundlagen und therapeutische Konzepte der Anthroposophischen Medizin. 2nd ed. Salumed, Berlin.

Girke, M. (Editor). 2014. Geriatrie: Grundlagen und therapeutische Konzepte der Anthroposophischen Medizin. Salumed, Berlin.

Gisin, N. 2009. Quantum nonlocality: How does nature do it? Science, 326: 1357–1358.

Glöckler, M. (Editor). 1998. Gesundheit und Schule. Edition Persephone. Medizinische Sektion am Goetheanum, Dornach.

Glöckler, M. (Editor). 2002. Spirtuelle Ethik. Situationsgerechtes, selbstverantwortliches Handeln. Verlag am Goetheanum, Dornach.

Glöckler, M. (Editor). 2005. Anthroposophische Arzneitherapie für Ärzte und Apotheker. Wissenschaftliche Verlagsgesellschaft, Stuttgart.

Glöckler, M.; Goebel, W. 2007. A Guide to Child Health. Floris Books, Edinburgh.

Glöckler, M.; Girke, M.; Matthes, H. 2011. In: Uhlenhoff, R. (Editor), Anthroposophie in Geschichte und Gegenwart. Berliner Wissenschaftsverlag, Berlin, pp. 514–611.

Goerttler, K. 1972. Morphologische Sonderstellung des Menschen im Reich der Lebensformen auf der Erde. In: Gadamer, H.-G.; Vogler, P. (Editors), Neue Anthropologie. Thieme, Stuttgart, pp. 215–257.

Goethe, J.W. 1848. The Autobiography of Johann Goethe: Truth and Poetry from my own Life [E-book]. Henry G. Bohn, London. [Viewed 20/10/2015]. URL: https://archive.org/stream/autobiographyofg00goet#page/282/mode/2up.

Goethe, J.W. 1915. Goethe, with special consideration of his philosophy [E-book]. Open Court Publishing Library, London. [Viewed 14/10/2015]. URL: https://archive.org/stream/goethewithspeci00carugoog#page/n276/mode/2up.

Goethe, J.W. 1950b. Analyse und Synthese. In: Beutler, E. (Editor), Johann Wolfgang Goethe. Sämtliche Werke. Vol. 16. Artemis, Zürich, pp. 887–890.

Goethe, J.W. 1950c. Aphorismen und Fragmente. In: Beutler, E. (Editor), Johann Wolfgang Goethe. Sämtliche Werke. Vol. 17. Artemis, Zürich.

Goethe, J.W. 1950e. Beiträge zur Optik. §14. In: Beutler, E. (Editor), Johann Wolfgang Goethe. Sämtliche Werke. Vol. 16. Artemis, Zürich, p. 770.

Goethe, J.W. 1950f. Beobachtung und Denken. In: Beutler, E. (Editor), Johann Wolfgang Goethe. Sämtliche Werke. Vol. 17. Artemis, Zürich, pp. 856–857.

Goethe, J.W. 1950g. Der Versuch als Vermittler von Objekt und Subjekt. In: Beutler, E. (Editor), Johann Wolfgang Goethe. Sämtliche Werke. Vol. 17. Artemis, Zürich, pp. 844–855.

Goethe, J.W. 1950i. Erfahrung und Wissenschaft. In: Beutler, E. (Editor), Johann Wolfgang Goethe. Sämtliche Werke. Vol. 17. Artemis, Zürich, pp. 869–871.

Goethe, J.W. 1950j. Erster Entwurf einer allgemeinen Einleitung in die vergleichende Anatomie, ausgehend von der Osteologie. In: Beutler, E. (Editor), Johann Wolfgang Goethe. Sämtliche Werke. Vol. 17. Artemis, Zürich, pp. 231–269.

Goethe, J.W. 1950k. Faust. Eine Tragödie. In: Beutler, E. (Editor), Johann Wolfgang Goethe. Sämtliche Werke. Vol. 5. Artemis, Zürich, pp. 139–526.

Goethe, J.W. 1950l. Fragmente zur vergleichenden Anatomie. In: Beutler, E. (Editor), Johann Wolfgang Goethe. Sämtliche Werke. Vol. 17. Artemis, Zürich, pp. 414–435.

Goethe, J.W. 1950m. Naturwissenschaftliche Schriften: Vergleichende Anatomie, Zoologie. In: Beutler, E. (Editor), Johann Wolfgang Goethe. Sämtliche Werke. Vol. 17. Artemis, Zürich, pp. 217–436.

Goethe, J.W. 1950n. Naturwissenschaftliche Schriften: Botanik. In: Beutler, E. (Editor), Johann Wolfgang Goethe. Sämtliche Werke. Vol. 17. Artemis, Zürich, pp. 9–216.

Goethe, J.W. 1950o. Schriften zur Farbenlehre. In: Beutler, E. (Editor), Johann Wolfgang Goethe. Sämtliche Werke. Vol. 16. Artemis, Zürich, pp. 7–838.

Goethe, J.W. 1950p. Über den Zwischenkiefer des Menschen und der Tiere. In: Beutler, E. (Editor), Johann Wolfgang Goethe. Sämtliche Werke. Vol. 17. Artemis, Zürich, pp. 288–328.

Goethe, J.W. 1950q. Vorarbeiten zu einer Physiologie der Pflanzen. In: Beutler, E. (Editor), Johann Wolfgang Goethe. Sämtliche Werke. Vol. 17. Artemis, Zürich, pp. 119–133.

Goethe, J.W. 1950r. Wenige Bemerkungen. In: Beutler, E. (Editor), Johann Wolfgang Goethe. Sämtliche Werke. Vol. 17. Artemis, Zürich, pp. 101–102.

Goethe, J.W. 1958a. Brief an Knebel, 17. November 1784. In: Bach, R. (Editor), Johann Wolfgang Goethe. Briefe. Carl Hanser, München, pp. 208–209.

Goethe, J.W. 1958b. Brief an Sömmering, 28. Aug. 1796. In: Bach, R. (Editor), Johann Wolfgang Goethe. Briefe. Carl Hanser, München, pp. 349–351.

Goethe, J.W. 1975a. Sprüche in Prosa. In: Steiner, R. (Editor), Goethes Naturwissenschaftliche Schriften. Vol. 5. Rudolf Steiner Verlag, Dornach, pp. 349–537.

Goethe, J.W. 1975b. Wirkung meiner Schrift. In: Steiner, R. (Editor), Goethes Naturwissenschaftliche Schriften. Vol. 1. Rudolf Steiner Verlag, Dornach, pp. 194–217.

Goethe, J.W. 1988. Brief an H.W.F. Wackenroder vom 21. Januar 1832. Goethes Briefe und Briefe an Goethe. Hamburger Ausgabe. Beck, München.

Goethe, J.W. 1994. True Enough: To the physicist (1820). Collected Works, Selected Poems. Edited by Christopher Middleton. Princeton University Press, Princeton, NJ.

Goethe, J.W. 2005. Faust [E-book]. The World Publishing Company, Cleveland, New York. [Viewed 14/10/2015]. URL: http://www.gutenberg.org/files/14591/14591.txt.

Goethe, J.W. 2009. The Metamorphosis of Plants. Massachusetts Institute of Technology, Cambridge, MA.

Goleman, D. 1996. Emotional Intelligence. Bloomsbury Publishing PLC, London.

Golombek, E. 2000. Plastisch-therapeutisches Gestalten. Anthroposophische Kunsttherapie. Vol. 1. Urachhaus, Stuttgart.

Görnitz, T.; Görnitz, B. 2008. Die Evolution des Geistigen. Quantenphysik – Bewusstsein – Religion. Vandenhoeck & Ruprecht, Göttingen.

Görnitz, T.; Görnitz, B. 2016. Von der Quantenphysik zum Bewusstsein. Kosmos, Geist und Materie. Springer, Berlin.

Goschke, T. 2006. Der bedingte Wille. Willensfreiheit und Selbststeuerung aus der Sicht der kognitiven Neurowissenschaft. In: Roth, G.; Grün, K.-J. (Editors), Das Gehirn und seine Freiheit. Beiträge zur neurowissenschaftlichen Grundlegung der Philosophie. Vandenhoeck & Ruprecht, Göttingen, pp. 107–156.

Gottschling, S.; Gronwald, B.; Schmitt, S.; Schmitt, C.; Längler, A.; Leidig, E.; Meyer, S.; Baan, A.; Shamdeen, M.G.; Berrang, J.; Graf, N. 2013. Use of complementary and alternative medicine in healthy children and children with chronic medical conditions in Germany. Complement Ther Med 2013; (Suppl. 1): 61–69. doi:10.1016/j.ctim.2011.06.001.

Gould, J.L.; Gould, C.G. 1994. The Animal Mind. Scientific American Library, New York.

Goyert, A. 2015. Anthroposophische Medizin und die Prozesse des Stoffwechsels. Verlag Freies Geistesleben, Stuttgart.

Grandpré, T.; Strittmatter, S.M. 2001. Nogo: A Molecular Determinant of Axonal Growth and Regeneration. Neuroscientist 7: 377–386.

Gregor, T.; Wieschaus, E.F.; McGregor, A.P.; Bialek, W.; Tank, D.W. 2007. Stability and nuclear dynamics of the Bicoid morphogen gradient. Cell 130: 141–152.

Grimm, R.; Kaschubowski, G. (Editors). 2008. Kompendium der anthroposophischen Heilpädagogik. Ernst Reinhardt, München.

Gross, J.R. 1974. Regeneration. Thieme, Stuttgart.

Gründemann, C.; Papagiannopoulos, M.; Lamy, E.; Mersch-Sundermann, V.; Huber, R. 2011. Immunomodulatory properties of a lemon-quince preparation (Gencydo®) as an indicator of anti-allergic potency. Phytomedicine 18(8–9): 760–768. doi:10.1016/j.phymed.2010.11.016.

Gründemann, C.; Diegel, C.; Sauer, B.; Garcia-Käufer, M.; Huber, R. 2015. Immunomodulatory effects of preparations from Anthroposophical Medicine for parenteral use. BMC Complement Altern Med 15: 219. doi:10.1186/s12906-015-0757-6.

Guggisberg, A.G.; Baumgartner, S.; Tschopp, C.M.; Heusser, P. 2005. Replication study concerning the effects of homeopathic dilutions of histamine on human basophil degranulation in vitro. Complement Ther Med 13: 91–100.

Gumustekin, K.; Seven, B.; Karabulut, N.; Aktas, O.; Gursan, N.; Aslan, S.; Keles, M.; Varoglu, E.; Dane, S. 2004. Effects of sleep deprivation, nicotine, and selenium on wound healing in rats. Neuroscience 114: 1433–1442.

Gurdon, J.B.; Bourillot, P.Y. 2001. Morphogen gradient interpretation. Nature 413: 797–803.

Gurley, K.A.; Rink, J.C.; Sánchez Alvarado, A. 2008. ▢-Catenin defines head versus tail identity during planarian regeneration and homeostasis. Science 319: 323–327.

Gurwitsch, A. 1922. Über den Begriff des Embryonalen Feldes. Wilhelm Roux Arch Entwickl Mech Org 51: 383–415.

Gutenbrunner, C.; Hildebrandt, G.; Moog, R. (Editors). 1987. Chronobiology and Chronomedicine. Basic Research and Applications. Proceedings of the 2nd Annual Meeting of the European Society for Chronobiology. Peter Lang, Frankfurt a. M.

Gutenbrunner, C.; Hildebrandt, G.; Moog, R. (Editors). 1993. Chronobiology and Chronomedicine. Basic Research and Applications. Proceedings of the 7th Annual Meeting of the European Society for Chronobiology. Peter Lang, Frankfurt a. M.

Hachtel, B.; Gäch, A. 2007. Bibliographie Heileurythmie 1920–2007. Natur-Kunst-Medizin Verlags GmbH, Bad Boll.

Hacking, I. 1996. Einführung in die Philosophie der Naturwissenschaften. Reclam, Stuttgart.

Hadorn, E. 1981. Experimentelle Entwicklungsforschung – im Besonderen an Amphibien. Springer, Berlin, Heidelberg, New York.

Haggard, P.; Eimer, M. 1999. On the relation between brain potentials and the awareness of voluntary movements. Exp Brain Res 126: 28–133.

Halusková, J. 2010. Epigenetic studies in human diseases. Folia Biol. (Praha) 56(3): 83–96.

Hamre, H.J.; Mittag, I.; Glockmann, A.; Kiene, H.; Tröger, W. 2011. Pulpa dentis D30 for acute reversible pulpitis: A prospective cohort study in routine dental practice. Altern Ther Health Med 17(1): 16–21.

Hamre, H.J.; Kiene, H.; Glockmann, A.; Ziegler, R.; Kienle, G.S. 2013. Long-term outcomes of anthroposophic treatment for chronic disease: a four-year follow-up analysis of 1510 patients from a prospective observational study in routine outpatient settings. BMC Res Notes 2013; 6: 269. doi:10.1186/1756-0500-6-269.

Hamre, H.J.; Kiene, H.; Ziegler, R.; Tröger, W.; Meinecke, C.; Schnürer, C.; Vögler, H.; Glockmann, A.; Kienle, G.S. 2014a. Overview of the Publications from the Anthroposophic Medicine Outcomes Study (AMOS): A Whole System Evaluation Study. Glob Adv Health Med 2014; 3(1): 54–70. doi:10.7453/gahmj.2013.010.

Hamre, H.J.; Glockmann, A.; Schwarz, R.; Riley, D.S.; Baars, E.W.; Kiene, H.; Kienle, G.S. 2014b. Antibiotic Use in Children with Acute Respiratory or Ear Infections: Prospective Observational Comparison of Anthroposophic and Conventional Treatment under Routine Primary Care Conditions. Evid Based Complement Alternat Med 2014: 243801. doi:10.1155/2014/243801.

Handwerker, H.O. 1995. Allgemeine Sinnesphysiologie. In: Schmidt, R.F.; Thews, G. (Editors), Physiologie des Menschen. Springer, Berlin, Heidelberg, New York, p. 195.

Harten, H.-U. 1974. Physik für Mediziner. Springer, Berlin, Heidelberg, New York.

Hartmann, N. 1964. Der Aufbau der realen Welt. Grundriss der allgemeinen Kategorienlehre. 3rd ed. Walter de Gruyter, Berlin.

Hauschka, M. 1971. Rhythmische Massage nach Dr. Ita Wegman. Menschenkundliche Grundlagen. Schule für Künstlerische Therapie und Massage, Boll über Göppingen.

Hauschka, R. 1974. Heilmittellehre. Ein Beitrag zu einer zeitgemäßen Heilmittelerkenntnis. 2nd ed. Vittorio Klostermann, Frankfurt a. M.

Hauschka, R. 2002. The Nature of Substance: Spirit and Matter. Translated by Marjorie Spock and M.T. Richards. Sophia Books, Forest Row, East Sussex.

Hawks, J.; Wang, E.T.; Cochran, G.M.; Harpending, H.C.; Moyzis, R.K. 2007. Recent acceleration of human adaptive evolution. Proc. Natl. Acad. Sci. U.S.A. 104(52): 20753–20758.

Hay, E.D. 1968. Dedifferentiation and Metaplasia in Vertebrate and Invertebrate Regeneration. In: Ursprung, H. (Editor), The Stability of the Differentiated State. Springer, Berlin, Heidelberg, New York, pp. 85–108.

Haynes, J.D.; Rees, G. 2006. Decoding mental states from brain activity in humans. Nat. Rev. Neurosci. 7: 523–534.

Head, J.F.; Inouye, S.; Teranishi, K.; Shimomura, O. 2000. The crystal structure of the photoprotein aequorin at 2.3. A resolution. Nature 405: 372–376.

Hegel, G.W.F. 1969. Enzyklopädie der philosophischen Wissenschaften im Grundrisse (1830). Felix Meiner, Hamburg.

Hegel, G.W.F. 1977. Phenomenology of Spirit. Oxford University Press, Oxford.

Hegel, G.W.F. 1979. Phänomenologie des Geistes. Suhrkamp, Frankfurt a. M.

Hegel, G.W.F. 2001a. Philosophy of Right. Translated by S.W. Dyde [E-book]. Batoche Books, Kitchener. [Viewed 15/10/2015]. URL: http://socserv2.socsci.mcmaster.ca/econ/ugcm/3ll3/hegel/right.pdf.

Hegel, G.W.F. 2001b. Encyclopaedia of the Philosophical Sciences Part One [E-book]. Blackmask Online. [Viewed 15/10/2015]. URL: http://www.hegel.net/en/pdf/Hegel-Enc-1.pdf.

Hegel, G.W.F. 2010. The Science of Logic. Cambridge University Press, Cambridge.

Heine, R.; Bay, F. (Editors). 1995. Pflege als Gestaltungsaufgabe. Anregungen aus der Anthroposophie für die Praxis. Hippokrates, Stuttgart.

Heinz, J.; Fiori, W.; Heusser, P.; Ostermann, T. 2013. Cost Analysis of Integrative Inpatient Treatment Based on DRG Data: The Example of Anthroposophic Medicine. Evid Based Complement Alternat Med 2013: 748932. doi:10.1155/2013/748932.

Heisenberg, W. 1959. Physics and Philosophy [E-book]. Penguin Books, London. [Viewed 13/10/2015]. URL: http://www.naturalthinker.net/trl/texts/Heisenberg,Werner/Heisenberg,%20Werner%20-%20Physics%20and%20philosophy.pdf.

Heitler, W. 1972. Naturwissenschaft ist Geisteswissenschaft. Die Waage, Zürich.

Hensel, H. 1966. Allgemeine Sinnesphysiologie. Hautsinne, Geschmack, Geruch. Springer, Berlin, Heidelberg, New York.

Herder, J.G. 2014. Outlines of a Philosophy of the History of Man. Translated by T.O. Churchill. Reprint. BiblioBazaar, Charleston, SC.

Heusser, P. 1984. Der Schweizer Arzt Ignaz Paul Vital Troxler (1780–1866). Seine Philosophie, Anthropologie und Medizintheorie. Schwabe, Basel, Stuttgart.

Heusser, P. 1989. Das zentrale Dogma nach Watson und Crick und seine Widerlegung durch die moderne Genetik. Verhandlungen der Naturforschenden Gesellschaft in Basel, Basel, pp. 1–14.

Heusser, P. 1997. Führt die Molekularbiologie zu einem neuen Konzept des Organismus? Forsch Komplementärmed 4: 106–111.

Heusser, P. (Editor). 1999a. Akademische Forschung in der Anthroposophischen Medizin. Beispiel Hygiogenese: Natur- und geisteswissenschaftliche Zugänge zur Selbstheilungskraft des Menschen. Peter Lang, Bern.

Heusser, P. 1999b. Anthroposophie als Geisteswissenschaft und die naturwissenschaftliche Medizin. In: Heusser, P. (Editor), Akademische Forschung in der Anthroposophischen Medizin. Peter Lang, Bern, pp. 21–38.

Heusser, P. 1999c. Intuition. Die innere Basis von Wissenschaft und Ethik. Anthroposophische und konventionelle Medizin. In: Ausfeld-Hafter, B. (Editor), Intuition in der Medizin. Peter Lang, Bern, pp. 77–96.

Heusser, P. 1999d. Probleme von Studiendesigns mit Randomisation, Verblindung und Placebogabe. Forsch Komplementärmed 6: 89–102.

Heusser, P. 2000a. Goethe, die moderne Sinnesphysiologie und das Leib/Seele-Problem. In: Heusser, P. (Editor), Goethes Beitrag zur Erneuerung der Naturwissenschaften. Paul Haupt, Bern, Stuttgart, Wien, pp. 435–456.

Heusser, P. (Editor). 2000b. Goethes Beitrag zur Erneuerung der Naturwissenschaften. Paul Haupt, Bern, Stuttgart, Wien.

Heusser, P. 2000c. Über die Notwendigkeit einer Erneuerung heutiger Wissenschaften – Der Beitrag von Goethes Wissenschaftsmethode. In: Heusser, P. (Editor), Goethes Beitrag zur Erneuerung der Naturwissenschaften. Paul Haupt, Bern, Stuttgart, Wien, pp. 13–41.

Heusser, P. 2001a. Erklärung der Evolution durch Selbstorganisation: nominalistisch oder realistisch? Bull. Soc. Frib. Sc. Nat. 90: 33–48.

Heusser, P. 2001b. Kriterien zur Beurteilung des Nutzens von komplementärmedizinischen Methoden. Forsch Komplementärmed Klass Naturheilkd 8: 14–23.

Heusser, P. 2002a. Das materialistische Weltbild stösst an seine Grenzen. In: Stauffacher, W.; Bircher, J. (Editors), Zukunft Medizin Schweiz. Das Projekt „Neu-Orientierung der Medizin" geht weiter. EMH Schweizerischer Ärzteverlag und Schweizerische Akademie der Medizinischen Wissenschaften, Basel, pp. 106–113.

Heusser, P. (Editor). 2002b. Gesundheitsförderung – Eine neue Zeitforderung. Peter Lang, Bern.

Heusser, P. 2002c. Physiologische Grundlagen der Gesundheitsförderung und das anthroposophisch-medizinische Konzept. In: Heusser, P. (Editor),

Gesundheitsförderung, eine neue Zeitforderung. Peter Lang, Bern, pp. 101–130.

Heusser, P. 2003. Sterben, Leben nach dem Tod und wiederholte Erdenleben. In: Heusser, P. (Editor), Sterbebegleitung, Sterbehilfe, Euthanasie. Haupt, Bern, Stuttgart, Wien, pp. 127–156.

Heusser, P. 2006a. Holistic Thinking in Basic Sciences and Medicine. A New Elective at the University of Bern. Overall Concept and Evaluation Strategy, Specific Needs Assessment, Pilot Course with an Emphasis on Role-Play, and Feasibility Study. Thesis, Master of Medical Education. Medical Faculty. University of Bern, Bern.

Heusser, P. 2006b. Spiritualität in der modernen Medizin. Peter Lang, Bern.

Heusser, P. 2007. Anthroposophic Concepts of Pain. In: Ernst, E.; Pittler, M.H.; Wider, B.; Boddy, K. (Editors), Complementary Therapies for Pain Management. An Evidence-Based Approach. Elsevier, Mosby, Edinburgh, pp. 57–61.

Heusser, P. 2013. Emergenz und Kausalität: Systemische Interaktion von Körper, Leben, Seele und Geist des Menschen. In: Heusser, P.; Weinzirl, J. (Editors), Medizin und die Frage nach dem Menschen. Königshausen und Neumann, Würzburg, pp. 35–50.

Heusser, P.; Riggenbach, B. (Editors). 2003. Sterbebegleitung – Sterbehilfe – Euthanasie. Die aktuelle Problematik aus anthroposophisch-medizinischer Sicht. Paul Haupt, Bern.

Heusser, P.; Stutz, M.; Haeberli, A. 2004. Does homeopathically potentized antimony stimulate coagulation? A summary of previous findings and results of an in vitro pilot study by means of thrombelastography. J Altern Complement Med 10: 829–834.

Heusser, P.; Berger Braun, S.; Bertschy, M.; Burkhard, R.; Ziegler, R.; Helwig, S.; van Wegberg, B.; Cerny, T. 2006a. Palliative In-Patient Cancer Treatment in an Anthroposophic Hospital: II Quality of Life during and after Stationary Treatment, and Subjective Treatment Benefits. Forsch Komplementärmed 13: 156–166.

Heusser, P.; Berger Braun, S.; Ziegler, R.; Bertschy, M.; Helwig, S.; van Wegberg, B.; Cerny, T. 2006b. Palliative In-Patient Cancer in an Anthroposophic Hospital: I Treatment Patterns and Compliance with Anthroposophic Medicine. Forsch Komplementärmed 13: 94–100.

Heusser, P.; Berger, S.; Stutz, M.; Hüsler, A.; Haeberli, A.; Wolf, U. 2009a. Efficacy of Homeopathically Potentized Antimony on Blood Coagulation. A Randomized Placebo Controlled Crossover Trial. Forsch Komplementärmed 16: 14–18.

Heusser, P.; Bertschy, M.; Burkhard, R.; Ziegler, R.; Cerny, T.; Wolf, U. 2009b. Langzeiterhaltung der Lebensqualität bei fortgeschrittener Krebskrankheit.

Prospektive Studie über zwölf Monate während und nach stationärer Behandlung in einer anthroposophischen Klinik. In: Scheer, R.; Alban, S.; Becker, H.; Holzgrabe, U.; Kemper, F.H.; Kreis, W.; Matthes, H. (Editors). Die Mistel in der Tumortherapie 2. Aktueller Stand und klinische Forschung. KVC-Verlag, Essen, pp. 477–493.

Heusser, P.; Scheffer, C.; Neumann, M.; Tauschel, D.; Edelhäuser, F. 2012. Towards non-reductionistic medical anthropology, medical education and practitioner-patient-interaction: the example of Anthroposophic Medicine. Patient Educ Couns 89(3): 455–460. doi:10.1016/j.pec.2012.01.004.

Heusser, P.; Kienle, G.S. 2014a. Anthroposophic Medicine, Integrative Oncology, and Mistletoe Therapy of Cancer. In: Abrams, D.I.; Weil, A.T. (Editors), Integrative Oncology. 2nd ed. Oxford University Press, Oxford, New York, pp. 560–588.

Heusser, P.; Eberhard, S.; Berger, B.; Weinzirl, J.; Orlow, P. 2014b. The subjectively perceived quality of postgraduate medical training in integrative medicine within the public healthcare systems of Germany and Switzerland: the example of anthroposophic hospitals. BMC Complement Altern Med 2014; 14: 191. doi:10.1186/1472-6882-14-191.

Heusser, P.; Eberhard, S.; Weinzirl, J.; Orlow, P.; Berger, B. 2014c. Problems in integrative postgraduate medical training of physicians at anthroposophic hospitals in Germany and Switzerland. Forsch Komplementärmed 21(4): 223–230. doi:10.1159/000366187.

Hildebrandt, G. 1998. Therapeutische Physiologie. In: Gutenbrunner, C.; Hildebrandt, G. (Editors), Balneologie und medizinische Klimatologie. Springer, Berlin, Heidelberg, New York, pp. 5–84.

Hildebrandt, G. 1999. Physiologische Grundlagen der Hygiogenese. In: Heusser, P. (Editor), Akademische Forschung in der Anthroposophischen Medizin. Beispiel Hygiogenese: Natur- und geisteswissenschaftliche Zugänge zur Selbstheilungskraft des Menschen. Peter Lang, Bern, pp. 57–82.

Hildebrandt, G.; Moser, M.; Lehofer, M. 1998. Chronobiologie und Chronomedizin. Biologische Rhythmen. Medizinische Konsequenzen. Hippokrates, Stuttgart.

Hinrichsen, K.V. (Editor). 1990. Humanembryologie. Lehrbuch und Atlas der vorgeburtlichen Entwicklung des Menschen. Springer, Berlin, Heidelberg, New York.

Hobert, O. 2008. Gene Regulation by Transcription Factors and MicroRNAs. Science 319: 1785–1786.

Hoerner, W. 1991. Zeit und Rhythmus. Die Ordnungsgesetze der Erde und des Menschen. 2nd ed. Urachhaus, Stuttgart.

Holdrege, C. 1999. Der vergessene Kontext. Entwurf einer ganzheitlichen Genetik. Verlag Freies Geistesleben, Stuttgart.

Holliday, R. 1990. DNA Methylation and Epigenetic Inheritance. Philos Trans R Soc Lond B Biol Sci 326(1235): 329–338.

Honegger, R. 1925. Goethe und Hegel. Jahrbuch der Goethegesellschaft, Weimar, pp. 38–111.

Hösle, V. 2005. Objective Idealism and Darwinism. In: Hösle, V.; Illies, C. (Editors), Darwinism and Philosophy. University of Notre Dame Press, Notre Dame, IN, pp. 216–242.

Hösle, V.; Illies, C. (Editors). 2005. Darwinism and Philosophy. University of Notre Dame Press, Notre Dame, IN.

Hosomi, J.K.; Ghelman, R.; Quintino, M.P.; de Souza, E.; Nakamura, M.; Moron, A.F. 2014. Effects of chronic Bryophyllum pinnatum administration on Wistar rat pregnancy. Forsch Komplementärmed 21(3): 184–189. doi:10.1159/000363709.

Houchmandzadeh, B.; Wieschaus, E.; Leibler, S. 2002. Establishment of developmental precision and proportions in the early Drosophila embryo. Nature 415: 798–802.

Hügli, A.; Lübcke, P. (Editors). 2003. Philosophielexikon. Personen und Begriffe der abendländischen Philosophie von der Antike bis zur Gegenwart. Rowohlt, Reinbek bei Hamburg.

Hürny, C.; Heusser, P.; Bernhard, J.; Castiglione, M.; Cerny, T. 1994. Verbessern nicht-konventionelle Zusatztherapien die Lebensqualität von Krebspatienten? Eine methodenkritische Übersicht. Schweiz Med Wochenschr Suppl 124(62): 55–63.

Husemann, A.J. 1994. The Harmony of the Human Body: Musical Principles in Human Physiology. Floris Books, Edinburgh.

Husemann, A.J. 1996. Der Zahnwechsel des Kindes. Ein Spiegel seiner seelischen Entwicklung. Verlag Freies Geistesleben, Stuttgart.

Husemann, A.J. 2007. Menschenwissenschaft durch Kunst: Die plastisch-musikalisch-sprachliche Menschenkunde. Einführung – Quellentexte – Dokumentation. Verlag Freies Geistesleben, Stuttgart.

Husemann, A.J. 2015. Form, Leben und Bewusstsein. Einführung in die Menschenkunde der Anthroposophischen Medizin. Verlag Freies Geistesleben, Stuttgart.

Husemann, F. 2009. Anthroposophische Medizin. Ein Weg zu den heilenden Kräften. Verlag am Goetheanum, Dornach.

Husemann, F.; Wolff, O. 1982, 1987, 1989. The Anthroposophical Approach to Medicine. 3 Vols. Anthroposophic Press, Hudson, NY.

Husemann, G. 1962. Erdengebärde und Menschengestalt. Das Zinn in Erde und Mensch. Verlag Freies Geistesleben, Stuttgart.

Jachens, L. 2012. Dermatologie: Grundlagen und therapeutische Konzepte der Anthroposophischen Medizin. Salumed, Berlin.

Jaeger, J.; Surkova, S.; Blagov, M.; Janssens, H.; Kosman, D.; Kozlov, K.N.; Manu; Myasnikova, E.; Vanario-Alonso, C.E.; Samsonova, M.; Sharp, D.H.; Reinitz, J. 2004. Dynamic control of positional information in the early Drosophila embryo. Nature 430: 368–371.

Jaenisch, R.; Bird, A. 2003. Epigenetic regulation of gene expression: how the genome integrates intrinsic and environmental signals. Nat. Genet. (Suppl.) 33: 245–254.

Jäger, T.; Scherr, C.; Shah, D.; Majewsky, V.; Betti, L.; Trebbi, G.; Bonamin, L.; Simões-Wüst, A.P.; Wolf, U.; Simon, M.; Heusser, P.; Baumgartner, S. 2011a. Use of homeopathic preparations in experimental studies with abiotically stressed plants. Homeopathy 2011; 100(4): 275–287.

Jäger, T.; Scherr, C.; Wolf, U.; Simon. M.; Heusser, P.; Baumgartner, S. 2011b. Investigation of arsenic-stressed yeast (Saccharomyces cerevisiae) as a bioassay in homeopathic basic research. ScientificWorldJournal 2011; 11: 568–583. doi:10.1100/tsw.2011.45.

Jäger, T.; Scherr, C.; Simon, M.; Heusser, P.; Baumgartner, S. 2011c. Development of a test system for homeopathic preparations using impaired duckweed (Lemna gibba L.). J Altern Complement Med 2011; 17(4): 315–323. doi:10.1089/acm.2010.0246.

Jenny, H. 2009. Kymatik. Wellenphänomene und Schwingungen. AT Verlag, AZ Fachverlage, Aarau.

Jeschke, E.; Ostermann, T.; Tabali, M.; Kröz, M.; Bockelbrink, A.; Witt, C.M.; Willich, S.N.; Matthes, H. 2011. Anthroposophic medicine in pediatric primary care: a prospective, multicenter observational study on prescribing patterns. Altern Ther Health Med 17(2): 18–28.

Jeschke, E.; Ostermann, T.; Vollmar, H.C.; Tabali, M.; Matthes, H. 2012. Depression, Comorbidities, and Prescriptions of Antidepressants in a German Network of GPs and Specialists with Subspecialisation in Anthroposophic Medicine: A Longitudinal Observational Study. Evid Based Complement Alternat Med 2012: 508623. doi:10.1155/2012/508623.

Jong, M.C.; Jong, M.U.; Baars, E.W. 2012. Adverse drug reactions to anthroposophic and homeopathic solutions for injection: a systematic evaluation of German pharmacovigilance databases. Pharmacoepidemiol Drug Saf 21(12): 1295–1301. doi:10.1002/pds.3298.

Jütte, R. 1996. Geschichte der Alternativen Medizin. C.H. Beck, München.

Kanitz, J.L.; Pretzer, K.; Reif, M.; Voss, A.; Brand, R.; Warschburger, P.; Längler, A.; Henze, G.; Seifert, G. 2011. The impact of eurythmy therapy on stress coping strategies and health-related quality of life in healthy, moderately stressed adults. Complement Ther Med 19(5): 247–255. doi:10.1016/j.ctim.2011.06.008.

Kanitz, J.L.; Pretzer, K.; Calaminus, G.; Wiener, A.; Längler, A.; Henze, G.; Driever, P.H.; Seifert, G. 2013. Eurythmy therapy in the aftercare of pediatric posterior fossa tumour survivors – a pilot study. Complement Ther Med 2013; 21 (Suppl. 1): 3–9. doi:10.1016/j.ctim.2012.02.007.

Kant, I. 1926. Kritik der reinen Vernunft. 13th ed. Felix Meiner, Leipzig.

Kant, I. 1998. Critique of Pure Reason. Cambridge University Press, Cambridge.

Karplus, M.; McCammon, J.A. 1986. Das dynamische Verhalten der Proteine. Spektrum der Wissenschaft, 6: 108–118.

Kauffmann, S. 1995. At Home in the Universe. The Search for Laws of Self-Organization and Complexity. Oxford University Press, New York, Oxford.

Keller, P.J.; Schmidt, A.D.; Wittbrodt, J.; Stelzer, E.H.K. 2008. Reconstruction of zebrafish early embryonic development by scanned light microscopy. Science 322: 1065–1069.

Kellert, S.H. 1993. In the Wake of Chaos. University of Chicago Press, Chicago, London.

Kiefer, B. 2007. Das Prinzip der Emergenz. Schweizerischer Nationalfonds. Horizonte, 33.

Kiene, H. 1984. Essentiale Wissenschaftstheorie. Die Erkenntnistheorie Rudolf Steiners im Spannungsfeld moderner Wissenschaftstheorien. Perspektiven essentialer Wissenschaft. Urachhaus, Stuttgart.

Kiene, H. 1994. Komplementärmedizin – Schulmedizin. Der Wissenschaftsstreit am Ende des 20. Jahrhunderts. Schattauer, Stuttgart.

Kiene, H. 2001. Komplementäre Methodenlehre. Cognition Based Medicine. Springer, Berlin, Heidelberg, New York.

Kienle, G. 1965. Die optischen Wahrnehmungsstörungen und die nichteuklidische Struktur des Sehraums. Habilitationsschrift. Johann Wolfgang von Goethe-Universität, Frankfurt.

Kienle, G. 1974. Arzneimittelsicherheit und Gesellschaft. Eine kritische Untersuchung. Schattauer, Stuttgart, New York.

Kienle, G. 1980. Das Formalisierungsproblem in der Medizin. Therapie der Gegenwart, 119: 1407–1421.

Kienle, G.S. 2005. Gibt es Gründe für pluralistische Evaluationsmodelle? Limitationen der randomisierten klinischen Studie. Z Ärztl Fortbild Qualalitätssich 99: 289–294.

Kienle, G.S. 2009. Bedeutet Evidence-based Medicine eine Abkehr von der ärztlichen Therapiefreiheit? In: Jütte, R. (Editor), Die Zukunft der Individualmedizin. Deutscher Ärzte-Verlag, Köln, pp. 69–77.

Kienle, G.; Burkhardt, R. 1983. Der Wirksamkeitsnachweis für Arzneimittel. Analyse einer Illusion. Urachhaus, Stuttgart.

Kienle, G.S.; Kiene, H. 2003. Die Mistel in der Onkologie. Fakten und konzeptionelle Grundlagen. Wissenschaftliche Ergebnisse und Diskussionen im Kontext moderner Tumorimmunologie, klinischer Methodologie und aktueller Krebskonzepte. Schattauer, Stuttgart, New York.

Kienle, G.S.; Kiene, H.; Albonico, H.-U. 2006. Anthroposophische Medizin in der klinischen Forschung. Wirksamkeit, Nutzen, Wirtschaftlichkeit, Sicherheit. Schattauer, Stuttgart, New York.

Kienle, G.S.; Kiene, H. 2010. Review article: Influence of Viscum album L (European mistletoe) extracts on quality of life in cancer patients: a systematic review of controlled clinical studies. Integr Cancer Ther 9(2): 142–157.

Kienle, G.S.; Hampton Schwab, J.; Murphy, J.B.; Andersson, P.; Lunde, G.; Kiene, H.; Hamre, H.J. 2011a. Eurythmy Therapy in anxiety. Altern Ther Health Med 17(4): 56–63.

Kienle, G.S.; Glockmann, A.; Grugel, R.; Hamre, H.J.; Kiene, H. 2011b. Clinical research on anthroposophic medicine: update of a health technology assessment report and status quo. Forsch Komplementärmed 2011; 18(5): 269–282.

Kienle, G.S.; Albonico, H.-U.; Baars, E.W.; Hamre, H.J.; Zimmermann, P.; Kiene, H. 2013. Anthroposophic medicine: an integrative medical system originating in Europe. Glob Adv Health Med 2(6): 20–31. doi:10.7453/gahmj.2012.087.

Kim, J. 1999. Making sense of emergence. Philos Stud 95: 3–36.

Kimelman, D.; Pyati, U.J. 2005. Bmp Signaling: Turning a Half into a Whole. Cell 123: 982–984.

Kipp, F. 1948. Höherentwicklung und Menschwerdung. Hippokrates, Stuttgart.

Kirchner-Bockholt, M. 1997. Grundelemente der Heileurythmie. Verlag am Goetheanum, Dornach.

Kirschner, M.; Gerhart, J.; Mitchison, T. 2000. Molecular "Vitalism". Cell 100: 79–88.

Klein, S.D.; Sandig, A.; Baumgartner, S.; Wolf, U. 2013. Differences in Median Ultraviolet Light Transmissions of Serial Homeopathic Dilutions of Copper Sulfate, Hypericum perforatum, and Sulfur. Evid Based Complement Alternat Med 2013: 370609. doi:10.1155/2013/370609.

Kolisko, E. 1922a. Hypothesenfreie Chemie im Sinne der Geisteswissenschaft. In: Kolisko, E. (Editor), Aenigmatisches aus Kunst und Wissenschaft. Der Kommende Tag AG, Stuttgart, pp. 165–230.

Kolisko, E. 1989. Auf der Suche nach neuen Wahrheiten. Goetheanistische Studien. Philosophisch-Anthroposophischer Verlag am Goetheanum, Dornach.

Kolisko, E. 2002. Vom therapeutischen Charakter der Waldorfschule. Edited and with a foreword by Peter Selg. Verlag am Goetheanum, Dornach.

Kolisko, L. 1922b. Milzfunktion und Plättchenfrage. Der Kommende Tag AG, Stuttgart.

Kolisko, L. 1961. Physiologischer und physikalischer Nachweis der Wirksamkeit kleinster Entitäten. 1923–1959. Hohenheimer Druck- und Verlagshaus, Gerabronn.

Kolisko, L. 1997. Physiologischer und physikalischer Nachweis der Wirksamkeit kleinster Entitäten. New edited and with contributions by G. und F. Husemann. Verlag am Goetheanum, Dornach.

Kolisko, E.; Rozumek, M. 2012. Hypothesenfreie Chemie: «Hypothesenfreie Chemie» im Sinne der Geisteswissenschaft, der Atomismusstreit 1922/23 und Rudolf Steiners Stellung zum Atomismus. Verlag am Goetheanum, Dornach.

König, K. 2000. Eugen Kolisko. In: Selg, P. (Editor), Anthroposophische Ärzte. Lebens- und Arbeitswege im 20. Jahrhundert. Verlag am Goetheanum, Dornach, pp. 164–170.

Koolhaas, J.M. 2008. Coping style and immunity in animals: making sense of individual variation. Brain Behav. Immun. 22: 662–667.

Kooreman, P.; Baars, E.W. 2012. Patients whose GP knows complementary medicine tend to have lower costs and live longer. Eur J Health Econ 13(6): 769–776. doi:10.1007/s10198-011-0330-2.

Koster, E.B.; Ong, R.R.; Heybroek, R.; Delnoij, D.M.; Baars, E.W. 2014. The consumer quality index anthroposophic healthcare: a construction and validation study. BMC Health Serv Res 2014; 14: 148. doi:10.1186/1472-6963-14-148.

Kragl, M.; Knapp, D.; Nacu, E.; Khattak, S.; Maden, M.; Epperlein, H.H.; Tanaka, E.M. 2009. Cells keep a memory of their tissue origin during axolotl limb regeneration. Nature 460: 60.

Kranich, E.-M. 2003. Der innere Mensch und sein Leib. Verlag Freies Geistesleben, Stuttgart.

Krischer, K. 2007. Von Glühwürmchen und Elektroden. Wie Strukturen aus ungeordneter Materie entstehen. In: Sandhoff, K.; Donner, W. (Editors). Vom Urknall zum Bewusstsein – Selbstorganisation der Materie. Verhandlungen der Gesellschaft Deutscher Naturforscher und Ärzte. Thieme, Stuttgart, pp. 133–149.

Kröz, M.; Reif, M.; Büssing, A.; Zerm, R.; Feder, G.; Bockelbrink, A.; von Laue, H.B.; Matthes, H.; Willich, S.N.; Girke, M. 2011. Does self-regulation and autonomic regulation have an influence on survival in breast and colon carcinoma patients? Results of a prospective outcome study. Health Qual Life Outcomes 2011; 9: 85. doi:10.1186/1477-7525-9-85.

Kröz, M.; Reif, M.; Zerm, R.; Winter, K.; Schad. F.; Gutenbrunner, C.; Girke, M.; Bartsch, C. 2015. Do we have predictors of therapy responsiveness for a multimodal therapy concept and aerobic training in breast cancer survivors with chronic cancer-related fatigue? Eur J Cancer Care (Engl) 2015; 24(5): 707–717. doi:10.1111/ecc.12278.

Kugler, W. (Editor). 1997. Rudolf Steiner und die Gründung der Weleda. Beiträge zur Rudolf Steiner Gesamtausgabe Nr. 118/119. Verlag der Rudolf Steiner-Nachlassverwaltung, Dornach.

Kühn, A. 1965. Vorlesungen über Entwicklungsphysiologie. 2nd ed. Springer, Berlin, Heidelberg, New York.

Küppers, G. 1996. Selbstorganisation: Selektion durch Schließung. In: Küppers, G. (Editor), Chaos und Ordnung. Formen der Selbstorganisation in Natur und Gesellschaft. Philipp Reclam jun., Stuttgart, pp. 123–148.

Küppers, G.; Paslack, R. 1996. Die natürlichen Ursachen von Ordnung und Organisation. In: Küppers, G. (Editor), Chaos und Ordnung. Formen der Selbstorganisation in Natur und Gesellschaft. Philipp Reclam jun., Stuttgart, pp. 44–60.

Lambrigger-Steiner, C.; Simões-Wüst, A.P.; Kuck, A.; Fürer, K.; Hamburger, M.; von Mandach, U. 2014. Sleep quality in pregnancy during treatment with Bryophyllum pinnatum: an observational study. Phytomedicine 21(5): 753–757. doi:10.1016/j.phymed.2013.11.003.

Lanctot, C.; Cheutin, T.; Cremer, M.; Cavalli, G.; Cremer, T. 2007. Dynamic genome architecture in the nuclear space: regulation of gene expression in three dimensions. Nat. Rev. Genet. 8: 104–115.

Längler, A.; Zuzak, T.J. 2013. Complementary and alternative medicine in paediatrics in daily practice – a European perspective. Complement Ther Med 2013; 21 (Suppl 1): 26–33. doi:10.1016/j.ctim.2012.01.005.

Largo, R. 2000. Kinderjahre. Die Individualität des Kindes als erzieherische Herausforderung. Piper, München.

Laubichler, M.D. 2005. Organismen. Systemtheoretische Organismuskonzeptionen. In: Krohs, U.; Toepfer, G. (Editors), Philosophie der Biologie. Suhrkamp, Frankfurt a. M., pp. 109–124.

Layer, M. (Editor). 2003. Praxishandbuch Rhythmische Einreibungen nach Wegman/Hauschka. Hans Huber, Bern.

Lecuit, T.; Le Goff, L. 2007. Orchestrating size and shape during morphogenesis. Nature 450: 189–192.

Lees, J. 2013. Facilitating self-healing in anthroposophic psychotherapy. Forsch Komplementärmed 20(4): 286–289. doi:10.1159/000354192.

Lehmann, J. 1991. Die vitalistische Grundfrage und ihr Verhältnis zur modernen Molekularbiologie. Dissertation. Medizinhistorisches Institut. Universität Bern, Bern.

Leopold, D.A. 2009. Pre-emptive blood flow. Nature 457: 387–388.

Leuenberger, P.; Longchamp, C. 2002. Was erwartet die Bevölkerung von der Medizin? Ergebnisse einer Umfrage des GfS-Forschungsinstitutes, Politik und Staat, Bern, im Auftrag der SAMW. In: Stauffacher, W.; Bircher, J. (Editors), Zukunft Medizin Schweiz. Das Projekt „Neu-Orientierung der Medizin" geht weiter. Schweizerischer Ärzteverlag AG, Basel, pp. 181–235.

Levin, M. 2012. Morphogenetic fields in embryogenesis, regeneration, and cancer: Non-local control of complex patterning. BioSystems 109(3): 243–261. doi:10.1016/j.biosystems.2012.04.005.

Levine, J. 1983. Materialism and qualia: The explanatory gap. Pac Philos Q 64: 354–361.

Lewin, B. 1988. Gene. Lehrbuch der molekularen Genetik. VHC-Verlag, Stuttgart.

Lewis, J. 2008. From signals to patterns: space, time and mathematics in developmental biology. Science 322: 399–403.

Libet, B. 1985. Unconscious cerebral initiative and the role of conscious will in voluntary action. Behav Brain Sci 8: 529–539.

Libet, B.; Gleason, C.A.; Wright, E.W.; Pearl, K. 1983. Time of conscious intention to act in relation to onset of cerebral activity (readiness potential). Brain 106: 623–642.

Linke, D.B.; Kurthen, M. 1988. Parallelität von Gehirn und Seele. Neurowissenschaft und Leib-Seele-Problem. Enke, Stuttgart.

Löffler, G.; Petrides, P.E. 1998. Biochemie und Pathobiochemie. 6th ed. Springer, Berlin, Heidelberg, New York.

Lötzke, D.; Heusser, P.; Büssing, A. 2015. A systematic literature review on the effectiveness of eurythmy therapy. J Integr Med 2015; 13(4): 217–230. doi:10.1016/S2095-4964(15)60163-7.

Lu, H.P. 2012. Enzymes in coherent motion. Science 335: 300–301.

Lutzker, P. 1996. Der Sprachsinn: Sprachwahrnehmung als Sinnesvorgang. Verlag Freies Geistesleben, Stuttgart.

Ma, D.K.; Jang, M.H.; Guo, J.U.; Kitabatake, Y.; Chang, M.L.; Pow-Anpongkul, N.; Flavell, R.A.; Lu, B.; Ming, G.L.; Song, H. 2009. Neuronal activity-induced Gadd45b promotes epigenetic DNA demethylation and adult neurogenesis. Science 323: 1074–1077.

Maddox, J. 1988. Crystals from first principles. Nature 335: 201.

Majewsky, V.; Arlt, S.P.; Shah, D.; Scherr, C.; Jäger, T.; Betti, L.; Trebbi, G.; Bonamin, L.; Klocke, P.; Baumgartner, S. 2009. Use of homeopathic preparations in experimental studies with healthy plants. Homeopathy 98(4): 228–243.

Majewsky, V.; Scherr, C.; Arlt, S.P.; Kiener, J.; Frrokaj, K.; Schindler, T.; Klocke, P.; Baumgartner, S. 2014. Reproducibility of effects of homeopathically potentised gibberellic acid on the growth of Lemna gibba L. in a randomised and blinded bioassay. Homeopathy 103(2): 113–126.

Majorek, M.; Tüchelmann, T.; Heusser, P. 2004. Therapeutic Eurythmy-movement therapy for children with attention deficit hyperactivity disorder (ADHD): a pilot study. Complement Ther Nurs Midwifery 10: 46–53.

Marti, E. 1994. Das Ätherische. Eine Erweiterung der Naturwissenschaft durch Anthroposophie. 2nd ed. Verlag die Pforte im Rudolf Steiner Verlag, Dornach.

Marti, T.; Heusser, P. 2009. Gesundheit vier- bis achtjähriger Kinder vor dem Hintergrund des familiären Lebensstils: Eine retrospektive Querschnittstudie an Kindern aus Schulen in der Stadt Bern und Umgebung. Peter Lang, Bern.

Martin, J. 2000. The idea is more important than the experiment. Lancet 356: 934–937.

Matthiessen, P.F. 2002. Perspektivität und Paradigmenpluralismus in der Medizin. In: Fuchs, B.; Kobler-Fumasoli, N. (Editors), Hilft der Glaube? Heilung auf dem Schnittpunkt zwischen Theologie und Medizin. Lit Verlag, Münster.

Matthiessen, P.F. 2008. Pluralität – auf dem Weg zu einer Integrativen Medizin? Forsch Komplementärmed 15: 248–250.

Matthiessen, P.F. 2009. Paradigmenpluralität und ärztliche Praxis. Autonomie des Arztes und Methodenpluralismus. In: Jütte, R. (Editor), Die Zukunft der Individualmedizin. Deutscher Ärzte-Verlag, Köln, pp. 37–53.

Matthiessen, P.F.; Rosslenbroich, B.; Schmidt, S. 1992. Unkonventionelle Medizinische Richtungen. Bestandsaufnahme zur Forschungssituation. Verlag für neue Wissenschaft GmbH, Bonn.

Mayr, E. 1984. Die Entwicklung der biologischen Gedankenwelt. Vielfalt, Evolution und Vererbung. Springer, Berlin, Heidelberg, New York.

Mayr, E. 1999. This is biology: the science of the living world. 7th ed. Harvard University Press, Cambridge, MA.

McKeen, T. 1996. Wesen und Gestalt des Menschen. Verlag Freies Geistesleben, Stuttgart.

McMahon, A.; Supatto, W.; Fraser, S.E.; Stathopoulos, A. 2008. Dynamic analyses of Drosophila gastrulation provide insights into collective cell migration. Science 322: 1546–1550.

Medicus, G. 2006. Grundlagen der Anthropologie – Eine interdisziplinäre Wissenschaft mit biologischen Wurzeln. Naturwiss Rundsch 159: 65–71.

Mees-Christeller, E.; Denzinger, I.; Altmaier, M.; Künstner, H.; Umfrid, H.; Frieling, E.; Auer, S. 2000. Therapeutisches Zeichnen und Malen. Anthroposophische Kunsttherapie. Vol. 2. Urachhaus, Stuttgart.

Melchart, D.; Mitscherlich, F.; Amiet, M.; Eichenberger, R.; Koch, P. 2005. Programm Evaluation Komplementärmedizin (PEK) Schlussbericht. In: BAG, Bundesamt für Gesundheit, Eidgenössisches Departement des Innern, Bern.

Messiah, A. 1961. Quantum Mechanics. North-Holland, Amsterdam.

Metzinger, T. (Editor), 1996. Bewusstsein. Beiträge aus der Gegenwartsphilosophie. Ferdinand Schöningh, Paderborn.

Meier, C.; Heusser, P. 1999. Stibium in der Blutstillung. Eine Ärzteumfrage. Der Merkurstab, 52: 99–104.

Michalopoulos, G.K. 2007. Liver regeneration. J. Cell. Physiol. 213: 286–300.

Mildenberger, M.; Schöll, A. 1977. Die Macht der süßen Worte. Transzendentale Meditation. Aussaat Verlag, Wuppertal.

Misquitta, A.J.; Welch, G.W.; Stone, A.J.; Price, S.L. A first principles prediction of the crystal structure of $C_6Br_2ClFH_2$. 2008. Chem Phys Lett 456: 105–109.

Mitchell, S. 2008. Komplexitäten. Warum wir erst anfangen, die Welt zu verstehen. Suhrkamp, Frankfurt a.M.

Mocek, R. 1996. Ganzheit und Selbstorganisation. Auf den Spuren eines biologischen Grundproblems. In: Küppers, G. (Editor), Chaos und Ordnung. Formen der Selbstorganisation in Natur und Gesellschaft. Philipp Reclam jun., Stuttgart, pp. 60–96.

Montazeri, A. 2008 Health-related quality of life in breast cancer patients: a bibliographic review of the literature from 1974 to 2007. J. Exp. Clin. Cancer Res. 27: 32.

Morrissey, M.J.; Duntley, S.P.; Anch, A.M.; Nonneman, R. 2004. Active sleep and its role in the prevention of apoptosis in the developing brain. Med. Hypotheses 62: 876–879.

Müller, O.L. 2015. Mehr Licht. Goethe mit Newton im Streit um die Farben. Fischer, Frankfurt a.M.

Nagel, T. 2012. Mind and Cosmos. Why the Materialist Neo-Darwinian Conception of Nature is Almost Certainly False. Oxford University Press, Oxford.

Nager, F. 1991. Der heilkundige Dichter. Goethe und die Medizin. Artemis, Zürich.

Nair, S.; Schilling, T.S. 2008. Chemokine signalling controls endodermal migration during Zebrafish gastrulation. Science 322: 89–92.

Nishida, K. 1992 An Inquiry into the Good, Yale University Press, New Haven, CT.

Novalis, 1960. Schriften. Die Werke Friedrich von Hardenbergs. Historisch-kritische Ausgabe. Edited by Paul Kluckhohn and Richard Samuel. 2nd ed. Kohlhammer, Stuttgart.

Novalis, 1997. Philosophical Writings. Translated and edited by Margaret Mahony Stoljar. State University of New York Press, Albany, NY.

Nüsslein-Volhard, C.; Wieschaus, E. 1980. Mutations affecting segment number and polarity in Drosophila. Nature 287: 795–801.

O'Dowd, H.; Gladwell, P.; Rogers, C.A.; Hollinghurst, S.; Gregory, A. 2006. Cognitive behavioural therapy in chronic fatigue syndrome: a randomised controlled trial of an outpatient group programme. Health Technol Assess 10: 1–121.

Oberheim, K.W.; Luther, W. 1958. Versuche über die Extremitätenregeneration von Salamanderlarven bei umgekehrter Polarität des Amputationsstumpfes. Wilhelm Roux Arch Entwickl Mech Org, 150: 373–382.

Olney, J.W. 1987. Glutamate. In: Adelman, G. (Editor), Encyclopedia of Neuroscience. Birkhäuser, Boston, Basel, Stuttgart, pp. 468–470.

Ostermann, T.; Matthiessen, P.F. (Editors). 2003. Einzelfallforschung in der Medizin. Bedeutung, Möglichkeiten, Grenzen. Verlag für Akademische Schriften, Frankfurt a. M.

Ostermann, T.; Raak, C.; Büssing, A. 2009. Survival of cancer patients treated with mistletoe extract (Iscador): a systematic literature review. BMC Cancer 2009; 9: 451.

Ostermann, T.; Büssing, A. 2012. Retrolective studies on the survival of cancer patients treated with mistletoe extracts: a meta-analysis. Explore 8(5): 277–281.

Ostwald, W. 1895. Die Überwindung des wissenschaftlichen Materialismus. Verhandlungen der Gesellschaft Deutscher Naturforscher und Ärzte 1895. 67. Versammlung, Leipzig.

Overstolz, A. 1995. Dokumentation anthroposophisch-medizinischer Zeitschriften 1926–1994. Verlag am Goetheanum, Dornach.

Overstolz, A. 2001. Dokumentation anthroposophisch-medizinischer Bücher: Gesamtübersicht über die weltweit erschienenen Bücher zur anthroposophischen Medizin und ihrer Grenzgebiete. Verlag am Goetheanum, Dornach.

Oxford University Press. 2014. Oxford Dictionaries. [Internet] 04/06/2014 [Viewed 26/12/2015]. URL: http://www.oxforddictionaries.com/de/definition/englisch/genetic-information?q=genetic+information.

Pampallona, S.; von Rohr, E.; van Wegberg, B.; Bernhard, J.; Helwig, S.; Heusser, P.; Hürny, C.; Schaad, R.; Cerny, T. 2002. Socio-Demographic and Medical Characteristics of Advanced Cancer Patients Using Conventional or Complementary Medicines. Onkologie 25: 165–170.

Pauen, M. 2001. Grundprobleme der Philosophie des Geistes und die Neuro wissenschaften. In: Pauen, M.; Roth, G. (Editors), Neurowissenschaften und Philosophie. Wilhelm Fink, München, pp. 83–122.

Peat, F.D. 1999. Active Information, Meaning and Form. Frontier Perspectives 8: 49–53.

Pelikan, W. 1997–2012. Healing Plants: Insights through Spiritual Science. 2 Vols. Mercury Press, Spring Valley, NY.

Pelikan, W. 2006. The Secrets of Metals. 2nd ed. Lindisfarne Books, Great Barrington, MA.

Pelikan, W.; Unger, G. 1965. Die Wirkung potenzierter Substanzen. Pflanzenwachstumsversuche mit statistischer Auswertung. Philosophisch-Anthroposophischer Verlag, Dornach.

Penn, D.C.; Holyoak, K.J.; Povinelli, D.J. 2008. Darwin's mistake: Explaining the discontinuity between human and nonhuman minds. Behav Brain Sci 31: 109–178.

Penzlin, H. 2016. Das Phänomen Leben: Grundfragen der Theoretischen Biologie. 2nd ed. Springer, Berlin, Heidelberg.

Petersen, C.P.; Reddien, P.W. 2008. Smed-βcatenin-1 is required for anteroposterior blastema polarity in planarian regeneration. Science 319: 327–330.

Peveling, M. 2015. Der Sprachsinn bei Rudolf Steiner. Eine kritische Würdigung im Lichte der modernen Sprachforschung und der sozialen Neurobiologie. Thesis. Witten/Herdecke University.

Plessner, H. 1975. Die Stufen des Organischen und der Mensch. 3rd ed. Walter de Gruyter, Berlin.

Polanyi, M. 1968. Life's irreducible structure. Live mechanisms and information in DNA are boundary conditions with a sequence of boundaries above them. Science 160: 1308–1312.

Poppelbaum, H. 1924. Der Bildekräfteleib der Lebewesen als Gegenstand wissenschaftlicher Erfahrung. Der Kommende Tag AG, Stuttgart.

Poppelbaum, H. 1981. Mensch und Tier. Fünf Einblicke in ihren Wesensunterschied. 2nd ed. Fischer, Hamburg.

Popper, K.R. 1972. Objective Knowledge, an Evolutionary Approach. Clarendon Press, Oxford.

Popper, K.R. 1976. Logik der Forschung. 6th ed. J. C. B. Mohr (Paul Siebeck), Tübingen.

Portmann, A. 1982. Um eine basale Anthropologie. In: Portmann, A. (Editor), Biologie und Geist. Suhrkamp, Frankfurt a.M., pp. 277–291.

Portmann, A. 2006. Die Innerlichkeit – Die Weltbeziehung und das Erleben (1960). In: Senn, D.G. (Editor), Adolf Portmann. Lebensforschung und Tiergestalt. Ausgewählte Texte. Schwabe, Basel, pp. 60–73.

Priever, W. 1999. Aspekte des Unbewussten. Verlag Freies Geistesleben, Stuttgart.

Prigogine, I. 1977. Nobel Lecture: Time, Structure and Fluctuations [E-book]. Nobel Media AB, Stockholm. [Viewed 15/04/2009]. URL: http://nobelprize.org/chemistry/laureates/1977/prigogine-lecture.pdf.

Primas, H. 1985a. Kann Chemie auf Physik reduziert werden? Erster Teil: Das Molekulare Programm. Chemie in unserer Zeit 19: 109–119.

Primas, H. 1985b. Kann Chemie auf Physik reduziert werden? Zweiter Teil: Die Chemie der Makrowelt. Chemie in unserer Zeit 19: 160–166.

Primas, H. 1991. Reductionism: Palaver without Precedent. In: Agazzi, E. (Editor), The Problem of Reductionism in Science. Kluwer Academic Publishers, Dordrecht, Boston, London, pp. 161–172.

Primas, H. 1992. Umdenken in der Naturwissenschaft. Gaia 1: 5–15.

Purves, W.K.; Sadava, D.; Orians, G.H.; Heller, H.C. 2003. Life: The Science of Biology. 7th revised ed. W.H. Freeman & Co Ltd, Sunderland, MA.

Purves, W.K.; Sadava, D.; Orians, G.H.; Heller, H.C. 2006. Biologie. Elsevier, München.

Raak, C.; Molsberger, F.; Heinrich, U.; Bertram, M.; Ostermann, T. 2014. Mesembryanthemum crystallinum L. as dermatologically effective medicinal plant – first results from 3 pilot studies. Forsch Komplementärmed 21(6): 366–373. doi:10.1159/000369909.

Ravagli, L. 2009. Zanders Erzählungen. Eine kritische Analyse des Werkes „Anthroposophie in Deutschland". Berliner Wissenschafts-Verlag, Berlin.

Reversade, B.; De Robertis, E.M. 2005. Regulation of ADMP and BMP2/4/7 at Opposite Embryonic Poles Generates a Self-Regulating Morphogenetic Field. Cell 123: 1147–1160.

Riedl, R. 2005. A Systems Theory of Evolution. In: Hösle, V.; Illies, C. (Editors), Darwinism and Philosophy. University of Notre Dame Press, Notre Dame, IN, pp. 121–142.

Righetti, M. 1988. Forschung in der Homöopathie. Grundlagen, Problematik und Ergebnisse. Ulrich Burgdorf, Göttingen.

Ringe, D.; Petsko, G.A. 2008. How Enzymes Work. Science 320: 1428–1429.

Roberts, L. 2009. Sleeping to reset overstimulated synapses. Science 324: 22–23.

Robinson, G.E.; Fernald, R.D.; Clayton, D.F. 2008. Genes and social behavior. Science 322: 896–900.

Rohen, J.W. 2008. Functional Morphology. The Dynamic Wholeness of the Human Organism. Adonis Press, Hillsdale, NY.

Rohen, J.W.; Lütjen-Drecoll, E. 2000. Funktionelle Histologie. 4th ed. Schattauer, Stuttgart.

Rohen, J.W.; Lütjen-Drecoll, E. 2006a. Funktionelle Anatomie des Menschen. Lehrbuch der makroskopischen Anatomie nach funktionellen Gesichtspunkten. 11th ed. Schattauer, Stuttgart.

Rohen, J.W.; Lütjen-Drecoll, E. 2006b. Funktionelle Embryologie. Die Entwicklung der Funktionssysteme des menschlichen Organismus. 3rd ed. Schattauer, Stuttgart.

Rohen, J.W.; Yokochi, C.; Lütjen-Drecoll, E. 1998. Color Atlas of Anatomy. 4th ed. Williams & Wilkins, Baltimore.

Rossi, E.; Vita, A.; Baccetti, S.; Di Stefano, M.; Voller, F.; Zanobini, A. 2015. Complementary and alternative medicine for cancer patients: results of the EPAAC survey on integrative oncology centres in Europe. Support Care Cancer 23(6): 1795–1806. doi:10.1007/s00520-014-2517-4.

Rosslenbroich, B. 1994. Die Rhythmische Organisation des Menschen. Aus der chronobiologischen Forschung. Verlag Freies Geistesleben, Stuttgart.

Rosslenbroich, B. 2006. The notion of progress in evolutionary biology – the unresolved problem and an empirical suggestion. Biol Philos 21: 41–70.

Rosslenbroich, B. 2007. Autonomiezunahme als Modus der Makroevolution. Galunder, Nümbrecht.

Rosslenbroich, B. 2014. On the origin of autonomy. A new look at the major transitions in evolution. Springer, Cham, Heidelberg, New York, Dordrecht, London.

Roth, G. 2001. Die neurobiologischen Grundlagen von Geist und Bewusstsein. In: Pauen, M.; Roth, G. (Editors), Neurowissenschaften und Philosophie. Eine Einführung. Wilhelm Fink, München, pp. 155–209.

Roth, G. 2004. Worüber dürfen Hirnforscher reden – und in welcher Weise? In: Geyer, C. (Editor), Hirnforschung und Willensfreiheit. Suhrkamp, Frankfurt a. M., pp. 66–85.

Roth, G. 2006. Willensfreiheit und Schuldfähigkeit aus Sicht der Hirnforschung. In: Roth, G.; Grün, K.-J. (Editors), Das Gehirn und seine Freiheit. Beiträge zur neurowissenschaftlichen Grundlegung der Philosophie. Vandenhoeck & Ruprecht, Göttingen, pp. 9–27.

Roth, G. 2007. Die Physik des Geistes. In: Sandhoff, K.; Donner, W.; et al. (Editors), Vom Urknall zum Bewusstsein – Selbstorganisation der Materie. Verhandlungen der Gesellschaft Deutscher Naturforscher und Ärzte. Thieme, Stuttgart, pp. 301–312.

Rozumek, M. 2008. Lebensvorgänge als Erkenntnismittel für chemische Zusammenhänge. In: Rozumek, M.; Buck, P. (Editors), Das Chemische und die Stoffe. Verlag am Goetheanum, Dornach, pp. 45–66.

Rozumek, M.; Buck, P. (Editors). 2008. Das Chemische und die Stoffe. Verlag am Goetheanum, Dornach.

Rutherford, S.L.; Lindquist, S. 1998. Hsp90 as a capacitor for morphological evolution. Nature 396: 336–342.

Sander, K. 1990. Von der Keimplasmatheorie zur synergetischen Musterbildung. Verh. Dtsch. Zool. Ges. 83: 133–177.

Saunders, J.W. 1968. Animal Morphogenesis. Macmillan, New York.

Schad, F.; Axtner, J.; Happe, A.; Breitkreuz, T.; Paxino, C.; Gutsch, J.; Matthes, B.; Debus, M.; Kröz, M.; Spahn, G.; Riess, H.; von Laue, H.B.; Matthes, H. 2013a. Network Oncology (NO) – a clinical cancer register for health services research and the evaluation of integrative therapeutic interventions in anthroposophic medicine. Forsch Komplementärmed 20(5): 353–360. doi:10.1159/000356204.

Schad, F.; Axtner, J.; Buchwald, D.; Happe, A.; Popp, S.; Kröz, M.; Matthes, H. 2013b. Intratumoral Mistletoe (Viscum album L) Therapy in Patients with Unresectable Pancreas Carcinoma: A Retrospective Analysis. Integr Cancer Ther 13(4): 332–340.

Schad, W. 1977. Man and Mammals: Toward a Biology of Form. Waldorf Press, Garden City, NY.

Schad, W. 1982a. Biologisches Denken. In: Schad, W. (Editor), Goetheanistische Naturwissenschaft. Allgemeine Biologie. Verlag Freies Geistesleben, Stuttgart, pp. 9–25.

Schad, W. 1982b. Die Vorgeburtlichkeit des Menschen. Der Entwicklungsgedanke in der Embryologie. Urachhaus, Stuttgart.

Schad, W. (Editor). 1982c. Goetheanistische Naturwissenschaft. Allgemeine Biologie. Verlag Freies Geistesleben, Stuttgart.

Schad, W. (Editor). 1982d. Goetheanistische Naturwissenschaft. Botanik. Verlag Freies Geistesleben, Stuttgart.

Schad, W. (Editor). 1983. Goetheanistische Naturwissenschaft. Zoologie. Verlag Freies Geistesleben, Stuttgart.

Schad, W. (Editor). 1985a. Goetheanistische Naturwissenschaft. Anthropologie. Verlag Freies Geistesleben, Stuttgart.

Schad, W. 1985b. Stauphänomene am menschlichen Knochenbau. In: Schad, W. (Editor), Goetheanistische Naturwissenschaft. Anthropologie. Verlag Freies Geistesleben, Stuttgart, pp. 9–29.

Schad, W. 1992. Die Heterochronien in der Evolution der Wirbeltierklassen. Naturwissenschaftliche Fakultät Witten/Herdecke, Witten.

Schad, W. 1993. Vom Verstehen der Zeit. In: Kniebe, G. (Editor), Was ist Zeit? Verlag Freies Geistesleben, Stuttgart, pp. 233–272.

Schad, W. 1998. Zeitgestalten der Natur. Goethe und die Evolutionsbiologie. In: Matussek, P. (Editor), Goethe und die Verzeitlichung der Natur. Beck, München, pp. 345–382.

Schad, W. 2003. Chronobiologie ist Ätherforschung. Tycho de Brahe Jahrbuch für Goetheanismus. Tycho Brahe Verlag, Niefern-Öschelbronn, pp. 20–36.

Schad, W. 2012. Säugetier und Mensch: Ihre Gestaltbiologie in Raum und Zeit. 1st ed. Verlag Freies Geistesleben, Stuttgart.

Schad, W. 2014. Die Doppelnatur des Ich: Der übersinnliche Mensch und seine Nervenorganisation. Verlag Freies Geistesleben, Stuttgart.

Schaefer, K.E.; Hensel, H.; Brady, R. (Editors). 1977. Toward a Man-Centered Medical Science. A New Image in Medicine. Vol. I. Futura Publishing Company, Mt. Kisco, NY.

Schaefer, K.E.; Hildebrandt, G.; Macbeth, N. (Editors). 1979a. Basis of an Individual Physiology. A New Image in Medicine. Vol. II. Futura Publishing Company, Mt. Kisco, NY.

Schaefer, K.E.; Stave, U.; Blankenburg, W. (Editors). 1979b. Individuation Process and Biographical Aspects of Disease. A New Image in Medicine. Vol. III. Futura Publishing Company, Mt. Kisco, NY.

Schark, M. 2005. Lebewesen als ontologische Kategorie. In: Krohs, U.; Toepfer, G. (Editors), Philosophie der Biologie. Suhrkamp, Frankfurt a. M., pp. 175–192.

Scheffer, C.; Tauschel, D.; Neumann, M.; Lutz, G.; Cysarz, D.; Heusser, P.; Edelhäuser, F. 2012. Integrative medical education: educational strategies and preliminary evaluation of the Integrated Curriculum for Anthroposophic Medicine (ICURAM). Patient Educ Couns 89(3): 447–454. doi:10.1016/j.pec.2012.04.006.

Scheffer, C.; Tauschel, D.; Neumann, M.; Lutz, G.; Valk-Draad, M.; Edelhäuser, F. 2013. Active student participation may enhance patient centeredness: patients' assessments of the clinical education ward for integrative medicine. Evid Based Complement Alternat Med 2013: 743832. doi:10.1155/2013/743832.

Scheffer, C.; Debus, M.; Heckmann, C.; Cysarz, D.; Girke, M. 2016. Colchicum autumnale in Patients with Goitre with Euthyroidism or Mild

Hyperthyroidism: Indications for a Therapeutic Regulative Effect – Results of an Observational Study. Evid Based Complement Alternat Med 2016: 2541912. doi:10.1155/2016/2541912.

Scheffers, T.; Weinzirl, J. 2015. Die Milz in der „Okkulten Physiologie". In: Selg, P. (Editor), „Okkulte Physiologie". Rudolf Steiners Prager Kurs (1911). Verlag des Ita Wegman Instituts, Arlesheim, pp. 37–58.

Scheler, M. 1947. Die Stellung des Menschen im Kosmos. Nymphenburger, München.

Scheler, M. 1973. Wesen und Formen der Sympathie. 2nd ed. Franke, Bern.

Schelling, F.J.W. 1980. The Unconditional in Human Knowledge. Four Early Essays. Translation and commentary by Fritz Marti. Associated University Presses, Inc., Plainsboro, NJ.

Scherr, C.; Baumgartner, S.; Spranger, J.; Simon, M. 2006. Effects of Potentised Substances on Growth Kinetics of Saccharomyces cerevisiae and Schizosaccharomyces pombe. Forsch Komplementärmed 13: 298–306.

Scherr, C.; Simon, M.; Spranger, J.; Baumgartner, S. 2007. Duckweed (Lemna gibba L.) as a Test Organism for Homeopathic Potencies. J Altern Complement Med 13: 931–938.

Schiller, F. 1935. Briefwechsel zwischen Schiller und Goethe. 2 Vols. Schiller an Goethe 19. Januar 1798. Reclam, Leipzig.

Schindewolf, O.H. 1972. Phylogenie und Anthropologie aus paläontologischer Sicht. In: Gadamer, H.-G.; Vogler, P. (Editors), Neue Anthropologie. Thieme, Stuttgart, pp. 230–292.

Schink, M.; Garcia-Käufer, M.; Bertrams, J.; Duckstein, S.M.; Müller, M.B.; Huber, R.; Stintzing, F.C.; Gründemann, C. 2015. Differential cytotoxic properties of Helleborus niger L. on tumour and immunocompetent cells. J Ethnopharmacol 159: 129–136. doi:10.1016/j.jep.2014.11.003.

Schmal, F.W.; von Weizsäcker, C.F. 2000. Moderne Physik und Grundfragen der Medizin. Dtsch Ärztebl 97: A 165–167.

Schmidt, G. 1975–1979. Dynamische Ernährungslehre. Die Anregung Rudolf Steiners für eine neue Ernährungshygiene. 2 Vols. Proteus Verlag, St. Gallen.

Schneider, N. 1998. Erkenntnistheorie des 20. Jahrhunderts. Reclam, Stuttgart.

Schneider, P. 1985. Einführung in die Waldorfpädagogik. 2nd ed. Klett-Cotta, Stuttgart.

Schöne, A. 1987. Goethes Farbentheologie. C.H. Beck, München.

Schramm, H. 1983. Heilmittelfibel zur Anthroposophischen Medizin. Novalis Verlag, Schaffhausen.

Schumacher, R. 2007. Gehirn und Bewusstsein aus philosophischer Sicht. In: Sandhoff, K.; Donner, W.; et al. (Editors), Vom Urknall zum Bewusstsein – Selbstorganisation der Materie. Verhandlungen der Gesellschaft Deutscher Naturforscher und Ärzte. Thieme, Stuttgart, pp. 315–328.

Schüpbach, W. 1947–1948. Über das Geometrische im menschlichen Skelett. Die Beziehungen zur Pflanzengeometrie, zum Kosmischen und zur menschlichen Individualität. 3 Vols. Troxler Verlag, Bern.

Schwegler, H. 2001. Reduktionismus und Physikalismen. In: Pauen, M.; Roth, G. (Editors), Neurowissenschaften und Philosophie. Eine Einführung. Wilhelm Fink, München, pp. 59–82.

Schwenk, T. 2014. Sensitive Chaos. The Creation of Flowing Forms in Water and Air. 2nd ed. Rudolf Steiner Press, London.

Searle, J.R. 1992. The Rediscovery of the Mind. MIT Press, Cambridge, MA, London.

Seifert, G.; Rutkowski, S.; Jesse, P.; Madeleyn, R.; Reif, M.; Henze, G.; Längler, A. 2011. Anthroposophic supportive treatment in children with medulloblastoma receiving first-line therapy. J. Pediatr. Hematol. Oncol. 33(3): e105–108. doi:10.1097/MPH.0b013e31820946d3.

Seifert, G.; Kanitz, J.L.; Pretzer, K.; Henze, G.; Witt, K.; Reulecke, S.; Voss, A. 2012. Improvement of heart rate variability by eurythmy therapy after a 6-week eurythmy therapy training. Integr Cancer Ther 11(2): 111–119. doi:10.1177/1534735411413263.

Seifert, G.; Kanitz, J.L.; Pretzer, K.; Henze, G.; Witt, K.; Reulecke, S.; Voss, A. 2013. Improvement of circadian rhythm of heart rate variability by eurythmy therapy training. Evid Based Complement Alternat Med 2013: 564340. doi:10.1155/2013/564340.

Selawry, A. 1985. Metall-Funktionstypen in Psychologie und Medizin. 2nd ed. Haug, Heidelberg.

Selg, P. 2000a. Anfänge anthroposophischer Heilkunst. Philosophisch-Anthroposophischer Verlag am Goetheanum, Dornach.

Selg, P. (Editor). 2000b. Anthroposophische Ärzte. Lebens- und Arbeitswege im 20. Jahrhundert. Nachrufe und Kurzbiographien. Verlag am Goetheanum, Dornach.

Selg, P. 2000c. Vom Logos menschlicher Physis. Die Entfaltung einer anthroposophischen Humanphysiologie im Werk Rudolf Steiners. Vols. I and II. 2nd ed. Verlag am Goetheanum, Dornach.

Selg, P. 2003. Gerhard Kienle. Leben und Werk. Vols. I and II. Verlag am Goetheanum, Dornach.

Selg, P. 2004. Krankheit, Heilung und Schicksal des Menschen. Über Rudolf Steiners geisteswissenschaftliches Pathologie- und Therapieverständnis. Verlag am Goetheanum, Dornach.

Selg, P. 2005. Christliche Medizin. Die ideellen Beziehungen des Christentums zur Heilkunde und die Anthroposophische Medizin. Verlag am Goetheanum, Dornach.

Selg, P. 2006. ‚Die Medizin muss Ernst machen mit dem geistigen Leben'. Rudolf Steiners Hochschulkurse für die „jungen Mediziner". Verlag am Goetheanum, Dornach.

Selg, P. 2013. Die «Wärme-Meditation»: Geschichtlicher Hintergrund und ideelle Beziehungen. Verlag am Goetheanum, Dornach.

Selg, P. (Editor). 2015. „Okkulte Physiologie". Rudolf Steiners Prager Kurs (1911). Verlag des Ita Wegman Instituts, Arlesheim.

Service, R.F. 2008. Problem solved (sort of). Science 321: 784–789.

Sheldrake, R. 2009. A New Science of Life. 3rd ed. Icon Books Ltd, London.

Sieweke, H. 1982. Anthroposophische Medizin. Studien zu ihren Grundlagen. Erster Teil. 2nd ed. Philosophisch-Anthroposophischer Verlag, Dornach.

Sieweke, H. 1994. Anthroposophische Medizin. Studien zu ihren Grundlagen. Zweiter Teil. Gesundheit und Krankheit als Verwirklichungsformen menschlichen Daseins. 2nd ed. Philosophisch-Anthroposophischer Verlag, Dornach.

Sieweke, H. 2008. Anthroposophische Medizin. 3rd ed. Verlag am Goetheanum, Dornach.

Sijmons, J. 2009. Phänomenologie und Idealismus. Struktur und Methode der Philosophie Rudolf Steiners. 1st ed. Schwabe, Basel.

Simões-Wüst, A.P.; Jeschke, E.; Mennet, M.; Schnelle, M.; Matthes, H.; von Mandach, U. 2012. Prescribing pattern of Bryophyllum preparations among a network of anthroposophic physicians. Forsch Komplementärmed 19(6): 293–301. doi:10.1159/000345841.

Simões-Wüst, A.P.; Kummeling, I.; Mommers, M.; Huber, M.A.; Rist, L.; van de Vijver, L.P.; Dagnelie, P.C.; Thijs, C. 2014. Influence of alternative lifestyles on self-reported body weight and health characteristics in women. Eur J Public Health 24(2): 321–327. doi:10.1093/eurpub/ckt045.

Simões-Wüst, A.P.; Hassani, T.A.; Müller-Hübenthal, B.; Pittl, S.; Kuck, A.; Meden, H.; Eberhard, J.; Decker, M.; Fürer, K.; von Mandach, U.; Bryophyllum Collaborative Group. 2015. Sleep Quality Improves During Treatment with Bryophyllum pinnatum: An Observational Study on Cancer Patients. Integr Cancer Ther 2015; 14(5): 452–459. doi:10.1177/1534735415580680.

Simonis, W.C. 1955. Die unbekannte Heilpflanze. Vittorio Klostermann, Frankfurt a. M.

Simonis, W.C. 1981. Medizinisch-botanische Wesensdarstellungen einzelner Heilpflanzen. Lehrbriefe nach Heilpflanzen-Unterricht an einer ärztlichen Fortbildungsstätte. 2 Vols. Novalis Verlag, Schaffhausen.

Singer, W. 2001. Vom Gehirn zum Bewusstsein. In: Lüer, G.; Elsner, N. (Editors), Das Gehirn und sein Geist. Wallstein, Göttingen, pp. 189–204.

Singer, W. 2004. Verschaltungen legen uns fest: Wir sollten aufhören, von Freiheit zu sprechen. In: Geyer, C. (Editor), Hirnforschung und Willensfreiheit. Suhrkamp, Frankfurt a. M., pp. 30–65.

Singer, W. 2006. Gekränkte Freiheit. Interview mit Wolf Singer. In: Roth, G.; Grün, K.-J. (Editors), Das Gehirn und seine Freiheit. Beiträge zur neurowissenschaftlichen Grundlegung der Philosophie. Vandenhoeck & Ruprecht, Göttingen, pp. 83–87.

Sirontin, Y.B.; Das, A. 2009. Anticipatory haemodynamic signals in sensory cortex not predicted by local neural activity. Nature 457: 475–479.

Soldner, G. (Editor). 2008. Vademecum Anthroposophischer Arzneimittel. Gesellschaft Anthroposophischer Ärzte in Deutschland und Medizinische Sektion der Freien Hochschule für Geisteswissenschaft, Dornach, Filderstadt.

Soldner, G.; Stellmann, H.M. 2011. Individuelle Pädiatrie. Leibliche, seelische und geistige Aspekte in Diagnostik und Beratung. Anthroposophisch-homöopathische Therapie. 4th ed. Wissenschaftliche Verlagsgesellschaft, Stuttgart.

Soto, A.M.; Sonnenschein, C. 2011. The tissue organization field theory of cancer: a testable replacement for the somatic mutation theory. Bioessays 33: 332–340.

Spaemann, R.; Löw, R. 1991. Die Frage Wozu? Geschichte und Wiederentdeckung des teleologischen Denkens. Piper, München.

Spemann, H. 1921. Die Erzeugung tierischer Chimären durch heteroplastische embryonale Transplantation zwischen Triton cristatus und taeniatus. Wilhelm Roux Arch Entw Mech Org 48: 533–570.

Spemann, H. 1968. Experimentelle Beiträge zu einer Theorie der Entwicklung. 2nd ed. Springer, Berlin.

Spratt, N.; Hass, H. 1960. Integrative mechanisms in development of the early chick blastoderm. I. Regulative potentiality of separated parts. J. Exp. Zool. 145: 97–138.

Stamm, M.; Edelmann, D. 2013. Handbuch frühkindlicher Entwicklung. Springer, Wiesbaden.

Steele, M.L.; Axtner, J.; Happe, A.; Kröz, M.; Matthes, H.; Schad, F. 2014a. Adverse Drug Reactions and Expected Effects to Therapy with Subcutaneous

Mistletoe Extracts (Viscum album L.) in Cancer Patients. Evid Based Complement Alternat Med 2014: 724258. doi:10.1155/2014/724258.

Steele, M.L.; Axtner, J.; Happe, A.; Kröz, M.; Matthes, H.; Schad, F. 2014b. Safety of Intravenous Application of Mistletoe (Viscum album L.) Preparations in Oncology: An Observational Study. Evid Based Complement Alternat Med 2014: 236310. doi:10.1155/2014/236310.

Steele, M.L.; Axtner, J.; Happe, A.; Kröz, M.; Matthes, H.; Schad, F. 2015. Use and safety of intratumoral application of European mistletoe (Viscum album L) preparations in Oncology. Integr Cancer Ther 2015;14(2):140–148. doi:10.1177/1534735414563977.

Stein, W.J.; Steiner, R. 1985. Dokumentation eines wegweisenden Zusammenwirkens. W. J. Steins Dissertation in ihrem Entstehungsprozess und in ihrer Aktualität. Philosophisch-Anthroposophischer Verlag am Goetheanum, Dornach.

Steiner, R. Unpublished. The Spiritual Development of Man, Lecture I. Translated by K. Wegener [E-text]. [Viewed 07/10/2015]. URL: http://wn.rsarchive.org/Lectures/GA084/English/LZ0290/SpiDev_index.html.

Steiner, R. 1917. The Mission of Spiritual Science and of Its Building at Dornach, Switzerland. Translated by A.M. Wilson [E-text]. H.J. Heywood-Smith, London. [Viewed 17/11/2015]. URL: http://wn.rsarchive.org/Lectures/GA035/English/HJHS1917/19160111p01.html.

Steiner, R. 1929. Philosophy and Anthroposophy [E-text]. Anthroposophical Publishing Company, London. [Viewed 17/11/2015]. URL: http://wn.rsarchive.org/GA/GA0035/PhlAnt_index.html.

Steiner, R. 1945. The Effect of Occult Development upon the Self and the Sheaths of Man [E-text]. Rudolf Steiner Publishing. [Viewed 14/10/2015]. URL: http://wn.rsarchive.org/Lectures/GA145/English/RSPC1945/OccDev_index.html.

Steiner, R. 1948. Spiritual Science and Medicine. Rudolf Steiner Publishing, London.

Steiner, R. 1950. Anthroposophie und Akademische Wissenschaften. 1st ed. Europa Verlag, Zürich, Wien, Stuttgart.

Steiner, R. 1951a. An Occult Physiology. Rudolf Steiner Publishing, London.

Steiner, R. 1951b. Anthroposophical Approach to Medicine. Translated by Charles Davy. Anthroposophical Publishing Company, London. (Contains part of Steiner, R. 1989.).

Steiner, R. 1961. Goethes Recht in der Naturwissenschaft. In: Steiner, R., Methodische Grundlagen der Anthroposophie 1884–1901. Gesammelte Aufsätze. 2nd ed. Verlag der Rudolf Steiner-Nachlassverwaltung, Dornach.

Steiner, R. 1965. Die Geisteswissenschaft als Anthroposophie und die zeitgenössische Erkenntnistheorie. In: Steiner, R., Philosophie und Anthroposophie. Gesammelte Aufsätze 1904–1918. 1st ed. Rudolf Steiner Verlag, Dornach, pp. 307–331.

Steiner, R. 1968. Die Rätsel der Philosophie in ihrer Geschichte als Umriss dargestellt. 8th ed. Verlag der Rudolf Steiner-Nachlassverwaltung, Dornach.

Steiner, R. 1969a. Knowledge of the Higher Worlds. How is it achieved? Rudolf Steiner Press, London.

Steiner, R. 1969b. Occult Science – An Outline. Translated by George and Mary Adams. Rudolf Steiner Press, London.

Steiner, R. 1970a. Theosophy. An introduction to the supersensible knowledge of the world and the destination of man. Rudolf Steiner Press, London.

Steiner, R. 1970b. Man as Symphony of the Creative Word. Translated by Judith Compton-Burnett. Rudolf Steiner Press, London.

Steiner, R. 1972. Karmic Relationships: Esoteric Studies. Vol. 1. Translated by George Adams [E-text]. Rudolf Steiner Press, London. [Viewed 03/11/2015]. URL: http://wn.rsarchive.org/Lectures/GA235/English/RSP1972/ Karm01_index.html.

Steiner, R. 1973a. The Riddles of Philosophy. Anthroposophic Press, Spring Valley, NY.

Steiner, R. 1973b. Anthroposophical Leading Thoughts. Rudolf Steiner Press, London.

Steiner, R. 1978a. The case for Anthroposophy. Translated by Owen Barfield. The Anthroposophic Press, Great Barrington, MA.

Steiner, R. 1978b. Die Philosophie der Freiheit. Grundzüge einer modernen Weltanschauung. Seelische Beobachtungsresultate nach naturwissenschaftlicher Methode. 14th ed. Rudolf Steiner Verlag, Dornach.

Steiner, R. 1979a. Grundlinien einer Erkenntnistheorie der Goetheschen Weltanschauung. Mit besonderer Rücksicht auf Schiller. 7th ed. Rudolf Steiner Verlag, Dornach.

Steiner, R. 1980a. Wahrheit und Wissenschaft. Vorspiel einer Philosophie der Freiheit. 5th ed. Rudolf Steiner Verlag, Dornach.

Steiner, R. 1980b. Second Scientific Lecture-Course: Warmth Course. Mercury Press, Spring Valley, NY.

Steiner, R. 1981a. Truth and Knowledge, Introduction to the Philosophy of Freedom. Translated by Rita Stebbing. SteinerBooks, Great Barrington, MA.

Steiner, R. 1981b. The Being of Man and his Future Evolution. Translated by Pauline Wehrle. Rudolf Steiner Press, London.

Steiner, R. 1983a. The Boundaries of Natural Science. Anthroposophic Press, Spring Valley, NY.

Steiner, R. 1983b. Curative Eurythmy. Translated by Kristina Krohn in collaboration with Dr. Anthony Degenaar. Rudolf Steiner Press, London.

Steiner, R. 1983c. Antworten der Geisteswissenschaft auf die großen Fragen des Daseins. 2nd ed. Rudolf Steiner Verlag, Dornach.

Steiner, R. 1985a. Mystery Centres: Lecture V [E-text]. Garber Communications, Inc. [Viewed 13/10/2015]. URL: http://wn.rsarchive.org/Lectures/GA232/English/GC1985/ MysCen_index.htm.

Steiner, R. 1985b. Goethe's World View. Mercury Press, Spring Valley, NY.

Steiner, R. 1985c. True and False Paths in Spiritual Investigation. Rudolf Steiner Press, London.

Steiner, R. 1986. Physiology and Therapeutics. Translated by Alice Wulsin and Gerald Karnow [E-text] (Contains part of Steiner, R. 1989). Mercury Press, Spring Valley, NY. [Viewed 17/11/2015]. URL: http://wn.rsarchive.org/Lectures/GA314/English/ MP1986a/PhysTh_index.html.

Steiner, R. 1987. Pastoral Medicine. The Collegial Working of Doctors and Priests. Anthroposophic Press, Hudson, NY.

Steiner, R. 1988a. The Science of Knowing, Outline of an Epistemology Implicit in the Goethean World View. Mercury Press, Spring Valley, NY.

Steiner, R. 1988b. Goethean Science. Mercury Press, Spring Valley, NY.

Steiner, R. 1988c. Warmth Course. Mercury Press, Spring Valley, NY.

Steiner, R. 1989. Physiologisch-Therapeutisches auf Grundlage der Geisteswissenschaft. Zur Therapie und Hygiene. 3rd ed. Rudolf Steiner Verlag, Dornach.

Steiner, R. 1990. The Riddle of Man. Translated by William Lindeman [E-text]. Mercury Press, Spring Valley, NY. [Viewed 15/11/2015]. URL: http://wn.rsarchive.org/Books/GA020/English/MP1990/GA020_c05.html.

Steiner, R. 1991. Anthroposophical Spiritual Science and Medical Therapy. Mercury Press, Spring Valley, NY.

Steiner, R. 1994a. Anthroposophische Menschenerkenntnis und Medizin. 3rd ed. Rudolf Steiner Verlag, Dornach.

Steiner, R. 1994b. Course for young doctors. Mercury Press, Spring Valley, NY.

Steiner, R. 1996. Riddles of the Soul. Mercury Press, Spring Valley, NY.

Steiner, R. 2001. The Light Course. First course in natural science: Light, Color, Sound— Mass, Electricity, Magnetism. Translated by Raoul Cansino. Anthroposophic Press, Great Barrington, MA.

Steiner, R. 2004a. Texte zur Medizin. Anthroposophie und Heilkunst. Teil 1: Physiologische Menschenkunde. Edited by Peter Selg. Rudolf Steiner Verlag, Dornach.

Steiner, R. 2004b. Texte zur Medizin. Anthroposophie und Heilkunst. Teil 2: Pathologie und Therapie. Edited by Peter Selg. Rudolf Steiner Verlag, Dornach.

Steiner, R. 2005. Fachwissenschaften und Anthroposophie. 1st ed. Rudolf Steiner Verlag, Dornach.

Steiner, R. 2006. The Mission of Spiritual Science and of its Building at Dornach Switzerland (1918). Reprint. Standard Publications, Incorporated, Huntington, WV.

Steiner, R. 2011. The Philosophy of Freedom. The Basis for a Modern World Conception. Translated by Michael Wilson. Rudolf Steiner Press, London.

Steiner, R. 2014. Education for Special Needs: The Curative Education Course. Rudolf Steiner Press, London.

Steiner, R.; Wegman, I. 1991. Grundlegendes für eine Erweiterung der Heilkunst nach geisteswissenschaftlichen Erkenntnissen. 7th ed. Rudolf Steiner Verlag, Dornach.

Steiner, R.; Wegman, I. 1996. Extending Practical Medicine. Fundamental Principles based on the Science of the Spirit. New translation by A.R. Meuss. Rudolf Steiner Press, London.

Stent, G.S. 1981. Programmatic phenomena, hermeneutics and complex network. In: Steinberg, C.M.; Lefkovits, I. (Editors), The Immune System. Vol. I. Karger, Basel, pp. 6–13.

Stephan, A. 2001. Emergenz in kognitionsfähigen Systemen. In: Pauen, M.; Roth, G. (Editors), Neurowissenschaften und Philosophie. Wilhelm Fink, München, pp. 123–154.

Stern, C.M. 1954. Two or three bristles. Am. Sci. 42: 213–247.

Stern, E. 2006. Was Hänschen nicht lernt, lernt Hans hinterher. Erwerb geistiger Kompetenzen bei Kindern und Erwachsenen aus kognitionspsychologischer Perspektive. In: Nuissl, E. (Editor), Vom Lernen zum Lehren. Lern- und Lehrforschung für die Weiterbildung. Bertelsmann, Bielefeld, pp. 93–106.

Stöckler, M. 1991. A short history of emergence and reductionism. In: Agazzi, E. (Editor), The Problem of Reductionism in Science. Kluwer Academic Publishers, Dordrecht, Boston, London, pp. 71–90.

Stone, A.J. 2008. Intermolecular potentials. Science 321: 787–789.

Stotz, K. 2005. Organismen als Entwicklungssysteme. In: Krohs, U.; Toepfer, G. (Editors), Philosophie der Biologie. Suhrkamp, Frankfurt a.M., pp. 125–143.

Stratmann, F. 1988. Zum Einfluss der Anthroposophie in der Medizin. Zuckschwerdt, München, Bern, Wien, San Francisco.

Streubel, T. 2008. Gehirn und Geist. Peter Lang, Frankfurt a. M., Berlin, Bern.

Strohman, R.C. 1997. The coming revolution in biology. Nat. Biotechnol. 15: 194–200.

Stuart, H.A.; Klages, G. 1984. Kurzes Lehrbuch der Physik. 10th ed. Springer, Berlin, Heidelberg, New York, Tokyo.

Suchantke, A. 2009. Metamorphosis. Evolution in Action. Translated by Norman Skillen. Adonis Press, Hillsdale, NY.

Suchantke A. 2012. Lesen im Buch der Natur: Wege zum Erfahren des Ätherischen. Verlag am Goetheanum, Dornach.

Suddendorf, T. 2013. The Gap. The Science of what separates us from other animals. Basic Books, New York.

Superbielle, E.; Sanchez, P.E.; Kravitz, A.V.; Wang, X.; Ho, K.; Eilertson, K.; Devidze, N.; Kreitzer, A.C.; Mucke, L. 2013. Physiologic brain activity causes DNA double-strand breaks in neurons, with exacerbation by amyloid-. Nat. Neurosci. 16: 613–621.

Süsskind, M.; Thürmann, P.A.; Lüke, C.; Jeschke, E.; Tabali, M.; Matthes, H.; Ostermann, T. 2012. Adverse drug reactions in a complementary medicine hospital: a prospective, intensified surveillance study. Evid Based Complement Alternat Med 2012: 320760. doi:10.1155/2012/320760.

Tabali, M.; Ostermann, T.; Jeschke, E.; Witt, C.M.; Matthes, H. 2012. Adverse drug reactions for CAM and conventional drugs detected in a network of physicians certified to prescribe CAM drugs. J Manag Care Pharm 18(6): 427–438.

Thiel, M.T.; Findeisen, B.; Längler, A. 2011. Music therapy as part of integrative neonatology: 20 years of experience – 3 case reports and a review. Forsch Komplementärmed 18(1): 31–35. doi:10.1159/000323714.

Toepfer, G. 2005. Teleologie. In: Krohs, U.; Toepfer, G. (Editors), Philosophie der Biologie. Suhrkamp, Frankfurt a. M., pp. 36–70.

Tokuriki, N.; Tawfik, D.S. 2009. Protein Dynamism and Evolvability. Science 324: 203–208.

Tomasello, M. 1999. The Cultural Origins of Human Cognition. Harvard University Press, Cambridge, MA.

Tomasello, M. 2014. A Natural History of Human Thinking. Harvard University Press, Cambridge, MA.

Tonegawa, S. 1983. Somatic generation of antibody diversity. Nature 302: 575–581.

Torregrosa Paredes, P.; Gutzeit, C.; Johansson, S.; Admyre, C.; Stenius, F.; Alm, J.; Scheynius, A.; Gabrielsson, S. 2014. Differences in exosome populations in human breast milk in relation to allergic sensitization and lifestyle. Allergy 69(4): 463–471. doi:10.1111/all.12357.

Touitou, Y.; Haus, E. (Editors). 1992. Biologic Rhythms in Clinical and Laboratory Medicine. Springer, Berlin, Heidelberg, New York.

Treichler, R. 1981a. Der schizophrene Prozess. Beiträge zu einer erweiterten Pathologie und Therapie. Verlag Freies Geistesleben, Stuttgart.

Treichler, R. 1981b. Die Entwicklung der Seele im Lebenslauf. Verlag Freies Geistesleben, Stuttgart.

Tretter, F. 2007. Gehirn-Geist-Debatte. Wissenschaftsphilosophische Probleme im Hinblick auf die Psychiatrie. Nervenarzt 78: 498–504.

Tröger, W.; Ždrale, Z.; Stanković, N.; Matijašević, M. 2012. Five-year follow-up of patients with early stage breast cancer after a randomized study comparing additional treatment with viscum album (L.) extract to chemotherapy alone. Breast Cancer (Auckl) 6: 173–180. doi:10.4137/BCBCR.S10558.

Tröger, W.; Galun, D.; Reif, M.; Schumann, A.; Stanković, N.; Milićević, M. 2013. Viscum album [L.] extract therapy in patients with locally advanced or metastatic pancreatic cancer: a randomised clinical trial on overall survival. Eur. J. Cancer 49(18): 3788–3797.

Tröger, W.; Ždrale, Z.; Tišma, N.; Matijašević, M. 2014a. Additional Therapy with a Mistletoe Product during Adjuvant Chemotherapy of Breast Cancer Patients Improves Quality of Life: An Open Randomized Clinical Pilot Trial. Evid Based Complement Alternat Med 2014: 430518. doi:10.1155/2014/430518.

Tröger, W.; Galun, D.; Reif, M.; Schumann, A.; Stanković, N.; Milićević, M. 2014b. Quality of life of patients with advanced pancreatic cancer during treatment with mistletoe: a randomized controlled trial. Dtsch Ärztebl Int 111(29–30): 493–502. doi:10.3238/arztebl.2014.0493.

Troxler, I.P.V. 1806. Einige Worte über die grassierende Krankheit und Arzneykunde im Kantone Luzern im Jahre 1806, Zug.

Troxler, I.P.V. 1808. Elemente der Biosophie. J.G. Feind, Leipzig.

Troxler, I.P.V. 1829. Logik. Die Wissenschaft des Denkens und Kritik aller Erkenntnis. 3 Vols. Cotta, Stuttgart, Tübingen.

Troxler, I.P.V. 1921. Blicke in das Wesen des Menschen. Nach der Erstausgabe Aarau 1812. New ed. by H.E. Lauer. Der Kommende Tag AG, Stuttgart.

Troxler, I.P.V. 1925. Über das Leben und sein Problem. Nach der Erstausgabe von 1806. Edited by H. Kern und C. Bernoulli. Niels Kampmann, Celle.

Troxler, I.P.V. 1936. Fragmente. Erstveröffentlichung aus seinem Nachlasse. Edited by Willi Aeppli. Troxler Verlag, Bern.

Troxler, I.P.V. 1942. Vorlesungen über Philosophie. Über Inhalt, Bildungsgang, Zweck und Anwendung derselben aufs Leben. Nach der Druckausgabe Bern 1835. New ed. By F. Eymann. Troxler Verlag, Bern.

Troxler, I.P.V. 1944. Naturlehre oder Metaphysik. Nach der Erstausgabe 1828. New ed. by Willi Aeppli. Troxler Verlag, Bern.

Valentini, A. 1991. Non-local correlations in quantum electrodynamics. Phys Lett A 153 (6–7): 321–325.

Van Cauter, E.; Leproult, R.; Plat, L. 2000. Age-related changes in slow-wave sleep and REM sleep and relationship with growth hormone and cortisol levels in healthy men. JAMA 284: 861–868.

van der Pompe, G.; Antoni, M.; Visser, A. 1996. Adjustment to breast cancer: the psychobiological effects of psychosocial interventions. Patient Educ Couns 28: 209–219.

van Deventer, M. 1992. Die anthroposophisch-medizinische Bewegung in den verschiedenen Etappen ihrer Entwicklung. Natura-Verlag, Arlesheim.

Vandercruysse, R. 1999. Die therapeutische Dimension des Denkens. Anthroposophische Aspekte zur Psychoanalyse. Edition Hardenberg. Verlag Freies Geistesleben, Stuttgart.

Vanzetta, I.; Grinvald, A. 1999. Increased cortical oxidative metabolism due to sensory stimulation: implications for functional brain imaging. Science 286: 1555–1558.

Varela, C.R. 2009. Science for Humanism. The recovery of human agency. Routledge, London, New York.

Verde, P.E.; Ohmann, C. 2015. Combining randomized and non-randomized evidence in clinical research: a review of methods and applications. Res Synth Methods 6(1): 45–62.

Verhulst, J. 1999. Der Erstgeborene. Mensch und höhere Tiere in der Evolution. Verlag Freies Geistesleben, Stuttgart.

Vogel, G. 1999. Many modes of transport for an embryo's signals. Science 285: 100–105.

Vogel, L. 1979. Der dreigliedrige Mensch. Morphologische Grundlagen einer allgemeinen Menschenkunde. Philosophisch-Anthroposophischer Verlag, Dornach.

Volkamer, K.; Streicher, C.; Walton, K.G. 1991. Intuition, Kreativität und ganzheitliches Denken. Sauer Verlag, Heidelberg.

Vollmer, G. 1987. Evolutionäre Erkenntnistheorie. Erkenntnisstrukturen im Kontext von Biologie, Psychologie, Linguistik, Philosophie und Wissenschaftstheorie. 4th ed. Hirzel, Stuttgart.

von Bertalanffy, L. 1970. Gesetz oder Zufall: Systemtheorie und Selektion. In: Koestler, A.; Smythies, J.R. (Editors), Das neue Menschenbild. Die

Revolutionierung der Wissenschaften vom Leben. Ein internationales Symposion. Fritz Molden, Wien, München, Zürich, pp. 71–95.

von Bonin, D.; Grote, V.; Buri, C.; Cysarz, D.; Heusser, P.; Moser, M.; Wolf, U.; Laederach, K. 2014. Adaption of cardio-respiratory balance during dayrest compared to deep sleep – an indicator for quality of life? Psychiatry Res 219 (3): 638–644.

von Hagens, C.; Schiller, P.; Godbillon, B.; Osburg, J.; Klose, C.; Limprecht, R.; Strowitzki, T. 2012. Treating menopausal symptoms with a complex remedy or placebo: a randomized controlled trial. Climacteric 15(4): 358–367. doi:10.3109/13697137.2011.597895.

von Helmholtz, H. 1995. On Goethe's Scientific Researches. In: Cahan, D. (Editor), Science and Culture. Popular and Philosophical Essays [E-book]. University of Chicago Press, Chicago. [Viewed 13/10/2015]. URL: https://books.google.it/books?id=xTxcgGX9dicC&printsec=frontcover#v=onepage&q&f=false.

von Mackensen, M.; Schoppmann, R. 2001. Prozesschemie aus spirituellem Ansatz. Pädagogische Forschungsstelle Kassel, Kassel.

von Rohr, E.; Pampallona, S.; van Wegberg, B.; Cerny, T.; Hürny, C.; Bernhard, J.; Helwig, S.; Heusser, P. 2000a. Attitudes and beliefs towards disease and treatment in patients with advanced cancer using anthroposophical medicine. Onkologie 23: 558–563.

von Rohr, E.; Pampallona, S.; van Wegberg, B.; Hürny, C.; Bernhard, J.; Heusser, P.; Cerny, T. 2000b. Experiences in the realisation of a research project in anthroposophical medicine in patients with advanced cancer. Schweiz Med Wochenschr 130: 1173–1184.

von Uexküll, T.; Wesiak, W. 1998. Theorie der Humanmedizin. Grundlagen ärztlichen Denkens und Handelns. 3rd ed. Urban & Schwarzenberg, München, Wien, Baltimore.

von Weizsäcker, C.F. 1986. Naturgesetz und Theodizee. In: Dürr, H.-P. (Editor), Physik und Transzendenz. Scherz, Bern, München, Wien, pp. 250–261.

von Zabern, B. 2002. Kompendium der ärztlichen Behandlung seelenpflegebedürftiger Kinder, Jugendlicher und Erwachsener. Erfahrungen und Hinweise aus der anthroposophischen Therapie. Edition Persephone. Medizinische Sektion am Goetheanum, Dornach.

Wachsmuth, G. 1924. Die ätherischen Bildekräfte in Kosmos, Erde und Mensch. Der Kommende Tag AG, Stuttgart.

Wachsmuth, G. (Editor). 1926–1932. Gäa Sophia. Jahrbuch der Naturwissenschaftlichen Sektion der Freien Hochschule am Goetheanum. Vols. I–VI. Rudolf Geering Verlag, Dornach, Stuttgart, Basel.

Wachsmuth, G. 1932. The Etheric Formative Forces in Cosmos, Earth and Man; A Path of Investigation into the World of the Living. Translated from the 2nd German ed. by Olin D. Wannamaker. Anthroposophic Press, London.

Wachsmuth, G. 1965. Erde und Mensch. Ihre Bildekräfte, Rhythmen und Lebensprozesse. 3rd ed. Philosophisch-Anthroposophischer Verlag, Dornach.

Walach, H. 2011. Placebo controls: historical, methodological and general aspects. Philos. Trans. R. Soc. Lond., B, Biol. Sci. 366(1572): 1870–1878. doi:10.1098/rstb.2010.0401.

Walach, H.; Falkenberg, T.; Fønnebø, V.; Lewith, G.; Jonas, W.B. 2006. Circular instead of hierarchical: methodological principles for the evaluation of complex interventions. BMC Med Res Methodol 2006; 6: 29.

Wälchli, C.; Baumgartner, S.; Bastide, M. 2006. Effect of low doses and high homeopathic potencies in normal and cancerous human lymphocytes: an in vitro isopathic study. J Altern Complement Med 12: 421–427.

Walter, H. 1955. Abnormitäten der geistig-seelischen Entwicklung in ihren Krankheitserscheinungen und deren Behandlungsmöglichkeiten. Verlag des Klinisch-Therapeutischen Instituts, Arlesheim.

Walter, H. 1966. Die sieben Hauptmetalle. Ihre Beziehungen zu Welt, Erde und Mensch. Philosophisch-Anthroposophischer Verlag am Goetheanum, Dornach.

Walter, H. 1971. Die Pflanzenwelt. Ihre Verwandtschaft zur Erden- und Menschheitsentwicklung. Versuch einer Pflanzensystematik als Verständigungsgrundlage für die Therapie. Natura-Verlag, Arlesheim.

Wandschneider, D. 2005. The Problem of Direction and Goal in Biological Evolution. In: Hösle, V.; Illies, C. (Editors), Darwinism and Philosophy. University of Notre Dame Press, Notre Dame, IN, pp. 196–215.

Weaver, W.; Shannon, C.E. 1949. The Mathematical Theory of Communication. University of Illinois Press, Urbana, IL.

Wegman, I. (Editor). 1981. Natura. Eine Zeitschrift zur Erweiterung der Heilkunde nach geisteswissenschaftlicher Menschenkunde. Photomechanischer Nachdruck der 8 Bände von 1926–1940. Natura-Verlag, Arlesheim.

Wegman, I. (Editor). 2000. Beiblätter der Zeitschrift Natura. Zusammenstellung der 1926–1936 erschienenen Einzelhefte «für die Mitglieder der medizinischen Sektion». Natura-Verlag, Arlesheim.

Weihs, T.J. 1971. Das entwicklungsgestörte Kind. Heilpädagogische Erfahrungen in der therapeutischen Gemeinschaft. Verlag Freies Geistesleben, Stuttgart.

Weiss, P. 1927. Potenzprüfung am Regenerationsblastem. I. Extremitätenbildung aus Schwanzblastem im Extremitätenfeld bei Triton. Wilhelm Roux Arch Entwickl Mech Org 111: 317–340.

Weiss, P.A. 1970. Das lebende System: Ein Beispiel für den Schichtendeterminismus. In: Koestler, A.; Smythies, J.R. (Editors), Das neue Menschenbild. Die

Revolutionierung der Wissenschaften vom Leben. Ein internationales Symposion. Fritz Molden, Wien, München, Zürich, pp. 13–70.

Werthmann, P.G.; Sträter, G.; Friesland, H.; Kienle, G.S. 2013. Durable response of cutaneous squamous cell carcinoma following high-dose perilesional injections of Viscum album extracts – a case report. Phytomedicine 20(3–4): 324–327. doi:10.1016/j.phymed.2012.11.001.

Werthmann, P.G.; Helling, D.; Heusser, P.; Kienle, G.S. 2014. Tumour response following high-dose intratumoural application of Viscum album on a patient with adenoid cystic carcinoma. BMJ Case Rep 2014; pii:bcr2013203180. doi:10.1136/bcr-2013-203180.

Whitelegg, M. 2003. Goethean science: An alternative approach. J Altern Complement Med 9: 311–320.

Widmer, M.; Lauer, H.E. 1980. Ignaz Paul Vital Troxler. Walter Kugler, Zug.

Wiener, N. 1948. Cybernetics. Wiley, New York.

Willmann, O. 1907. Geschichte des Idealismus. Vol. 2. Friedrich Vieweg, Braunschweig.

Wimmenauer, W. 1992. Zwischen Feuer und Wasser. Gestalten und Prozesse im Mineralreich. Urachhaus, Stuttgart.

Wirz, J. 1996. Schritte zur Komplementarität in der Genetik. Elemente der Naturwissenschaft, 64: 37–52.

Wirz, J. 2000. Typusidee und Genetik. In: Heusser, P. (Editor), Goethes Beitrag zur Erneuerung der Naturwissenschaften. Paul Haupt, Bern, Stuttgart, Wien, pp. 313–336.

Wirz, J. 2008. Nicht Baukasten, sondern Netzwerk – die Idee des Organismus in Genetik und Epigenetik. Elemente der Naturwissenschaft 88: 5–21.

Witt, C.M. 2009. Efficacy, Effectiveness, Pragmatic Trials – Guidance on Terminology and the Advantages of Pragmatic Trials. Forsch Komplementärmed 16: 292–294.

Witt, C.M.; Bluth, M.; Albrecht, H.; Weisshuhn, T.E.; Baumgartner, S.; Willich, S.N. 2007. The in vitro evidence for an effect of high homeopathic potencies – A systematic review of the literature. Complement Ther Med 15: 128–138.

Wolf, U.; Wolf, M.; Heusser, P.; Thurneysen, A.; Baumgartner, S. 2011. Homeopathic Preparations of Quartz, Sulfur and Copper Sulfate Assessed by UV-Spectroscopy. Evid Based Complement Alternat Med 2011: 692798. doi:10.1093/ecam/nep036.

Wolff, O. 1990. Heilmittel für typische Krankheiten. Zu den von Rudolf Steiner methodisch neu konzipierten Heilmitteln. Verlag Freies Geistesleben, Stuttgart.

Wolff, O. 2013. Grundlagen einer geisteswissenschaftlich erweiterten Biochemie. 2nd ed. Verlag Freies Geistesleben, Stuttgart.

Wolff, O.; Schürholz, J.; McKeen, T. 1990. Anthroposophische Medizin. Der Merkurstab Sonderheft: 24–27.

Wuketits, F.M. 1981. Biologie und Kausalität. Parey, Berlin, Hamburg.

Xie, Q.; Cvekl, A. 2011. The orchestration of mammalian tissue morphogenesis through a series of coherent feed-forward loops. J. Biol. Chem. 286(50): 43259–43271.

Zander, H. 2007. Anthroposophie in Deutschland. 2 Vols. Vandenhoeck & Ruprecht, Göttingen.

Zeller, E. 1908. Grundriss der Geschichte der griechischen Philosophie. 9th ed. O. R. Reisland, Leipzig.

Zerm, R.; Lutnæs-Mast, F.; Mast, H.; Girke, M.; Kröz, M. 2013. Effects of eurythmy therapy in the treatment of essential arterial hypertension: a pilot study. Glob Adv Health Med 2013; 2(1): 24–30. doi:10.7453/gahmj.2013.2.1.006. Erratum in: Glob Adv Health Med 2013; 2(2): 8. Zerm, Ronald.

Ziegler, R. 2000. Goethe und die Mathematik als Kulturfaktoren. In: Heusser, P. (Editor), Goethes Beitrag zur Erneuerung der Naturwissenschaften. Paul Haupt, Bern, Stuttgart, Wien, pp. 457–486.

Ziegler, R. 2014. Vorbedingungen und Konsequenzen der Fragefähigkeit des Menschen: Erkenntniswissenschaft als Grundlage von Natur- und Geisteswissenschaft. In: Heusser, P.; Weinzirl, J. (Editors), Rudolf Steiner. Seine Bedeutung für Wissenschaft und Leben heute. Schattauer, Stuttgart, pp. 42–50.

Ziegler, R.; Richter, R.; Spengler Neff, A.; Wirz, J. 2015. Biologische Evolution als Erscheinungsentwicklung. Elemente der Naturwissenschaft 103: 41–115.

Zinke, J.F. 2003. Luftlautformen sichtbar gemacht. Sprache als plastische Gestaltung der Luft. Edited by Rainer Patzlaff. Verlag Freies Geistesleben, Stuttgart.

Zürcher, E.; Schlaepfer, R.; Conedera, M.; Guidici, F. 2010. Looking for differences in wood properties as a function of the felling date; lunar phase-correlated variations in the drying behavior of Norway spruce (Picea abies Karst.) and sweet chestnut (Castanea sativa Mill.). Trees 24: 31–41.